Methods in Molecular Biology

Series Editor
John M. Walker
School of Life and Medical Sciences
University of Hertfordshire
Hatfield, Hertfordshire, AL10 9AB, UK

For further volumes:
http://www.springer.com/series/7651

RNA Methylation

Methods and Protocols

Edited by

Alexandra Lusser

Division of Molecular Biology, Biocenter, Medical University of Innsbruck, Innsbruck, Austria

Humana Press

Editor
Alexandra Lusser
Division of Molecular Biology
Biocenter, Medical University of Innsbruck
Innsbruck, Austria

ISSN 1064-3745 ISSN 1940-6029 (electronic)
Methods in Molecular Biology
ISBN 978-1-4939-6805-3 ISBN 978-1-4939-6807-7 (eBook)
DOI 10.1007/978-1-4939-6807-7

Library of Congress Control Number: 2017933288

Printed on acid-free paper

This Humana Press imprint is published by Springer Nature
The registered company is Springer Science+Business Media LLC
The registered company address is: 233 Spring Street, New York, NY 10013, U.S.A.

Preface

The aim of this book is to provide a comprehensive compendium of current methods and protocols to study posttranscriptional base modifications, in particular methylation, in RNA. The study of RNA methylation has emerged as an exciting new research area over the past few years not least due to the discovery that internal base methylation, such as N6-methyladenosine (m6A) and 5-methylcytosine (m5C), is widespread not only in abundant RNAs (tRNAs, rRNA) but also in mRNAs and in long noncoding RNAs (lncRNAs). Thus, it might represent an additional layer of gene regulation, leading to the coining of the terms "RNA epigenetics" and "epitranscriptomics." Although RNA methylation research is still in its early stages, disruption of RNA methylation has been linked to a number of disease conditions, including obesity, type 2 diabetes mellitus, Alzheimer's disease, cancer, and cardiovascular disease. Moreover, at least m6A is a reversible modification, which further underscores the regulatory potential of RNA methylation.

The protocols collected in this book represent the state-of-the-art methodology for investigating diverse aspects of RNA methylation on different types of RNA. Protocols include wet lab techniques for the detection of methylation, instructions for bioinformatics analyses of transcriptome-scale data sets as well as protocols for the functional examination of RNA modifications and enzymes. Thus, this laboratory manual may constitute a valuable source for biochemists and molecular biologists from different fields who wish to include the investigation of different types of RNA methylation in their studies.

Innsbruck, Austria *Alexandra Lusser*

Contents

Contributors

THOMAS AMORT • *Division of Molecular Biology, Biocenter, Medical University of Innsbruck, Innsbruck, Austria*
ZACHARY ANDERSON • *Department of Biology, University of Pennsylvania, Philadelphia, PA, USA*
LILIA AYADI • *IMoPA UMR7365 CNRS-UL, BioPole Lorraine University, Vandoeuvre-les-Nancy, France*
ULF BIRKEDAL • *Department of Cellular and Molecular Medicine, The Panum Institute, University of Copenhagen, Copenhagen, Denmark*
FLORENCE BLANLOEIL-OILLO • *IMoPA UMR7365 CNRS-UL, BioPole Lorraine University, Vandoeuvre-les-Nancy, France; Next-Generation Sequencing Core Facility, FR3209 BMCT CNRS-UL, BioPole Lorraine University, Vandoeuvre-les-Nancy, France*
ZSUZSANNA BODI • *School of Biosciences, The University of Nottingham, Plant Sciences, Leicestershire, UK*
MARKUS T. BOHNSACK • *Centre for Biochemistry and Molecular Cell Biology, Institute for Molecular Biology, Georg-August-University, Göttingen, Germany; Göttingen Centre for Molecular Biosciences, Georg-August-University, Göttingen, Germany*
AARON E. COZEN • *Department of Biomolecular Engineering, University of California Santa Cruz, Santa Cruz, CA, USA*
DARIA KHOKHLOVA-CUBBERLEY • *Zymo Research Corp., Irvine, CA, USA*
MATTHIAS D. ERLACHER • *Division of Genomics and RNomics, Biocenter Innsbruck, Medical University of Innsbruck, Innsbruck, Austria*
FRANCESCA FINOTELLO • *Division of Bioinformatics, Biocenter, Medical University of Innsbruck, Innsbruck, Austria*
RUPERT G. FRAY • *School of Biosciences, The University of Nottingham, Plant Sciences, Leicestershire, UK*
HARRY GEORGE • *Department of Biology and Biochemistry, University of Bath, Bath, UK*
COSMAS C. GIALLOURAKIS • *Gastrointestinal Unit, Massachusetts General Hospital, Harvard Medical School, Boston, MA, USA*
BRIAN D. GREGORY • *Department of Biology, University of Pennsylvania, Philadelphia, PA, USA*
ANYA V. GROZHIK • *Department of Pharmacology, Weill Medical College, Cornell University, New York, NY, USA*
SARA HAAG • *Centre for Biochemistry and Molecular Cell Biology, Institute for Molecular Biology, Georg-August-University, Göttingen, Germany*
ASEEL EL HAJJ • *IMoPA UMR7365 CNRS-UL, BioPole Lorraine University, Vandoeuvre-les-Nancy, France*
MARK HELM • *Institute of Pharmacy and Biochemistry, Johannes Gutenberg-University of Mainz, Mainz, Germany*
CLAUDIA HÖBARTNER • *Institute for Organic and Biomolecular Chemistry, Georg-August-University, Göttingen, Germany*
THOMAS P. HOERNES • *Division of Genomics and RNomics, Biocenter Innsbruck, Medical University of Innsbruck, Innsbruck, Austria*

EVA HRABETA-ROBINSON • *Department of Biomolecular Engineering, University of California Santa Cruz, Santa Cruz, CA, USA*
SHOBBIR HUSSAIN • *Department of Biology and Biochemistry, University of Bath, Bath, UK*
SAMIE R. JAFFREY • *Department of Pharmacology, Weill Medical College, Cornell University, New York, NY, USA*
PATRICK KELLER • *Institute of Pharmacy and Biochemistry, Johannes Gutenberg-University of Mainz, Mainz, Germany*
JENS KRETSCHMER • *Centre for Biochemistry and Molecular Cell Biology, Institute for Molecular Biology, Georg-August-University, Göttingen, Germany*
NICOLAI KROGH • *Department of Cellular and Molecular Medicine, The Panum Institute, University of Copenhagen, Copenhagen, Denmark*
PAVEL P. KUKSA • *Department of Pathology and Laboratory Medicine, University of Pennsylvania, Philadelphia, PA, USA; Penn Institute for Biomedical Informatics, University of Pennsylvania, Philadelphia, PA, USA*
YUK YEE LEUNG • *Department of Pathology and Laboratory Medicine, University of Pennsylvania, Philadelphia, PA, USA; Penn Institute for Biomedical Informatics, University of Pennsylvania, Philadelphia, PA, USA*
XIAOYU LI • *State Key Laboratory of Protein and Plant Gene Research, School of Life Sciences, and Peking-Tsinghua Center for Life Sciences, Peking University, Beijing, China*
PATRICK A. LIMBACH • *Rieveschl Laboratories for Mass Spectrometry, Department of Chemistry, University of Cincinnati, Cincinnati, OH, USA*
BASTIAN LINDNER • *Department of Pharmacology, Weill Medical College, Cornell University, New York, NY, USA*
TODD M. LOWE • *Department of Biomolecular Engineering, University of California Santa Cruz, Santa Cruz, CA, USA*
ALEXANDRA LUSSER • *Division of Molecular Biology, Biocenter, Medical University of Innsbruck, Innsbruck, Austria*
VIRGINIE MARCHAND • *IMoPA UMR7365 CNRS-UL, BioPole Lorraine University, Vandoeuvre-les-Nancy, France; Next-Generation Sequencing Core Facility, FR3209 BMCT CNRS-UL, BioPole Lorraine University, Vandoeuvre-les-Nancy, France*
ERIN MARCUS • *Department of Biochemistry & Biophysics, Center for RNA Biology, University of Rochester School of Medicine, Rochester, NY, USA*
RONALD MICURA • *Institute of Organic Chemistry and Center for Molecular Biosciences, CMBI, Leopold-Franzens University, Innsbruck, Austria*
BENOIT MOLINIE • *Gastrointestinal Unit, Massachusetts General Hospital, Harvard Medical School, Boston, MA, USA*
YURI MOTORIN • *IMoPA UMR7365 CNRS-UL, BioPole Lorraine University, Vandoeuvre-les-Nancy, France; Next-Generation Sequencing Core Facility, FR3209 BMCT CNRS-UL, Lorraine University, Vandoeuvre-les-Nancy, France*
HENRIK NIELSEN • *Department of Cellular and Molecular Medicine, The Panum Institute, University of Copenhagen, Copenhagen, Denmark*
ANTHONY O. OLARERIN-GEORGE • *Department of Pharmacology, Weill Medical College, Cornell University, New York, NY, USA*
BRIAN J. PARKER • *Department of Biology, New York University, New York, NY, USA*
MELLIE JUNE PAULINES • *Rieveschl Laboratories for Mass Spectrometry, Department of Chemistry, University of Cincinnati, Cincinnati, OH, USA*
JINYING PENG • *State Key Laboratory of Protein and Plant Gene Research, School of Life Sciences, and Peking-Tsinghua Center for Life Sciences, Peking University, Beijing, China*

ERIC M. PHIZICKY • *Department of Biochemistry & Biophysics, Center for RNA Biology, University of Rochester School of Medicine, Rochester, NY, USA*
DIETMAR RIEDER • *Division of Bioinformatics, Biocenter, Medical University of Innsbruck, Innsbruck, Austria*
CHRISTIAN RIML • *Institute of Organic Chemistry and Center for Molecular Biosciences, CMBI, Leopold-Franzens University, Innsbruck, Austria*
KATHARINA SCHMID • *Institute of Pharmacy and Biochemistry, Johannes Gutenberg-University of Mainz, Mainz, Germany*
KATHERINE E. SLOAN • *Centre for Biochemistry and Molecular Cell Biology, Institute for Molecular Biology, Georg-August-University, Göttingen, Germany*
XUEGUANG SUN • *Zymo Research Corp., Irvine, CA, USA*
KATHRIN THÜRING • *Institute of Pharmacy and Biochemistry, Johannes Gutenberg-University of Mainz, Mainz, Germany*
JERNEJ ULE • *Department of Molecular Neuroscience, University College London, London, UK*
LEE E. VANDIVIER • *Cell and Molecular Biology Graduate Group, University of Pennsylvania, Philadelphia, PA, USA; Department of Biology, University of Pennsylvania, Philadelphia, PA, USA*
LI-SAN WANG • *Department of Pathology and Laboratory Medicine, University of Pennsylvania, Philadelphia, PA, USA; Penn Institute for Biomedical Informatics, University of Pennsylvania, Philadelphia, PA, USA; Institute of Aging, University of Pennsylvania, Philadelphia, PA, USA*
CHENGQI YI • *State Key Laboratory of Protein and Plant Gene Research, School of Life Sciences, Peking-Tsinghua Center for Life Sciences, Peking University, Beijing, China; Department of Chemical Biology, Synthetic and Functional Biomolecules Center, College of Chemistry and Molecular Engineering, Peking University, Beijing, China*
BI-FENG YUAN • *Key Laboratory of Analytical Chemistry for Biology and Medicine (Ministry of Education), Department of Chemistry, Wuhan University, Wuhan, China*

Part I

Modifications in Abundant RNA Species

Chapter 1

LC-MS Analysis of Methylated RNA

Kathrin Thüring*, Katharina Schmid*, Patrick Keller, and Mark Helm

Abstract

The detection and quantification of methylated RNA can be beneficial to understand certain cellular regulation processes such as transcriptional modulation of gene expression, immune response, or epigenetic alterations. Therefore, it is necessary to have methods available, which are extremely sensitive and accurate, for instance liquid chromatography-tandem mass spectrometry (LC-MS/MS). Here, we describe the preparation of RNA samples by enzymatic hydrolysis and the subsequent analysis of ribonucleosides by LC-MS/MS via NLS (Neutral loss scan) and DMRM (Dynamic multiple reaction monitoring). Also, we provide variations of these methods including chromatographic techniques and different kinds of quantification.

Key words RNA methylation, LC-MS, DMRM, NLS, SIL-IS, External calibration

1 Introduction

Methylation of RNA occurs in all three kingdoms of life and has multiple functions in, e.g., gene regulation, intracellular trafficking, stress/immune response, and RNA stabilization [1]. To analyze and thus further understand these functions, mass spectrometry is the tool of choice due to its remarkable sensitivity and accuracy in the detection and quantification of modified nucleosides. For instance, LC-MS/MS techniques that utilize a triple quadrupole-based detection allow limits of quantification in the low femtomol range [2, 3]. In this case, relative or even absolute quantification of the modification content is possible, when the sample signal can be related to an absolute amount of the corresponding modification. Therefore, spike-in measurements [4] or external calibrations [5] are required which in turn require the availability of weighable amounts of material. To overcome this, the use of an internal standard generated by stable isotope labeling (SIL-IS) is one alternative [6].

* These authors contributed equally to this work.

Alexandra Lusser (ed.), *RNA Methylation: Methods and Protocols*, Methods in Molecular Biology, vol. 1562,
DOI 10.1007/978-1-4939-6807-7_1, © Springer Science+Business Media LLC 2017

Here, we provide all necessary information about the preparation of RNA samples and their subsequent analysis via LC-MS/MS. For this purpose, RNA samples are digested to nucleosides by means of nucleases. The single modified nucleosides are then separated by using different gradients considering the type of sample and the desired quantification/detection method. The complete workflow is viewed in Fig. 1.

For the mass spectrometric analysis of the separated nucleosides, the most common techniques are the dynamic multiple reaction monitoring (DMRM) and the neutral loss scan (NLS). During DMRM, one or multiple desired precursor ions are chosen and fragmented in the collision cell. These fragmented nucleosides are then detected with high sensitivity due to the fact that only expected fragments (product ions) get monitored, which reduces background noise and enables quantification in the low femtomol range. The NLS method, on the other hand, is a broader screening and not limited to the analysis of known modifications, but can also detect unknown nucleosides with a certain fragmentation pattern. In this case, the nucleoside detection is just carried out if the precursor ions undergo a particular and adjusted neutral loss (loss of ribose (-132 Dalton) or 2′-O-methylated ribose (-146 Dalton) moiety) in the collision cell. Both methods are described in detail in the following sections by providing mass spectrometric parameters and guidelines for the quantification via internal standard/external calibration and qualification of modified nucleosides.

2 Materials

Prepare all solutions and buffers in ultrapure type I water.

Prefer salts and reagents of LC-MS grade, they should at least have HPLC grade.

2.1 Enzymes for RNA Digestion

1. Nuclease P1: Dissolve lyophilized nuclease P1 in MilliQ water to a final concentration of 0.3 U/μL, calculated by the manufacturer's quantification. Aliquote in 20 μL batches to limit the number of freeze-thaw cycles.
2. 10× Nuclease P1 buffer: 90 % 250 mM NH_4OAc pH 5.0, 10 % 2 mM $ZnCl_2$. Use glacial acetic acid for pH adjustment.
3. Snake venom phosphodiesterase (lyophilized): Dissolve in MilliQ water to a final concentration of 0.1 U/μL. Aliquot as described for nuclease P1.
4. Fast alkaline phosphatase: Dissolve lyophilized enzyme to 1 U/μL using MilliQ water.
5. 10× Fast alkaline phosphatase buffer: 100 mM Tris HCl pH 8, 50 mM $MgCl_2$, 1 M KCl, 0.2 % Triton X-100, 1 mg/mL BSA [7].

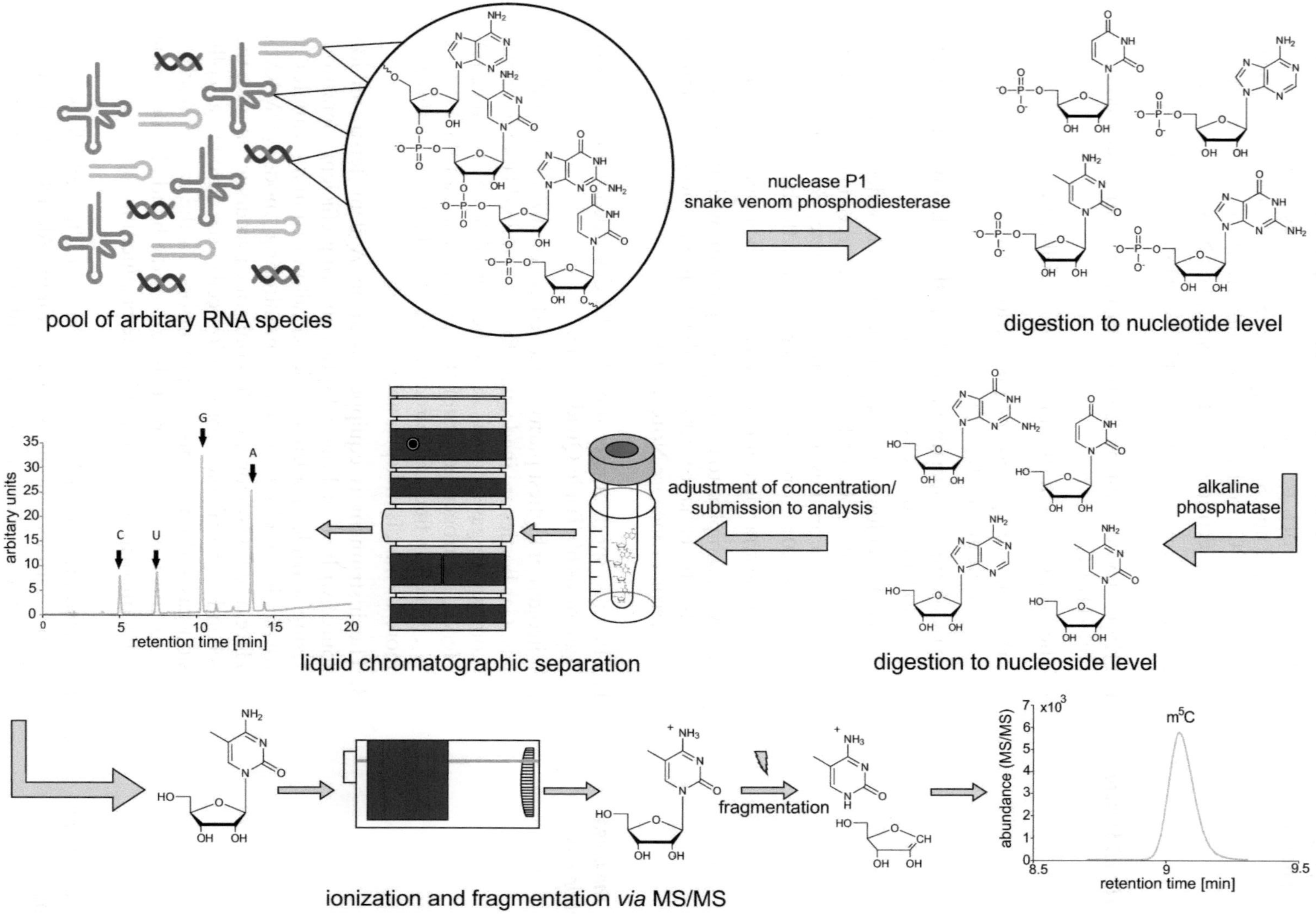

Fig. 1 LC-MS/MS analysis workflow. First RNAs are digested to nucleoside level by using Nuclease P1, snake venom phosphodiesterase and fast alkaline phosphatase. Then these samples are injected into the LC-MS, where they get separated by liquid chromatography and measured by the mass spectrometer. The obtained data is viewed in the corresponding mass chromatogram

2.2 Preparation of LC Running Buffers

1. Eluent A: 5 mM ammonium acetate pH 5.3, 0.1 % acetonitrile. Adjust the pH to 5.3 using glacial acetic acid, every pH between 5.2 and 5.4 is acceptable. Sterile filtration is optional. Store at 4 °C until use and equilibrate to room temperature before connecting to the LC-system (*see* **Note 1**).
2. Eluent B: commercially available LC-MS grade acetonitrile.

2.3 The LC System

1. HPLC device: Agilent 1200 series LC system or any comparable HPLC system.
2. The HPLC system should include the following devices:
   (a) Binary or quaternary pump, for analyte separation by a solvent gradient.
   (b) Column compartment, to keep the column temperature and with it the selectivity constant.
   (c) HPLC column: For separation of nucleosides a RP18 column with polar embedded groups is recommendable, e.g., the Synergi Fusion, 4 μm particle size, 80 Å pore size, 250 × 2.0 mm from Phenomenex (Aschaffenburg, Germany).
   (d) Diode array detector, for detection of the main nucleosides.
   (e) Auto sampler, to enable overnight runs with many samples and to enhance reproducibility as injection by hand is error-prone (*see* **Note 2**)[6].

2.4 The Mass Spectrometer

1. The mass spectrometer used for the described workflow is an Agilent 6460 Triple Quadrupole.
2. Nitrogen of highest purity is needed as nebulizing and collision gas. Gas bottles are inadequate for the required large gas flows (compare to gas amounts in Subheading 3.6. ESI parameters). Hence, a nitrogen generator that produces at least 95.0 % pure N_2, which is free of hydrocarbons, at a flow rate of three or more L/min, is indispensable.
3. The instrument is equipped with an Agilent Jetstream ion source with an orthogonal spray, ensuring efficient ionization. Using a triple quadrupole device allows running samples in (dynamic) multiple reaction monitoring mode (DMRM). Here, the first quadrupole lets only one precursor ion pass, the second "quadrupole" (or collision cell) uses nitrogen as collision gas and fragments the precursors. In the case of nucleosides this typically occurs at the glycosidic bond. Through the third quadrupole, again, only one product ion may pass [8].

2.5 Nucleoside Standards

1. Nucleoside standards: They are commercially available as solids from various suppliers, e.g., Sigma-Aldrich (Steinheim, Germany) or Berry & Associates Inc. (Dexter, USA). Dissolve substances to a stock concentration of 10 mM in type I water.

3 Methods

3.1 Sample Preparation by Nuclease Digestion

1. RNA samples for LC-MS/MS analysis should be prepared in MilliQ water, avoid the presence of salts/buffers (*see* **Note 3**).
2. For the digestion of up to 10 μg RNA to nucleosides (volume as needed), add 1/10 of sample volume 10× nuclease P1 buffer, 0.3 U of nuclease P1, and 0.1 U of snake venom phosphodiesterase and incubate at 37 °C for 2 h. Subsequently, add 1/10 volume of 10× fast alkaline phosphatase buffer and 1 U of fast alkaline phosphatase and incubate for additional 60 min. There is no further purification of the sample required. If desired, the sample volume can be adjusted with MilliQ water as needed and stored at −20 °C.
3. If available, add equal amounts of internal standard (^{13}C–labeled total RNA of *Escherichia coli* or *Saccharomyces cerevisiae*) to each sample (do not add for NLS analysis). The absolute amount of internal standard added to the samples is dependent on the abundance of the modification of interest in the standard and thus has to be determined individually for each modification analysis (*see* **Note 6**).

3.2 Preparation of Stable Isotope Labeled Internal Standard from E. Coli

1. Prepare M9 minimal growth medium with ^{13}C glucose. Grow *E. coli* (e.g., strain MC4100) in M9 medium to an OD_{600} of 1.4 (*see* **Note 4**) [6].
2. Pellet cells by centrifugation for 15′ at 4400 × *g*. Wash cell pellet two times with 1× PBS.
3. Extract RNA using TriReagent (Sigma-Aldrich) according to the manufacturer's instructions [9].
4. Measure RNA concentration in ng/μL by UV absorbance.
5. Hydrolyze RNA to nucleosides, according to Subheading 3.1.
6. Dilute concentration to 100 ng/μL.
7. Dilute further if abundance of your nucleoside of interest leads to a saturated MS/MS signal in (D)MRM mode.

3.3 Preparation of Stable Isotope Labeled Internal Standard from S. cerevisiae

1. Grow *S. cerevisiae* (strain: BY4741) in commercially available ^{13}C OD-2 rich growth medium to an OD_{600} of 1.4.
2. Proceed according to Subheading 3.2, **steps 2–7**.

3.4 Preparation of Calibration Series for Quantitative Measurements

All calibration solutions are prepared by diluting stock solutions of the respective nucleoside in MilliQ water.

1. For UV calibration of the canonical nucleosides prepare calibration solutions containing 0.1, 1, 10, and 100 μM of the main nucleosides. Inject 2, 5, and 10 μL of each solution (only 10 μL of the 0.1 μM solution) into the LC-MS/MS-system

and record the UV chromatogram using the DAD at 254 nm. Thus, a calibration curve in the range of 1–1000 pmol main nucleoside is received.

2. For external calibration of a modified nucleoside (if no internal standard is available, *see* **Note 5**), prepare calibration solutions in the range of the expected amount. Be sure to avoid saturation of the mass signal to stay in the linear calibration range.
3. For calibration with internal standard (response factor determination), prepare calibration solutions in the range of 0.1 nM to 1 μM, each containing equal and sufficient amounts of ^{13}C–labeled total RNA of either *E. coli* or *S. cerevisiae* (*see* Subheadings 3.2 or 3.3 respectively, and **Note 6**).

3.5 HPLC Methods and UV Detection

1. All HPLC separations are performed on a Synergi 4 μm Fusion-RP 80 Å column at 35 °C column temperature. The separation of nucleosides is performed using different gradients of eluent A and pure acetonitrile (eluent B).
2. Use one of the following gradients (*see* also Fig. 2) for nucleoside separation, depending on sample type and modification of interest. Be sure to equilibrate the column in 100 % eluent A before starting analysis (*see* **Note 7**).
   (a) Gradient A: use in standard RNA modification analysis for known nucleosides (Table 1).
   (b) Gradient B: use in standard DNA modification analysis (Table 2).
   (c) Gradient C: use in Neutral Loss Scans (NLS) for searching nonstandard nucleosides (Table 3).
3. Use a diode array detector (DAD) for UV detection of the main nucleosides (C, U, G, A) at 254 nm. This avoids saturation of the mass signals and helps to preserve the electron multiplier.

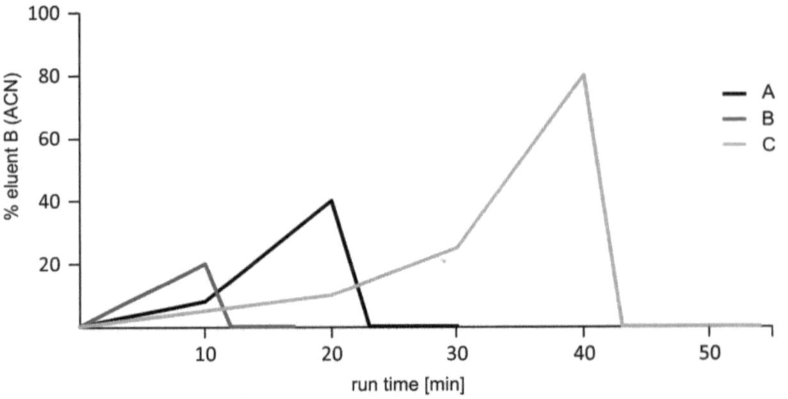

Fig. 2 Utilized gradients. Gradients *A*, *B*, and *C* show the percentages of eluent B over time during one complete chromatographic run

Table 1
Gradient A

Time [min]	% Eluent A	% Eluent B	Flow rate [mL/min]
0	100	0	0.35
10	92	8	
20	60	40	
23	100	0	
30	100	0	

Table 2
Gradient B

Time [min]	% Eluent A	% Eluent B	Flow rate [mL/min]
0	100	0	0.5
10	80	20	
12	100	0	
17	100	0	

Table 3
Gradient C

Time [min]	% Eluent A	% Eluent B	Flow rate [mL/min]
0	100	0	0.35
20	90	10	
30	75	25	
40	20	80	
43	100	0	
54	100	0	

3.6 Mass Spectrometric Parameters for MS/MS Analysis

1. Tandem mass analysis (MS/MS) is performed on a triple quadruple mass spectrometer, equipped with an electrospray ionization source in the positive ionization mode. For required RNA sample amounts, *see* **Note 8**.
2. Depending on the modification of interest, the following types of MS/MS analysis are used:
   (a) Use dynamic multiple reaction monitoring (DMRM mode) for qualitative and quantitative analysis and to reach a very high sensitivity.

Prerequisites: Fragmentation pattern (precursor and product ion) and retention time of the modification of interest is known (*see* **Note 9**).

Gradient to be used: A, B

ESI parameters (Table 4):

For each modification analyzed, fragment or voltage, collision energy and cell accelerator voltage have to be optimized separately. Follow the parameters for 5-methylcytidine analysis as an example (Table 5):

(b) Use Neutral Loss Scan (NLS mode) for investigating RNA modifications without the need of reference substances. It allows the identification of novel modifications, but has a lower sensitivity.

Prerequisites: modification of interest shows typical fragmentation at the glycosidic bond (loss of (2′-O-methyl)-ribose).

Exclusion of the main nucleosides from analysis might improve analysis (*see* **Note 10**).

Table 4
ESI parameters for DMRM mode

Parameter	Setting	Comments
Gas temperature	350 °C	N_2
Gas flow	8 L/min	N_2
Nebulizer pressure	50 psi	
Sheath gas temperature	350 °C	N_2
Sheath gas flow	12 L/min	N_2
Capillary voltage	3000 V	

Table 5
Parameters for DMRM-analysis of 5-methylcytidine

Parameter	Setting
Precursor ion	*m/z* 258
Product ion	*m/z* 126
Fragment or voltage	40 V
Collision energy	9 eV
Cell accelerator voltage	2 V
Retention time	9.4 min
Δ retention time	2 min

Gradient to be used: C

ESI parameters (Table 6):

Example settings for the loss of ribose (−132 Da) (Table 7):

3.7 Qualitative Data Analysis

All mass chromatograms are analyzed employing Agilent MassHunter Qualitative Analysis software (Version B.05.00). Make sure to inspect all peaks manually to avoid errors in peak identification.

1. For the analysis of main nucleosides (*see* **Note 10**) using MassHunter Qualitative Analysis software (or equivalent software of choice), extract the UV chromatogram of each sample. Manually integrate the peak of interest to receive the area of the peak (AUC) (*see* Fig. 3).
2. For the analysis of modified nucleosides (and their respective internal standards if using) use one of the following methods:
   (a) DMRM method: extract mass chromatograms for each modified nucleoside (*see* Fig. 4). Peak integration is performed automatically, but make sure to inspect the peak and its retention time to avoid errors in peak determination (*see* **Note 11**).

Table 6
ESI parameters for NLS mode

Parameter	Setting	Comments
Gas temperature	350 °C	N_2
Gas flow	8 L/min	N_2
Nebulizer pressure	50 psi	
Sheath gas temperature	350 °C	N_2
Sheath gas flow	12 L/min	N_2
Capillary voltage	3000 V	

Table 7
Parameters for neutral loss of ribose

Parameter	Setting
Fragment or voltage	80 V
Collision energy	15 eV
Cell accelerator voltage	2 V
Δ retention time	2 min

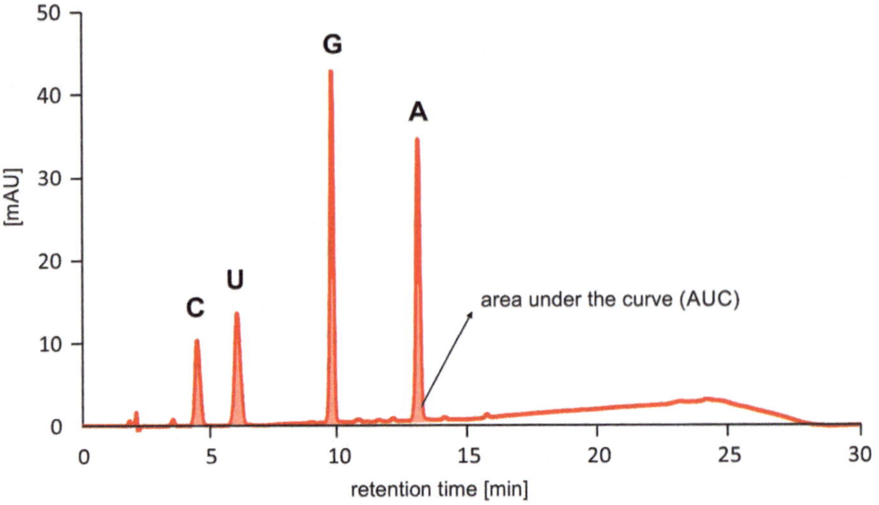

Fig. 3 UV chromatogram of the four canonical nucleosides using gradient A. The main nucleosides and the area under the curve (AUC) are marked

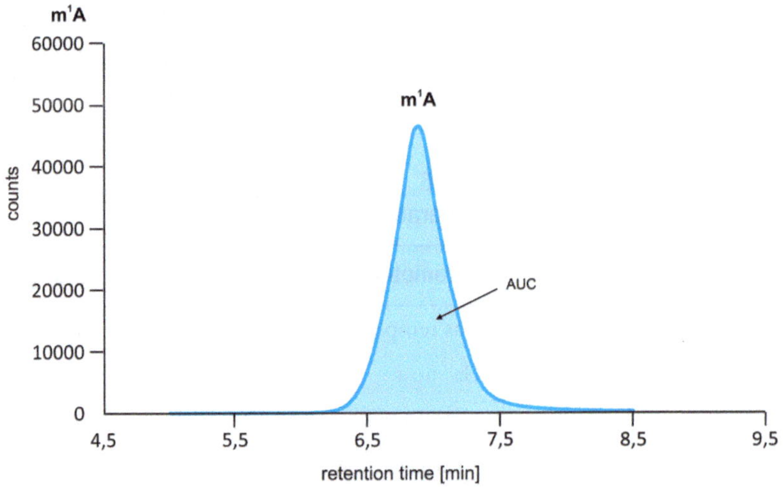

Fig. 4 Mass chromatogram of m^{1}A using gradient A. The AUC is indicated

(b) NLS method: use "extract ion chromatogram" to extract the chromatogram of the modification of interest by entering its precursor ion mass, integrate peaks manually. Alternatively, inspect the mass spectra of the peaks in the total ion current (TIC) to identify modifications of interest.

3. Normalize the modification peak area to the respective peak area of the ^{13}C–labeled internal standard if using (*see* Fig. 5).
4. To qualitatively compare samples in their modification content, normalize peak areas of detected modifications to the UV-peak area of one main nucleoside to counterbalance differences in the injected RNA amount.

3.8 Quantitative Data Analysis

Extract all peak areas (samples and calibration solutions) as described in Subheading 3.7.

1. For UV calibration of the main nucleosides, plot detected peak areas vs. the injected amount to receive the calibration curve (Fig. 6). Calculate the amount of one (or more) main nucleoside in each sample.
   Example for UV calibration:
2. For external calibration of modified nucleosides, plot detected peak areas vs. the injected amount to receive the calibration curve (Fig. 7). Calculate the amount of the modified nucleoside

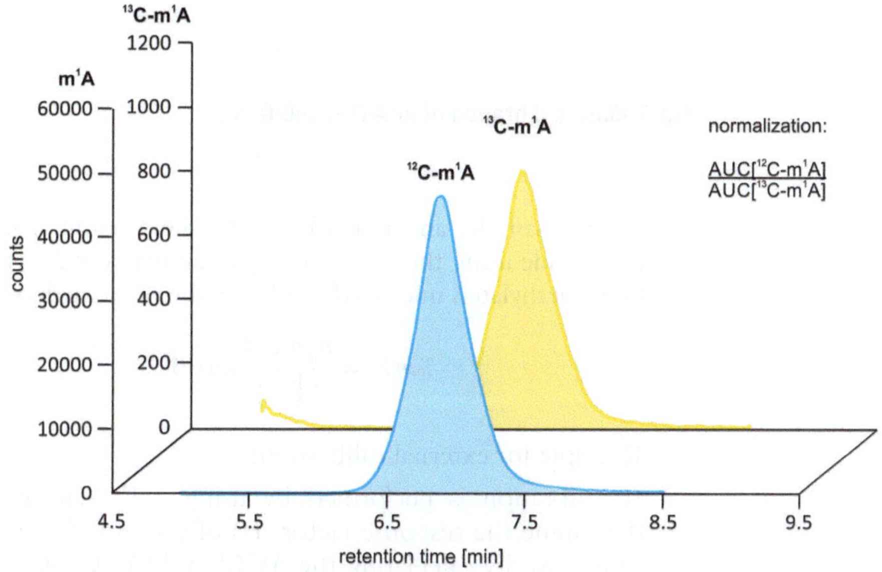

Fig. 5 Mass chromatograms of ^{12}C- and ^{13}C- m^{1}A using gradient A. The peak area of ^{12}C–m^{1}A is normalized to the respective ^{13}C peak area for quantification

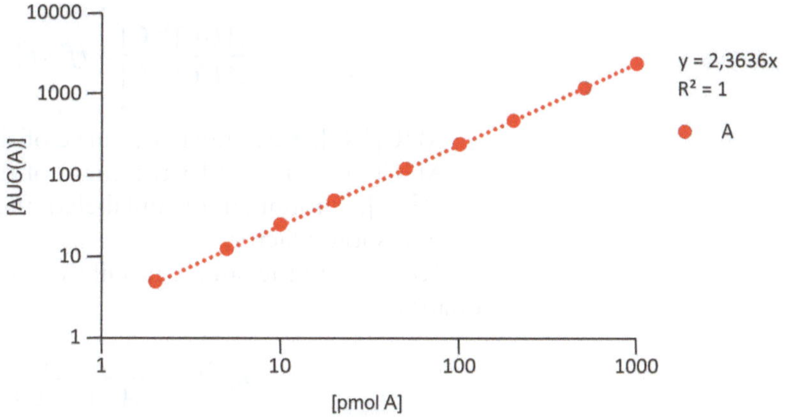

Fig. 6 UV calibration curve of adenosine (1–200 pmol)

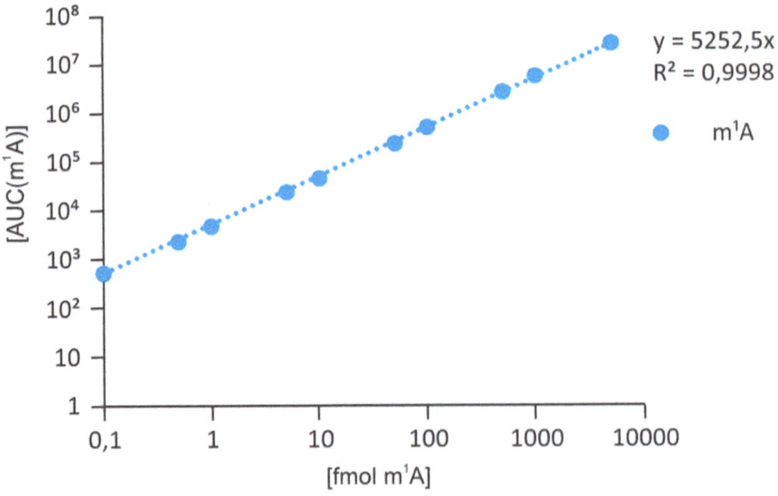

Fig. 7 Mass calibration of m¹A (1–5000 fmol)

in each sample, and normalize it to the amount of one main nucleoside using the following equation where *mN* is the modified/methylated nucleoside and *N* is the main nucleoside.

$$\%mN = \frac{n[mN]}{n[N]} \times 100$$

Example for external calibration:

3. If calibration is performed by using a ^{13}C–labeled SIL-IS, determine the response factor (rf) of the modified nucleoside of interest by calculating the AUC[^{12}C]/AUC[^{13}C] peak area ratio for each calibration sample and plotting it against the respective amount of ^{12}C nucleoside (*see* Fig. 8).

   The slope of the linear equation is the response factor for the modification of interest:

$$\frac{AUC[^{12}C]}{AUC[^{13}C]} = rf \times n[^{12}C]$$

AUC[^{12}C]: area under the curve of the unlabeled nucleoside.
AUC[^{13}C]: area under the curve of the labeled nucleoside.
n[^{12}C]: amount [fmol] unlabeled nucleoside.
rf: response factor.

To calculate the sample modification content use the following equation:

$$n[^{12}C] = \frac{AUC[^{12}C]}{AUC[^{13}C] \times rf}$$

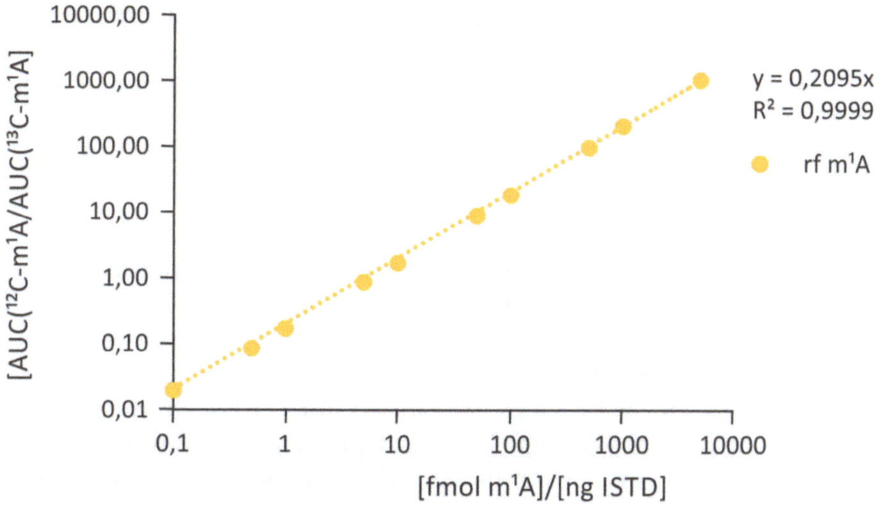

Fig. 8 **Calibration of m**[1]**A using 10 ng SIL-IS**

For normalization to a main nucleoside, first subtract the AUC of the ^{13}C peak of the main nucleoside (determined by measuring a control sample only including the SIL-IS) from the AUC of the respective peak in the sample. Proceed as described above (**step 1**). For analysis of specific RNA molecules instead of total RNA and the quantification of the modification content, *see* **Note 12**.

4 Notes

1. Buffer exchange/spoilage: To limit bacterial growth in the ammonium acetate buffer, add 1 mL of acetonitrile after adjustment of the pH. Exchange the buffer regularly, at least once a week, and remove the buffer if the device is not in use. Be sure to rinse the respective line with pure water to prevent spoilage of solvent filter and line. If the column is not in use for more than 1–2 days, store it in pure water/acetonitrile (50 % each).
2. Consumables for HPLC analysis: Sample vials are needed to use the auto sampler. For sample volumes above 200 μL use 1.5 mL (11.6 × 32 mm) glass vials with screw thread and appropriate screw caps with rubber/PTFE septum. Applying tight screw caps is necessary to avoid sample evaporation and hereby a change in the sample's concentration. Alternatively, one can crimp aluminum caps to the vial. If your sample volume is smaller use bottles with conical glass insert or extra inserts for the 1.5 mL vials.
3. Carefully avoid any salt contamination of your samples as they can form adducts with the nucleosides during ionization, which are not detected at the original mass of the protonated

nucleosides. If performing an ethanol precipitation step to purify the samples, use 0.5 M ammonium acetate as salt.

4. Composition of M9 minimal growth medium (Table 8):
   Prepare 5× M9 salts:
   Bring to 1 L with type 1 water.
   For 1 L medium (Table 9):
   Bring to 1 L with type 1 water and autoclave.
5. External calibration may yield inaccurate results due to ion suppression effects of the sample matrix, which yield decreased sample signals and an underestimation of the real modification content.
6. As SIL-IS, ^{13}C–labeled total RNA from either *E. coli* or *S. cerevisiae* is used, depending on the modifications of interest. Usage of the SIL-IS method is only possible when the modification of interest is present in sufficient amounts in one of the labeled RNA pools. The amount of SIL-IS which is sufficient for analysis is determined by the peak of the ^{13}C peak of the respective modification, which should have a signal-to-noise ratio of at least 10 to avoid detection problems due to signal suppression by the sample matrix.
7. To avoid shifts in retention time, always equilibrate the column in 100 % buffer A until the pump pressure is constant. Additionally, run at least one blank sample (pure water) before starting measuring actual samples.
8. The required injection amount of digested RNA depends on the modification itself (the limit of detection varies from the single-digit attomol to the double-digit fmol range using

Table 8
Amounts of M9 salts needed

Na_2HPO_4	33.9 g
KH_2PO_4	15 g
NaCl	2.5 g
NH_4Cl	5 g

Table 9
Composition of 1 L M9 medium

200 mL	5× M9 salts
1 mL	100 mM $CaCl_2$
20 mL	100 mM $MgSO_4$
40 mL	10 % glucose

multiple reaction monitoring), the abundance of the modification in the RNA sample and the scan type chosen.

9. Identification of nucleosides is performed by their characteristic fragmentation pattern. Most common modified nucleosides show fragmentation at the glycosidic bond; however, untypical fragmentation patterns are possible (e.g., in case of pseudourine, queuosine). For modifications with the same precursor and product ion mass-to-charge ratios (e.g., m^1A, m^6A, m^8A), the retention time is crucial for identification. Keep in mind that retention times might shift over time due to column aging.
10. As the abundances of the four main nucleosides are considerably higher than those of modified nucleosides, their mass signals might easily be saturated when injecting sample amounts allowing detection of the modified nucleosides. Thus, the main nucleosides should be excluded from mass analysis and instead quantified using UV detection. This additionally helps to preserve the electron multiplier.
11. Manual inspection of the peaks is crucial to avoid errors in the automatically performed peak identification. Such errors can occur if there are nucleosides with the same fragmentation pattern eluting close to each other, yielding in the detection of the wrong peak (e.g., m^1G and m^2G).
12. Analysis of specific RNA molecules requires their isolation from total RNA, as the digestion to nucleosides leads to a loss of sequence information. If the RNA of interest is isolated, the number of modified nucleosides per RNA molecule can be determined by calculating the amount of injected RNA molecules using the detected amount of one main nucleoside:

$$n[RNA] = \frac{n[\text{main nucleoside}]}{\text{number main nucleoside } \textit{per } RNA \text{ molecule}} \quad (1)$$

$$\text{number modified nucleoside } \textit{per } RNA = \frac{n[\text{mod.}]}{n[RNA]} \quad (2)$$

References

1. Motorin Y, Helm M (2011) RNA nucleotide methylation. Wiley Interdiscip Rev RNA 2:611–631
2. Nikcevic I, Wyrzykiewicz TK, Limbach PA (2011) Detecting low-level synthesis impurities in modified phosphorothioate oligonucleotides using liquid chromatography-high resolution mass spectrometry. Int J Mass Spectrom 304:98–104
3. Schmid K, Thuring K, Keller P, Ochel A, Kellner S, Helm M (2015) Variable presence of 5-methylcytosine in commercial RNA and DNA. RNA Biol 12:1152–1158
4. Contreras-Sanz A, Scott-Ward TS, Gill HS, Jacoby JC, Birch RE, Malone-Lee J, Taylor KM, Peppiatt-Wildman CM, Wildman SS (2012) Simultaneous quantification of 12 different nucleotides and nucleosides released from renal epithelium and in human urine samples using ion-pair reversed-phase HPLC. Purinergic Signal 8:741–751

5. Chen ML, Shen F, Huang W, Qi JH, Wang Y, Feng YQ, Liu SM, Yuan BF (2013) Quantification of 5-methylcytosine and 5-hydroxymethylcytosine in genomic DNA from hepatocellular carcinoma tissues by capillary hydrophilic-interaction liquid chromatography/quadrupole TOF mass spectrometry. Clin Chem 59:824–832
6. Kellner S, Ochel A, Thuring K, Spenkuch F, Neumann J, Sharma S, Entian KD, Schneider D, Helm M (2014) Absolute and relative quantification of RNA modifications via biosynthetic isotopomers. Nucleic Acids Res 42:e142
7. Crain PF (1990) Preparation and enzymatic hydrolysis of DNA and RNA for mass spectrometry. Methods Enzymol 193:782–790
8. Pomerantz SC, McCloskey JA (1990) Analysis of RNA hydrolyzates by liquid chromatography-mass spectrometry. Methods Enzymol 193:796–824
9. Chomczynski P, Sacchi N (1987) Single-step method of RNA isolation by acid guanidinium thiocyanate-phenol-chloroform extraction. Anal Biochem 162:156–159

Chapter 2

Comparative Analysis of Ribonucleic Acid Digests (CARD) by Mass Spectrometry

Mellie June Paulines and Patrick A. Limbach

Abstract

We describe the comparative analysis of ribonucleic acid digests (CARD) approach for RNA modification analysis. This approach employs isotope labeling during RNase digestion, which allows the direct comparison of a tRNA of unknown modification status against a reference tRNA, whose sequence or modification status is known. The reference sample is labeled with ^{18}O during RNase digestion while the candidate (unknown) sample is labeled with ^{16}O. These RNase digestion products are combined and analyzed by mass spectrometry. Identical RNase digestion products will appear in the mass spectrum as characteristic doublets, separated by 2 Da due to the $^{16}O/^{18}O$ mass difference. Singlets arise in the mass spectrum when the sequence or modification status of a particular RNase digestion product from the reference is not matched in the candidate (unknown) sample. This CARD approach for RNA modification analysis simplifies the determination of differences between reference and candidate samples, providing a route for higher throughput screening of samples for modification profiles, including determination of tRNA methylation patterns.

Key words Modified nucleosides, RNA sequencing, tRNA, rRNA, Modified bases, Tandem mass spectrometry, LC-MS/MS, MALDI-MS/MS, Isotope labeling, Epitranscriptome

1 Introduction

Transfer RNAs (tRNAs) are adapter molecules that decipher the genetic code via codon-anticodon interaction. In eukaryotes, tRNA is transcribed as a precursor molecule (pre-tRNA) by RNA polymerase III, which undergoes processing to generate the mature tRNA. Among the processing steps, the 5′ leader sequence is cleaved by RNAse P while the 3′ trailer sequence is removed by RNAse Z. A conserved CCA sequence is attached to the 3′ end, which serves as a site of amino acid attachment [1]. In addition, tRNAs are chemically modified by enzyme-mediated processes either on the nucleobase or on ribose [2, 3]. Modifications can be a simple methylation or can utilize complex multistep pathways involving many enzymes. Each individual tRNA will typically

Alexandra Lusser (ed.), *RNA Methylation: Methods and Protocols*, Methods in Molecular Biology, vol. 1562,
DOI 10.1007/978-1-4939-6807-7_2, © Springer Science+Business Media LLC 2017

contain multiple modified nucleosides. Depending on where modified nucleosides are located in the sequence of the tRNA, each modification could impact decoding accuracy, help in the proper folding of the tRNA, or serve as determinants of aminoacyl synthetase [4–6]. Recent evidence has also shown that tRNA modifications are dynamic, responding to environmental stress and impacting cellular translation [7, 8].

Several methods have been developed for the characterization of posttranscriptional chemical modifications on tRNAs. These methods include, but are not limited to, reverse transcriptase polymerase chain reaction (RT-PCR) [9, 10], electrophoresis [11], next-generation sequencing (NGS) [12], and mass spectrometry (MS) [13, 14]. Mass spectrometry-based RNA characterization has been shown to be a powerful analytical approach for identifying modified nucleosides and placing each modification in the appropriate tRNA sequence context. Each modification (with the exception of pseudouridine) increases the mass of the canonical nucleoside being modified, allowing mass spectrometry to easily identify the presence and type of such modifications.

The most commonly used mass spectrometry method for placing modified nucleosides onto tRNA sequences is an RNA modification mapping approach [15, 16]. The tRNA is digested to smaller oligonucleotides using a base-specific ribonuclease (RNase), such as RNase T1, which cleaves after unmodified guanosine residues. These oligonucleotides can be separated by reversed-phase high-performance liquid chromatography (RP-HPLC) and detected by mass spectrometry or tandem mass spectrometry (MS/MS). We recently introduced a variant of this RNA modification mapping approach that takes advantage of sequence conservation to readily identify differences in tRNA modification profiles [17–19]. This approach, referred to as the comparative analysis of RNA digest (CARD) method, employs isotope labeling during RNase digestion to differentiate known and unknown tRNA sequences and modification profiles.

A general overview of the CARD method is illustrated in Fig. 1. The RNA endonuclease cleavage mechanism generates a cyclic phosphate intermediate where oxygen from water can be incorporated onto the 3′ phosphate. If the RNase digestion is conducted using isotopically labeled water ($H_2{}^{18}O$), the ^{18}O label gets incorporated to the 3′ phosphate, thereby increasing the mass of the digestion product by 2 Da as compared with digestions conducted in normal (light) water ($H_2{}^{16}O$). This comparative approach requires a reference tRNA of known modification status [20–22] or an unmodified version of the tRNA of interest (e.g., an in vitro transcript) [23].

The tRNA to be analyzed can be isolated from tissue, bacterial cell pellets, or cultured cells. Typically, the reference tRNA (or tRNA mixture) is labeled with ^{16}O while the sample tRNA (or tRNA mixture) is labeled with ^{18}O. If the digestion product

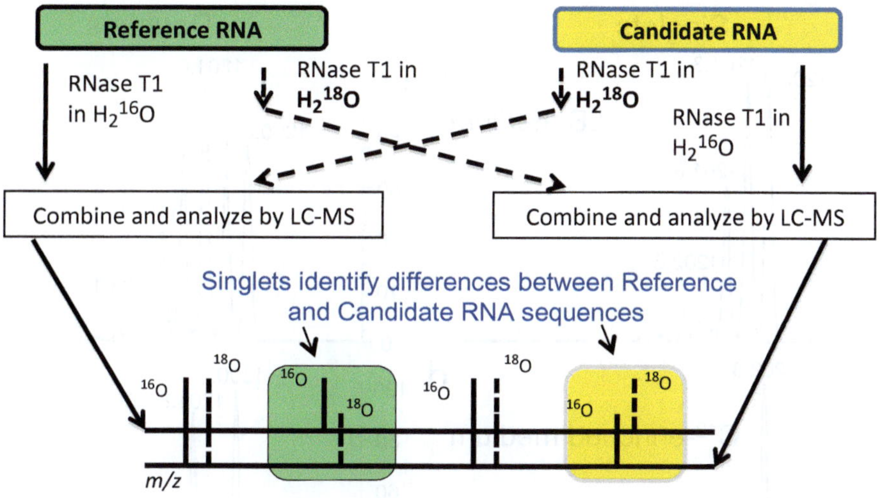

Fig. 1 Schematic outline of comparative sequencing by isotope labeling and LC-MS using a known reference RNA and a candidate (unknown) RNA to be comparatively sequenced. RNA endonuclease digestion products that are equivalent between the two samples will appear as doublets (separated by 2 Da) in the mass spectral data; digestion products that are different between the two samples will appear as a singlet. Reprinted with permission from "Method for comparative analysis of ribonucleic acids using isotope labeling and mass spectrometry"(2012) AnalChem84: 8607–8613. [17] Copyright 2012 American Chemical Society

sequences are identical, those digestion products will have the same mass and appear in the mass spectrum as a characteristic doublet, separated by 2 Da due to the $^{16}O/^{18}O$ mass difference (Fig. 2). The doublet ratio also provides quantitative information relating to each tRNA. When a singlet is detected, then the sequence or posttranscriptional modification status of the sample tRNA differs from that of the reference tRNA.

As singlets often reveal useful information about the state of the sample tRNA, they can be further analyzed to characterize the nature of the singlet. A common step is to reverse the labeling steps, so that the reference is labeled with ^{18}O and the sample is labeled with ^{16}O. If the singlet arises from the reference, the mass will increase by 2 Da after reverse labeling, while singlets from the sample will decrease in mass by 2 Da (Fig. 3). Further, singlets can be characterized by MS/MS to confirm the sequence and location of any modification changes.

The CARD approach has been used to characterize total tRNA modification profiles for closely related organisms, using the well-characterized *Escherichia coli* total tRNA pool as the reference. Moreover, by using DNA hybridization probes, the CARD approach can be developed into a targeted assay that would identify populations that lack one or more tRNA modifications. In particular, the CARD approach is a useful tool for rapidly screening for methylation changes present in gene deletion mutants or for examining RNA methylation patterns across samples obtained

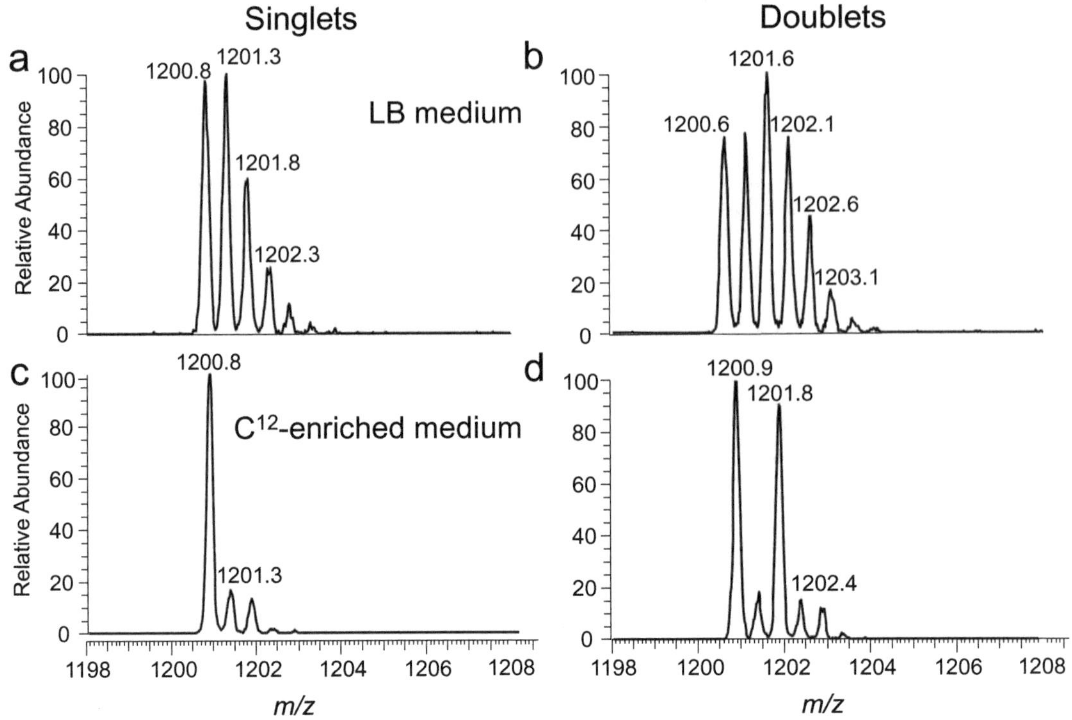

Fig. 2 Improvements in singlet and doublet identification using ^{12}C–enriched medium as illustrated with the doubly charged *E. coli* total tRNA RNase T1 digestion product A[ms^{2}i^{6}A]AACCGp (MW 2403.4 Da). (**a**) Mass spectrum from sample grown in LB medium and labeled with ^{16}O during RNase T1 digestion. (**b**) Same sample as in except labeled with both ^{16}O and ^{18}O during RNase T1 digestion. (**c**) Mass spectrum obtained when sample grown in ^{12}C–enriched medium and labeled with ^{16}Oduring RNase T1 digestion. (**d**) Same sample as in except labeled with both ^{16}O and ^{18}O during RNase T1 digestion. Singlet and doublet identifications are simplified in and d, respectively, by the use of ^{12}C–enriched medium. Reproduced from Ref. [19] with permission of Springer

from multiple populations. The protocol below describes the essential sample preparation steps required for successful CARD analysis of RNA methylation or complete modification profiles. Details on the mass spectrometry conditions can be obtained from a number of relevant publications [15, 17–19, 24].

2 Materials

All reagents used for RNA extraction and isolation should be reagent grade or higher. Glassware and water should be autoclaved to minimize RNase contamination. Gloves should be worn at all times.

2.1 Samples

CARD analysis is applicable to samples isolated from a variety of biological sources. The method is illustrated using both bacterial cell cultures and human tissue samples.

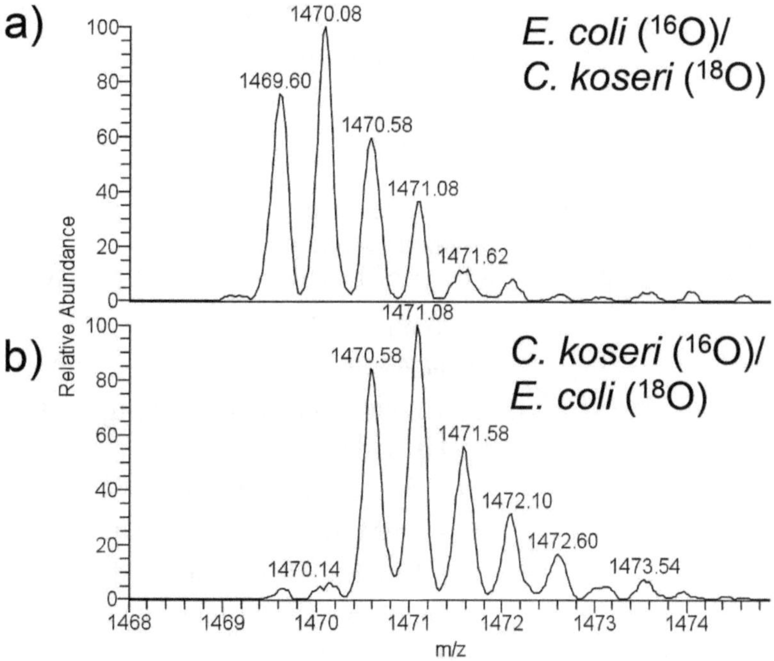

Fig. 3 Mass spectra corresponding to a detected singlet when (**a**) total tRNA from *Citrobacter koseri* is labeled with ^{18}O during RNase T1 digestion and (**b**) *E. coli* is labeled with ^{18}O. The singlet U[s^{4}U]AACAAAGp (*m/z* 1469.6, 2- charge) arises from *E. coli* tRNA-Cys(GCA) as confirmed by the +1 increase in the *m/z* isotopic envelope after ^{18}O–labeling of *E. coli*. Reproduced from Ref. [18] with permission from The Royal Society of Chemistry

2.2 RNA Extraction and Purification

1. TRI-reagent®.
2. Chloroform.
3. MonoExpress medium (^{12}C 99.95 %): 10× concentrate (Cambridge Isotope Laboratories).
4. Lysozyme buffer: 25 mM Tris–HCl, pH 7.5, 60 mM KCl, 10 mM $MgCl_2$.
5. 2 mm ceramic beads.
6. NucleoBond® PC 2000 column.
7. Equilibration buffer: 100 mM Tris–acetate, 15 % ethanol, 200 mM NaCl.
8. Wash buffer: 100 mM Tris–acetate, 15 % ethanol, 400 mM NaCl.
9. Elution buffer: 100 mM Tris–acetate, 15 % ethanol, 750 mM NaCl.
10. Streptavidin agarose beads.
11. 20× saline sodium citrate solution (SSC): 3 M NaCl, 0.3 M sodium citrate. 6× and 3× SSC is diluted from this stock.

12. Specific tRNA elution buffer: 0.1× SSC, 0.1 % SDS.
13. 7.5 M Ammonium acetate.
14. 95 % absolute ethanol.
15. Lysozyme.

2.3 CARD Reference Sample: In Vitro Transcribed tRNA

While reference tRNA samples can be generated from the previously characterized sample of known tRNAs [20–22], the method is described using in vitro transcribed tRNA samples.

1. Template DNA, 10 ng/μL.
2. Forward and reverse primer, 100 μM.
3. 10× PCR buffer: 15 mM $MgCl_2$, 500 mM KCl, 100 mM Tris–HCl, pH 8.3.
4. *PfuI* DNA polymerase: 2.5 U/μL.
5. 40 mM dNTPs mixture: 10 mM each dNTP.
6. QIAquick® PCR clean up kit.
7. Ampliscribe™ T7 high yield transcription kit, which includes: 10× reaction buffer, 100 mM rNTPs, 100 mM dithiothreitol, Riboguard® RNase inhibitor, T7 enzyme solution.
8. In vitro transcription cleanup (e.g., MEGAClear™ Transcription clean up kit).

2.4 RNA Sample Digestion Components

1. Ammonium acetate (NH_4OAC): 220 mM; mass spectrometry grade or higher.
2. Water-^{18}O, 97 %.
3. RNase T1 enzyme. The enzyme is purified by acetone precipitation overnight at −20 °C (300 μL acetone/700 μL enzyme solution). Remove the supernatant and resuspend the enzyme with 1 mL sterile water. Condition the Sep.-Pak C18 column with 3 mL acetonitrile. Equilibriate the column with 3 mL water. Slowly pass the enzyme solution through the column, collecting eluent. Repeat four times with eluent. Wash the column with 1 mL water. Elute the enzyme with 75 % acetonitrile. Aliquot the eluted enzyme (20 μL ~1000 U) in a microcentrifuge tube. Take into dryness and store at 4 °C.
4. Sample RNA (*see* Subheading 2.1).
5. Reference RNA (*see* Subheading 2.1 or 2.3).

2.5 HPLC Mobile and Stationary Phases for Mass Spectrometry

1. Solvent A: 200 mM Hexaflouro-2-propanol (HFIP), 8 mM triethyl amine (TEA).
2. Solvent B: 50 % solvent A (final concentration: 100 mM HFIP, 4 mM TEA), 50 % methanol.
3. C18 column, 5 μm pore size; 4.6 mm × 125 mm

2.6 Other Instrumentation

1. Mini Bead-Beater.
2. Vortex mixer.
3. Microfuge.
4. Vacuum drying apparatus.
5. UV-vis spectrophotometer.
6. Thermocycler.
7. Liquid chromatography tandem mass spectrometry system.

3 Methods

3.1 RNA Isolation

One key to successful CARD mass spectrometry is appropriate sample preparation and purification, due to the confounding nature of salts and buffers on the analysis [25]. The CARD method as described below can be used with RNA isolated from a variety of cell or tissue types. Representative examples of appropriate tRNA sample preparation are provided for bacterial cells and human tissue.

3.1.1 Bacterial Cell Cultures

Bacterial cells can be grown in Luria-Bertani (LB) media following a standard protocol [26]. However, a significant improvement in data analysis has been shown when cells are grown in a ^{12}C enriched/^{13}C depleted medium (MonoExpress Medium; Fig. 2). This approach reduces spectral interferences arising from naturally occurring ^{13}C isotopes, leading to significantly improved detection of singlets and doublets.

1. Pelletize the harvested bacterial cells and add 4.0 mL lysozyme buffer for every 200 mL culture.
2. Add lysozyme in a 5mg/mL buffer ratio. Incubate at 37 °C for 30 minutes.
3. For every 200 μL of suspension, add 1 mL of TRI-reagent®. Vortex the solution and incubate for 10 min at room temperature.
4. Add chloroform (200 μL per 1 mL Tri-Reagent®), invert by shaking and then incubate at room temperature for 10 min.
5. Spin the sample in a microcentrifuge for 15 min at 16,000 × *g*. The clear, upper layer contains total RNA.
6. Remove the aqueous phase and transfer to a clean, sterile microcentrifuge tube (*see* **Note 1**).
7. Add an equal volume of isopropyl alcohol.
8. Incubate at −20 °C from 1 h to overnight.
9. Precipitate the RNA by spinning the sample in a microcentrifuge for 5 min at 16,000 × *g*. Decant liquid and then wash the pellet with 1–2 volumes 75 % ethanol to remove residual salts.

10. Remove the ethanol and allow pellet to air dry.
11. Dissolve pellet in minimal amount of sterile water (~500 μL). Quantify the RNA isolated by checking the absorbance at A_{260} and A_{280} nm (*see* **Note 2**).

3.1.2 Tissue Samples (Placenta)

1. Remove the umbilical cord and other connective tissues. Slice into smaller pieces using sterile scalpel (*see* **Note 3**).
2. Wash with 100 mM $CaCl_2$ to remove blood clots.
3. For every ~2 mg placenta, add 600 μL of TRI-reagent®. Transfer in a tube containing 1.4 mm ceramic beads. Homogenize in a bead beater for 15 min at room temperature or below. Spin the sample in a microcentrifuge for 20 min at 16,000 × *g* and 4 °C.
4. Transfer the pinkish layer to a clean 1.5 mL microcentrifuge tube, add 200 μL chloroform and vortex to facilitate mixing. Let it stand for 20 min at room temperature.
5. Spin the sample in a microcentrifuge for 15 min at 16,000 × *g*. The clear, upper layer contains total RNA.
6. Remove the aqueous phase and transfer to a clean, sterile microcentrifuge (*see* **Note 1**).
7. Add an equal volume of isopropyl alcohol.
8. Incubate at −20 °C from 1 h to overnight to precipitate RNA.
9. Precipitate the RNA by spinning the sample in a microcentrifuge for 5 min at 16,000 × *g*. Decant liquid and then wash the pellet with 1–2 volumes 75 % ethanol to remove residual salts.
10. Remove the ethanol and allow pellet to air dry.
11. Dissolve pellet in minimal amount of sterile water (~1000 μL). Quantify the RNA isolated by checking the absorbance at A_{260} and A_{280} nm (*see* **Note 2**).

3.2 Transfer RNA Enrichment

The following enrichment procedure has been optimized for a Nucleobond® 2000 RX column, which has a binding capacity of ~2000 mg RNA.

1. Condition the column with 50 mL of equilibration buffer.
2. Dilute total RNA sample with two volumes of equilibration buffer (20 mL equilibration buffer/10 mL total RNA solution).
3. Load sample onto column and collect the flow through. Repeat this step twice to maximize binding and yield.
4. Add 2 × 50 mL of the wash buffer.
5. Elute transfer RNA from the column by adding 3 × 20 mL of a warm elution buffer (*see* **Note 4**).
6. Precipitate tRNA by adding an equal volume of isopropyl alcohol to the collected sample (column eluent) and incubate at −20 °C from 1 h to overnight.

7. Pelletize the tRNA by spinning the sample in a centrifuge for 20–30 min at 16,000 × *g*. Decant liquid and then wash the pellet with 1–2 volumes 75 % ethanol to remove residual salts.
8. Remove the ethanol and allow pellet to air dry. For CARD analysis using total tRNA samples, *see* **Note 5**.
9. Dissolve pellet in minimal amount of sterile water (~1000 μL). Quantify the RNA isolated by checking the absorbance at A_{260} and A_{280} nm (*see* **Note 2**).
10. Check the integrity of the sample by analyzing 2 μg of the tRNA on a 1.5 % agarose gel electrophoresis.

3.3 Isolation of a Single tRNA by Oligonucleotide Affinity Probe (Optional)

For CARD analyses focused on single tRNA species, an oligonucleotide affinity probe can be used to isolate the target tRNA from total RNA or total tRNA prepared above.

1. Wash 125 μL of streptavidin agarose beads with 750 μL of 6× SSC. Spin using a microfuge for 10 min and remove the supernatant. Repeat three times. Resuspend the beads in 375 μL 6× SSC.
2. Heat at 70 °C for 5 min a solution containing up to 10 μg of the 5′- biotinylated oligodeoxynucleotide probe, designed to be complimentary to a region of the target tRNA of interest (*see* **Note 6**).
3. Add the heated probe to the washed beads. Incubate at room temperature using a shaker for 1 h.
4. Remove the unbound probe by washing the beads twice with 6× SSC.
5. Prepare the sample RNA by adding sufficient 20× SSC to create a final concentration of 6× SSC. Warm the sample RNA to 85 °C for 10 min and then lower to 70 °C.
6. Add the beads from **step 4** to the RNA sample. Hold solution at 70 °C for 10 min.
7. Cool the mixture to 37 °C and incubate from 2 h to overnight.
8. After incubation, spin sample using a microfuge for 10 min and remove the supernatant containing unbound RNA.
9. Wash the beads with 750 μL of 3× SSC and 750 μL 1× SSC for increased stringency in hybridization.
10. Elute the bound tRNA by incubating the beads at 80 °C in 500 μL elution buffer for 5 min.
11. Remove the beads by spinning the sample in a microcentrifuge at 16,000 × *g* for 10 min. Decant 400 μL of the supernatant containing the elution buffer and probe-purified RNA (*see* **Note 7**). Save sample for further analysis.
12. (Optional) **Steps 10** and **11** can be repeated up to five times to increase the amount of probe-purified tRNA. If steps are repeated, add only 400 μL of the elution buffer before the 5 min incubation at 80 °C. Combine all samples.

13. Precipitate tRNA by adding 1/3rd volume 7.5 M ammonium acetate and two volumes absolute ethanol. Incubate overnight at −80 °C.
14. Spin the samples in a microcentrifuge for 20–30 min at 16,000 × *g*. Decant liquid and then wash the pellet with 1–2 volumes 75 % ethanol to remove residual salts.
15. Remove the ethanol and allow pellet to air dry.
16. Dissolve pellet in minimal amount of sterile water (~50 μL). Quantify the tRNA isolated by checking the absorbance at A_{260} and A_{280} nm (*see* **Note 2**). As necessary and if sufficient sample is available, check sample integrity by gel electrophoresis.

3.4 Reference tRNA from in Vitro Transcription (Optional)

The reference RNA sample for CARD analysis can be from a variety of sources. For applications examining the modification status of a particular tRNA, an in vitro transcript of the tRNA sequence serves as an appropriate unmodified reference (Fig. 4). The template used for the protocol described below is the PCR product from the amplification of a gene block. The gene of interest is downstream the T7 RNA polymerase promoter sequence.

1. Design a gene block with the following: forward primer sequence, T7 RNA polymerase promoter region, gene sequence of a specific RNA, and reverse primer.

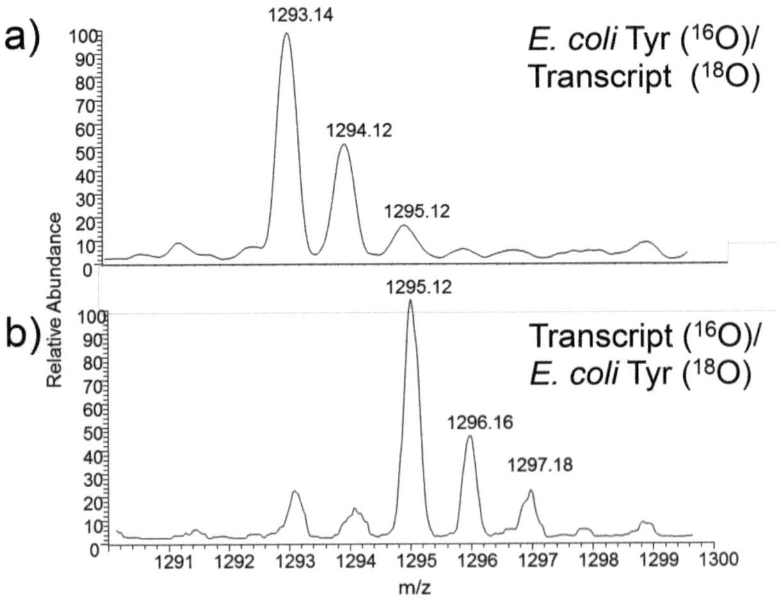

Fig. 4 A modified RNase T1 digestion product from *E. coli* tRNA Tyr detected as a singlet during CARD analysis using an in vitro transcript of *E. coli* tRNA Tyr as the reference. (**a**) Mass spectrum arising from the RNase T1 digestion product [m^{5}U][Ψ]CGp (*m/z* 1293.1 Da, −1 charge state) when labeled with ^{16}O. (**b**) The same singlet detected after reverse labeling using ^{18}O for the unknown. The +2 Da shifts confirm that this singlet is modified and arises from the *E. coli* tRNA Tyr

2. Prepare a PCR master mix from the following recipe (*see* **Note 8**).

   33 μL sterile water.

   4 μL 10× PCR buffer.

   0.8 μ L 40 mM dNTPs.

   0.4 μL Forward primer.

   0.4 μL Reverse primer.

   0.4 μL DNA polymerase (*PfuI*).

3. In one tube, add 1 μL of the gene block template. For the negative control, add 1 μL of sterile water.
4. Perform standard PCR reaction.
5. Run an aliquot of the PCR reaction on a 1.5 % agarose gel with standard DNA marker. Visualize the DNA with ethidium bromide staining under UV (*see* **Note 9**).
6. Purify the remaining PCR products using a commercial clean-up kit following manufacturer's instruction. The amplified gene block is used as the template for T7 RNA transcription.
7. Prepare the transcription reaction by assembling the components as listed. A single reaction has a total volume of 20 μL and is maintained by addition of sterile water.

   Sterile water.

   1 μg DNA template (PCR product).

   2 μL 10× reaction buffer.

   1.5 μL 100 mM ATP.

   1.5 μL 100 mM GTP.

   1.5 μL 100 mM CTP.

   1.5 μL 100 mM UTP.

   2 μL 100 mM dithiothreitol.

   0.5 μL RNase inhibitor.

   2 μL T7 RNA polymerase (2.5 U/μL).

8. Incubate the reaction at 37 °C for 2 h.
9. Purify and elute the transcripts using a commercial clean-up kit following manufacturer's instruction.
10. Resuspend the purified transcript in sterile water (~ 20 μL) and verify integrity of transcript using 1.5 % agarose gel.

3.5 CARD Sample Preparation and Enzymatic Digestion

1. Add 1 μg reference tRNA (if single tRNA) or 10 μg reference tRNA (if RNA mixture such as total tRNA) to a 500 μL microcentrifuge tube, add half total volume of 220 mM ammonium acetate, and take to dryness. Resuspend reference in 5 μL $H_2{}^{18}O$.

2. Add 1 μg sample tRNA (if single tRNA) or 10 μg sample tRNA (if RNA mixture such as total tRNA) to a 500 μL microcentrifuge tube, add half total volume of 220 mM ammonium acetate and take to dryness. Resuspend the sample in 5 μL $H_2{}^{16}O$.
3. Transfer the contents into separate tubes, each containing RNase T1 enzyme (50 U enzyme/μg RNA ratio) (*see* **Note 10**). Incubate for 2 h at 37 °C, take to dryness, and then reconstitute in 10 μL HPLC solvent A. Store at 4 °C until analysis by HPLC-MS/MS
4. The labeling approach (**steps 1** and **2**) can be reversed (reference in ^{16}O and sample in ^{18}O) as needed.

3.6 HPLC-MS/MS Analysis

The LC-MS/MS conditions are described for standard microbore LC connected in-line with a mass spectrometer. Any commercial LC-MS/MS system is compatible with CARD MS, although best results are obtained using linear ion trap, q-TOF, or FT mass spectrometers.

1. The flow rate of the HPLC should be at 30 μL/min.
2. Prepare the LC system by equilibrating the column in 95 % solvent A and 5 % solvent B.
3. The gradient is as follows: start with 5 % B and then increase to 20 % after 5 min, hold at 20 % B for 2 min, and then increase to 50 % B at 0.7 % B/min. Reequilibrate the column at 5 % B for at least 20 min before the next sample injection.
4. Combine 2.5 μL of reference tRNA digest with 2.5 μL sample tRNA digest and inject total amount (5 μL) on column.
5. Instrument appropriate mass spectrometer settings should be used for the negative polarity analysis of RNase T1 digestion products. When possible, data-dependent acquisition of MS/MS spectra should be acquired for confirmation of detected singlets and doublets (as needed) [24].
6. Vendor-specific software can be used for data processing and analysis. Singlets and doublets can be identified manually or using open source software (e.g., MZMine [27]).

4 Notes

1. Observe caution in removing the aqueous phase. It is best to remove only ~95 % of the upper layer so as not to introduce contamination of genomic DNA.
2. For a 1-cm path length, one A_{260} unit is equal to 40 μg/mL RNA. Pure RNA will have an A_{260}/A_{280} ~2.0. Protein or phenol contamination will have a higher A_{260}/A_{280} ratio.
3. It is best to work with fresh tissue as much as possible. However, if immediate extraction is not possible, the tissue can be “flash frozen” with liquid nitrogen and stored at −80 °C.

4. Long RNAs can be eluted from the column using a flush buffer (100 mM Tris–acetate, 15 % ethanol, 1500 mM NaCl). The column can also be reused several times. After the flush buffer, wash the column with distilled water and follow **step 1** again. Store the column in 50 % ethanol at 4 °C.
5. Salt adducts (i.e., Na^+ or K^+) confound mass spectrometry analysis. They can be minimized by re-precipitating the RNA solution with ammonium acetate (NH_4OAc). For every 100 μL volume of the solution, add 33.3 μL of 7.5 M NH_4OAc and 250 μL ethanol. Incubate the sample at −80 °C. Spin the sample in a microcentrifuge for 25 min at 16,000 × *g*. Repeat twice.
6. A 25–30 nucleotide long DNA probe is recommended for high specificity to the target tRNA.
7. Be very cautious in removing the supernatant so as not to disturb the agarose beads. If bead contamination is suspected, spin the sample in a microcentrifuge for 15 min at 16,000 × *g*. The beads will pelletize at the bottom. Carefully remove the supernatant and proceed to **step 13**.
8. The PCR buffer has to be warmed up at 37 °C for at least 5 min. All the other components should be on ice.
9. In some cases, there can be more than one product generated in the PCR or in vitro transcription reaction. If this happens, load the entire PCR reaction to an agarose gel and separate. The full-length product (based on the DNA marker) can be excised from the gel and eluted using commercial kits.
10. Back-exchange is typically not a problem during LC-MS/MS analysis of RNase T1 digestion products. However, if back-exchange is suspected, RNase T1 can be taken to dryness prior to the addition to the ^{18}O–containing solution. This is most easily accomplished by drying the required amount of RNase T1 in a microcentrifuge tube and adding the RNA and $H_2{}^{18}O$ into the tube containing dried RNase T1.

Acknowledgments

Financial support of this work was provided by the National Science Foundation (CHE1507357) and the University of Cincinnati.

References

1. Hopper A (2013) Transfer RNA post-transcriptional processing, turnover, and subcellular dynamics in the yeast. Genetics 194:43–67
2. Agris P (2015) The importance of being modified: an unrealized code to RNA structure and function. RNA 21:552–554
3. Cantara W, Murphy F 4th, Demirci H, Agris P (2013) Expanded use of sense codons is regulated by modified cytidines in tRNA. Proc Natl Acad Sci U S A 110:10964–10969
4. Weixlbaumer A, Murphy F 4th, Dziergowska A, Malkiewicz A, Vendeix F, Agris P, Ramakrishnan V (2007) Mechanism for

expanding the decoding capacity of transfer RNAs by modifications of uridines. Nat Struct Mol Biol 14:498–502

5. Murphy F 4th, Ramakrishnan V, Malkiewicz A, Agris P (2004) The role of modifications in codon discrimination by tRNA(Lys)UUU. Nat Struct Mol Biol 11:1186–1191
6. Helm M, Brulé H, Degoul F, Cepanec C, Leroux J, Giegé R, Florentz C (1998) The presence of modified nucleotide is required for the cloverleaf folding of a human mitochondrial. Nucleic Acids Res 26:1636–1643
7. Pang Y, Abo R, Levine S, Dedon P (2014) Diverse cell stresses induce unique patterns of tRNA up- and down-regulation: tRNA-seq for quantifying changes in tRNA copy number. Nucleic Acids Res 42:e170
8. Dedon PC, Begley TJ (2014) A System of RNA Modifications and Biased Codon Use Controls Cellular Stress Response at the Level of Translation. Chem Res Toxicol 27: 330–337
9. Ofengand J, Del Campo M, Kaya Y (2001) Mapping pseudouridine in RNA molecules. Methods 25:365–373
10. Motorin Y, Muller S, Behm-Ansmant I, Branlant C (2007) Identification of modified residues in RNA by reverse transcription-based methods. Methods Enzymol 425:21–53
11. Benedum-Wohlgamuth J, Rubio M, Paris Z (2009) Thiolation controls cytoplasmic tRNA stability and acts as a negative determinant for tRNA editing in mitochondria. J Biol Chem 284:23947–23953
12. Meyer K, Saletore Y, Zumbo P, Elemento O, Mason C, Jaffrey S (2012) Comprehensive analysis of mRNA methylation reveals enrichment in 3′ UTRs and near stop codons. Cell 149:1635–1646
13. Kowalak JA, Pomerantz SC, Crain PF, McCloskey JA (1993) A novel method for the determination of post-transcriptional modification in RNA by mass spectrometry. Nucleic Acids Res 21:4577–4585
14. Kowalak J, Bruenger E, McCloskey J (1995) Posttranscriptional modification of the central loop of domain V in *Escherichiacoli* 23 S ribosomal RNA. J Biol Chem 270:17758–17764
15. Wetzel C, Limbach P (2012) Global identification of transfer RNAs by liquid chromatography-mass spectrometry (LC-MS). J Proteomics 75:3450–3464
16. Hossain M, Limbach PA (2007) Mass spectrometry-based detection of transfer RNAs by their signature endonuclease digestion products. RNA 13:295–303
17. Li S, Limbach PA (2012) Method for comparative analysis of ribonucleic acids using isotope labeling and mass spectrometry. Anal Chem 84:8607–8613
18. Li S, Limbach PA (2013) Mass spectrometry sequencing of transfer ribonucleic acids by the comparative analysis of RNA digests (CARD) approach. Analyst 138:1386–1394
19. Wetzel C, Li S, Limbach PA (2014) Metabolic De-Isotoping for Improved LC-MS Characterization of Modified RNAs. J Am Soc Mass Spectrom 25:1114–1123
20. Machnicka MA, Milanowska K, Osman Oglou O, Purta E, Kurkowska M, Olchowik A, Januszewski W, Kalinowski S, Dunin-Horkawicz S, Rother KM, Helm M, Bujnicki JM, Grosjean H (2013) MODOMICS: a database of RNA modification pathways--2013 update. Nucleic Acids Res 41:D262–D267
21. Jühling F, Mörl M, Hartmann RK, Sprinzl M, Stadler PF, Pütz J (2009) tRNAdb 2009: compilation of tRNA sequences and tRNA genes. Nucleic Acids Res 37:D159–D162
22. Puri P, Wetzel C, Saffert P, Gaston KW, Russell SP, Varela JAC, van der Vlies P, Zhang G, Limbach PA, Ignatova Z, Poolman B (2014) Systematic identification of tRNAome and its dynamics in Lactococcus lactis. Mol Microbiol 93:944–956
23. Chan P, Lowe T (2009) GtRNAdb: A database of transfer RNA genes detected in genomic sequence. Nucleic Acids Res 37:D93–D97
24. Li S, Limbach PA (2015) Identification of RNA sequence isomer by isotope labeling and LC–MS/MS. J Mass Spectrom 49:1191–1198
25. Constantopoulos T, Jackson G, Enke C (1999) Effects of salt concentration on analyte response using electrospray ionization mass spectrometry. J Am Soc Mass Spectrom 10:625–634
26. Sambrook J, Fritch E, Maniatis T (1989) Molecular cloning: a laboratory manual, 2nd edn. Cold Spring Harbor Laboratory Press, Cold Spring Harbor, NY
27. Pluskal T, Castillo S, Villar-Briones A, Oresic M (2010) MZmine 2: modular framework for processing, visualizing, and analyzing mass spectrometry-based molecular profile data. BMC Bioinformatics 11:395

Chapter 3

Liquid Chromatography-Mass Spectrometry for Analysis of RNA Adenosine Methylation

Bi-Feng Yuan

Abstract

Dynamic RNA modifications recently were considered to constitute another realm for biological regulation in the form of "RNA epigenetics". *N*^{6}-methyladenosine (m^{6}A), one of the most important modifications on RNA, plays a fundamental role in epigenetic regulation of the mammalian transcriptome. We recently established various liquid chromatography-electrospray ionization-mass spectrometry (LC-ESI-MS/MS)-based methods for the sensitive and accurate determination of modified nucleosides in both DNA and RNA. Here, we describe a protocol to analyze m^{6}A in RNA by LC-ESI-MS/MS. And this protocol also can be extended to the analysis of other modified nucleosides in both DNA and RNA.

Key words RNA Methylation, *N*^{6}-Methyladenosine; Liquid Chromatography-Mass Spectrometry

1 Introduction

Given the rich layers of epigenetic regulation resulting from modifications of DNA and histones, reversible RNA modification has been proposed to represent another realm for biological regulation in the form of "RNA epigenetics" [1]. Naturally occurring RNA molecules contain various chemically modified nucleosides [2]. Up to date, more than 140 structurally distinct posttranscriptionally modified nucleosides have been identified in RNA molecules in all three kingdoms of life [2], among which *N*^{6}-methyladenosine (m^{6}A) is one of the most important modifications in RNA [3–5].

Several studies demonstrate that m^{6}A modification in RNA plays broad and critical roles in fundamental biological processes, including spermatogenesis, development, carcinogenesis, circadian rhythm, and stem cell renewal [6–10]. m^{6}A modification has been found in about 7000 mRNAs in both human and mouse cells with possible functions on the control and regulation of gene transcription and protein translation [11, 12]. The specific distribution of

Alexandra Lusser (ed.), *RNA Methylation: Methods and Protocols*, Methods in Molecular Biology, vol. 1562,
DOI 10.1007/978-1-4939-6807-7_3,

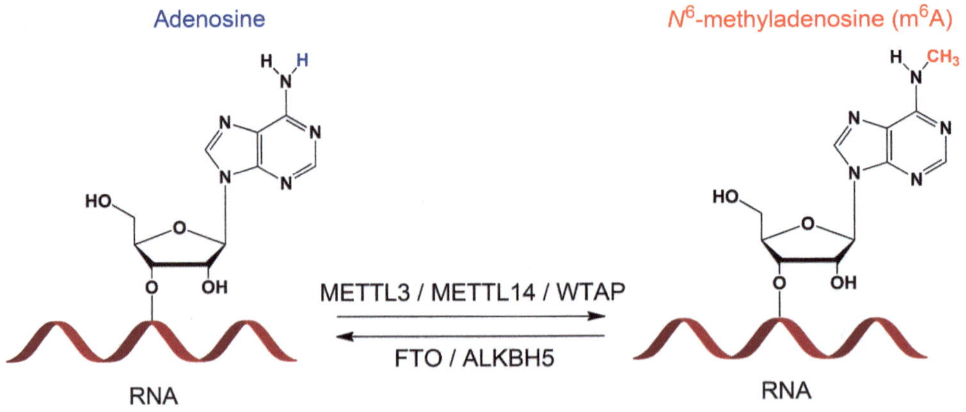

Fig. 1 Methylation of adenosine to m^6A by METTL3/METTL14/WTAP complex and demethylation of m^6A to adenosine by FTO or ALKBH5 in RNA

m^6A in different cells as well as in different cell cycles indicates that m^6A modification is highly regulated, which raises m^6A modification to a new layer of biological regulation at RNA level.

Previous studies showed that m^6A modification can be formed by a complex of the proteins METTL3/METTL14/WTAP [13] and m^6A can be demethylated by obesity-associated protein (FTO) [14] or ALKBH5 [15] (Fig. 1). In recent years, we established various liquid chromatography-electrospray ionization-mass spectrometry (LC-ESI-MS/MS)-based methods for the sensitive and accurate determination of modified nucleosides in both DNA and RNA [16–29]. In this chapter, we describe a routine procedure to analyze m^6A in RNA by LC-ESI-MS/MS established in our group. And this procedure can be extended to the analysis of other modified nucleosides in both DNA and RNA.

2 Materials

Prepare all solutions using Milli-Q water (Milli-Q apparatus, Millipore) and analytical grade reagents.

2.1 RNA Digestion

1. S1 nuclease: 180 units/μL.
2. Alkaline phosphatase (CIAP): 30 units/μL.
3. Venom phosphodiesterase I: 0.002 units/μL.

2.2 Nucleoside Standards

1. Cytidine (C): 5 mmol/L in water.
2. Guanosine (G): 5 mmol/L in water.
3. Denosine (A): 5 mmol/L in water.
4. Uridine (U): 5 mmol/L in water.
5. 2′-deoxycytidine (dC): 5 mmol/L in water.
6. 2′-deoxyguanosine (dG): 5 mmol/L in water.

7. 2′-deoxyadenosine (dA): 5 mmol/L in water.
8. Thymidine (T): 5 mmol/L in water.
9. N^6-methyladenosine (m^6A): 5 mmol/L in water.

2.3 Oligonucleotides

1. RNA standard 1 (10-mer RNA): 5′-AUCUAUAUGC-3′ (prepare 1 pmol/μL in water).
2. RNA standard 2 (15-mer RNA): 5′-UAAUAC(m^6A) GAGAAAUC-3′ (prepare 1 pmol/μL in water).
3. Quality control (QC) samples are prepared using different amounts of RNA standard 1 and RNA standard 2 (Table 1).

2.4 Chemicals and Reagents

1. Chloroform.
2. Formic acid.
3. Chromatographic grade methanol.
4. Solvent A: formic acid in water (0.1 %, v/v).
5. Solvent B: formic acid in methanol (0.1 %, v/v).
6. 10× S1 nuclease buffer: 300 mM CH_3COONa, pH 4.6, 2800 mM NaCl, 10 mM $ZnSO_4$.
7. 10× Alkaline phosphatase buffer: 500 mM Tris–HCl, 10 mM $MgCl_2$, pH 9.0.
8. Blood total RNA isolation kit.
9. Collect 250 μL of human whole blood samples from individuals according to standard clinical procedure.

2.5 Equipments

1. AB 3200 QTRAP mass spectrometer (Applied Biosystems) with an electrospray ionization source (Turbo Ionspray).
2. Shimadzu LC-20 AD HPLC with two LC-20 AD pumps, a SIL-20A autosampler, a CTO-20 AC thermostated column compartment, and a DGU-20A3 degasser.
3. B-500 spectrophotometer (Metash Instruments Co.).
4. Hisep C18-T column (150 mm×2.1 mm i.d., 5 μm, Weltech Co.).

Table 1
Amounts of the RNA standards used for the preparation of QC samples (reproduced from (20) with permission from Endocrine Society)

	Molar ratio (%)	
Methylation level m^6A/A (%)	RNA standard 1	RNA Standard 2
0.100	99.7(99.7 pmol)	0.3(0.3 pmol)
0.200	99.4(99.4 pmol)	0.6(0.6 pmol)
0.300	99.1(99.1 pmol)	0.9(0.9 pmol)

3 Methods

3.1 RNA Extraction

1. Total RNA is isolated from 200 μL of whole blood of human using the commercially available blood total RNA isolation kit according to the manufacturer's recommended protocol (*see* **Note 1**). Purified RNA should be dissolved in H_2O (*see* **Note 2**).
2. Determine the concentrations of the purified RNA using a B-500 spectrophotometer. The purity and yield of RNA is determined measuring absorbance at 260 and 280 nm.

3.2 Enzymatic Digestion of RNA

1. Denature specific amounts (20 ng–1 μg) of RNA (dissolved in 16 μL of H_2O) by heating at 95 °C for 5 min and then chilling on ice for 2 min (*see* **Note 3**).
2. Add 2 μL of 10× S1 nuclease buffer, 1 μL of S1 nuclease, and 1 μL of H_2O. Incubate the mixture (20 μL) at 37 °C for 4 h.
3. To the resulting solution add 10 μL of 10× alkaline phosphatase buffer, 5 μL of venom phosphodiesterase I, 1 μL of alkaline phosphatase, and 64 μL of H_2O.
4. Continue the incubation at 37 °C for an additional 1 h and subsequently add 100 μL H_2O (*see* **Note 4**). So the total volume of the mixture is 200 μL.
5. Extract the reaction mixture with 200 μL of chloroform twice (*see* **Note 5**).
6. Collect the resulting aqueous layer and lyophilize to dryness.
7. Redissolve the obtained digested nucleosides in 100 μL Milli-Q water for subsequent LC-ESI-MS/MS analysis. The sample can be analyzed immediately or stored at −20 °C until analysis.

3.3 Analysis of Nucleosides by LC-ESI-MS/ MS

1. For analysis of nucleosides by LC-ESI-MS/MS, we use a system consisting of an AB 3200 QTRAP mass spectrometer coupled with a Shimadzu LC-20 AD HPLC.
2. The HPLC separation of nucleosides is performed on a Hisep C18-T column with a flow rate of 0.2 mL/min at 35 °C (*see* **Note 6**).
3. Use solvent A and solvent B as mobile phases. Use a gradient of 5 min 5 % solvent B, 10 min 5–30 % solvent B, 5 min 30–50 % solvent B, 3 min 50 % -5 % solvent B and 17 min 5 % solvent B for the separation (*see* **Notes** 7 and **8**).
4. The mass spectrometry detection is performed under positive electrospray ionization mode.
5. Monitor the nucleosides by multiple reaction monitoring (MRM) mode using the mass transitions (precursor ions → product ions) of dC (228.1→ 112.1), T (243.1→ 127.1), dA (252.1→ 136.1), dG (268.1→ 152.1), C (244.1→ 112.1), U (245.1→ 113.0), A (268.1→ 136.1), G (284.1→ 152.1), m^6A (282.1→150.1).

Table 2
The MRM transitions and optimal parameters for the analysis of the nucleosides by LC-ESI-MS/MS

Analytes	Precursor ion	Product ion	DP (V)	EP (V)	CEP (V)	CE (V)	CXP (V)
A	268.1	136.1	15.0	5.0	16.6	23.0	2.0
C	244.1	112.1	20.0	8.0	10.0	21.3	2.4
G	284.1	152.1	40.0	5.0	9.0	23.0	6.0
U	245.1	113.0	24.7	6.0	13.0	13.7	3.8
dA	252.1	136.1	15.0	5.0	15.0	23.0	2.0
dC	228.1	112.1	11.0	5.0	15.0	23.0	3.0
dG	268.1	152.1	15.0	5.0	20.0	23.0	5.0
T	243.1	127.1	25.0	5.0	15.0	23.0	5.0
m^6A	282.1	150.1	32.0	8.0	10.0	27.0	2.6

6. The MRM parameters of all nucleosides are optimized to achieve maximal detection sensitivity. The optimized mass spectrometry conditions are listed in Table 2 (*see* **Note 9**).
7. The m^6A content is calculated using the following expression:

$$m^6A\% = \frac{M_{m^6A}}{M_A} \times 100\%$$

where M_{m6A} is the molar quantity of m^6A, and M_A is the molar quantity of adenosine determined in the RNA sample.
8. For data acquisition and processing, the AB SCIEX Analyst 1.5 Software can be applied.

3.4 Construction of Calibration Curve

1. Construct calibration curves by plotting the mean peak area ratio of m^6A/A versus the mean molar ratio of m^6A/A (ranging from 0.05 to 5 % of m^6A/A) based on data obtained from triplicate measurements. 40 pmol A standard supplemented with m^6A at different amounts ranging from 20 to 2000 fmol is used (Table 3).
2. Calculate the limit of detection (LOD) and the limit of quantification (LOQ) for m^6A as the amounts of the analyte at Signal/Noise ratios of 3 and 10, respectively. The LOD and LOQ are therefore determined to be 1.2 and 4.0 fmol for m^6A, respectively (Table 3).

3.5 Validation of the Analytical Method

1. Validation of the method is performed using the prepared QC samples by comparing the measured m^6A contents to the theoretical m^6A contents (Table 1).

Table 3
Linearity, LOD, and LOQ of m^6A by LC-ESI-MS/MS (reproduced from (20) with permission from Endocrine Society)

Analyte	Linear range (m^6A/A, %)	Regression line: Slope	Regression line: Intercept	R^2	LOD (fmol)	LOQ (fmol)
m^6A	0.05–5	0.0322 ±0.0003	0.0008 ±0.0001	0.9982	1.2	4.0

Table 4
Evaluation of the accuracy (by REs) and reproducibility (by RSDs) for the detection of m^6A (reproduced from (20) with permission from Endocrine Society)

									QC samples		
Nominal $[m^6A]/[A]\%$		0.050	0.100	0.200	0.500	1.000	2.000	5.000	0.100	0.200	0.300
Day 1 n=3	Measured mean $[m^6A]/[A]\%$	0.054	0.099	0.187	0.485	1.044	2.041	4.955	0.091	0.217	0.287
	RSD [a] (%)	7.8	3.7	2.4	7.7	4.7	1.2	1.3	3.7	4.7	0.4
	RE [b] (%)	8.0	−1.0	−6.5	-3.0	4.4	2.1	−0.9	−9.0	8.5	−4.3
Day 2 n=3	Measured mean $[m^6A]/[A]\%$	0.044	0.092	0.180	0.454	0.921	2.131	5.478	0.089	0.200	0.296
	RSD (%)	11.2	1.8	4.6	3.7	7.6	3.4	4.2	4.8	2.7	2.4
	RE (%)	−12.0	-8.0	−10.0	−9.2	−7.9	6.6	9.6	−11.0	0.0	−1.3
Day 3 n=3	Measured mean $[m^6A]/[A]\%$	0.047	0.103	0.219	0.545	1.094	1.985	4.668	0.106	0.191	0.282
	RSD (%)	6.7	3.1	4.0	3.3	5.8	2.2	1.6	7.1	0.8	2.7
	RE (%)	−6.0	3.0	9.5	9.0	9.4	−0.7	−6.6	6.0	−4.5	-6.0

[a]Relative standard deviation
[b]Relative error. *QC* quality control

2. The measured contents of m^6A by LC-ESI-MS/MS are calculated using the calibration curve (Table 3).
3. The accuracy of the method is evaluated by relative errors (*REs*) using the following expression:

$$RE\% = \frac{\text{Measured content} - \text{Theoretical content}}{\text{Theoretical content}} \times 100\%$$

4. Determine the reproducibility of the method by measuring intra- and inter-day precision. Calculate the intra- and inter-day relative standard deviations (RSDs) using QC samples. Three parallel analyses over a day give the intra-day RSDs, and the inter-day RSDs are determined by analysis of QC samples independently for three consecutive days.
5. Less than 15.0 % of RSD and RE are normally acceptable for quantification of m^6A by developed method. Shown in Table 4 is the typical validation data for the LC-ESI-MS/MS method.

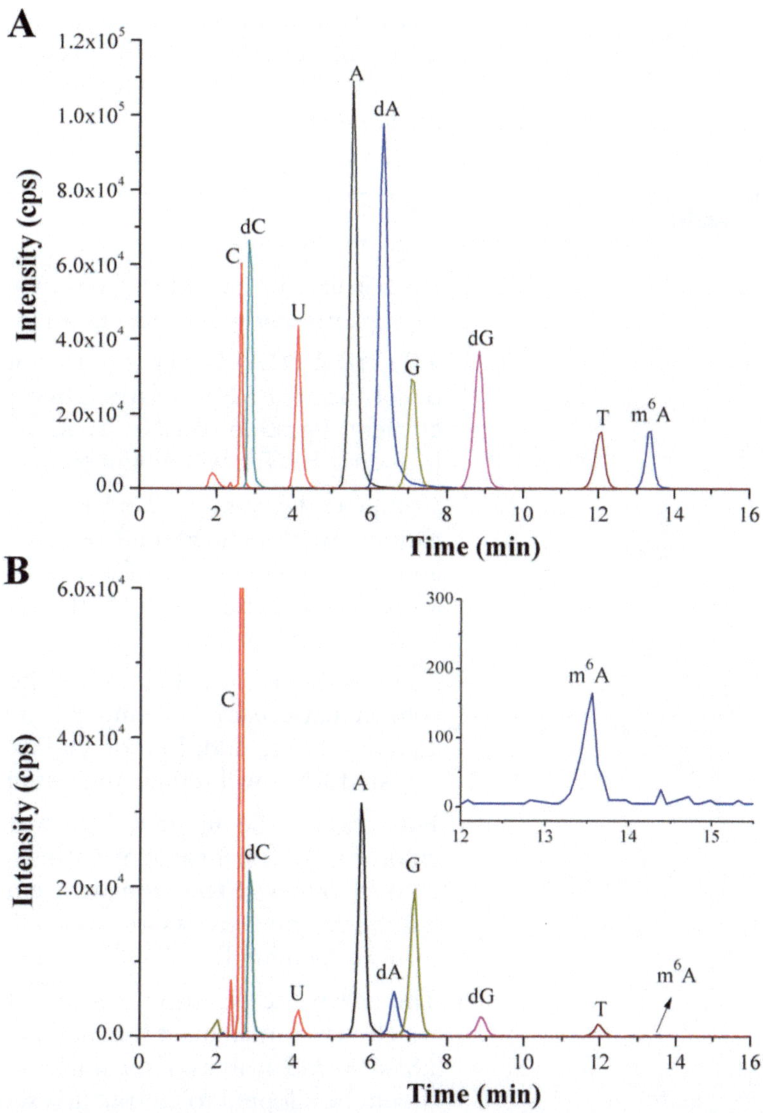

Fig. 2 The extracted ion chromatograms of nucleosides. (**a**) Nucleoside standards. (**b**) 20 ng RNA from a human whole blood sample. Shown in *inset* is the enlargement chromatogram of m^{6}A. Experimental conditions: separation column, Hisep C18-T column; temperature, 35 °C; flow rate, 0.2 mL/min; gradient elution, 5 min 5 % solvent B, 10 min 5–30 % solvent B, 5 min 30–50 % solvent B, 3 min 50 %–5 % solvent B and 17 min 5 % solvent B (reproduced from (20) with permission from Endocrine Society)

3.6 Analysis of m^{6}A Contents in RNA from Human Whole Blood Cells

1. With the established LC-ESI-MS/MS method, the m^{6}A contents in RNA from whole blood cells of type 2 diabetes mellitus (T2DM) patients can be analyzed. Figure 2a shows the typical extracted ion chromatograms of 9 nucleoside standards and Fig. 2b shows the typical extracted ion chromatograms of nucleosides from the enzymatic digestion product of 20 ng

RNA from a T2DM patient. The accurate quantification of m^6A is not interfered by the presence of high contents of normal nucleosides since m^6A can be well separated from other nucleosides.

4 Notes

1. RNA isolated from other types of samples (such as animal tissues, plants, and microbes) also can be used in this method.
2. TE (Tris–EDTA) buffer is not an appropriate buffer for RNA dissolving after RNA isolation. Because Zn^{2+} is necessary for the efficient digestion of RNA by S1 nuclease, chelating reagents, including EDTA, will inhibit the activity of S1 nuclease.
3. Complete digestion of RNA is necessary for accurate quantification of the modified nucleosides. Heating at 95 °C for 5 min and then chilling on ice for 2 min can efficiently destroy the secondary structure of RNA, thus facilitating the full digestion of RNA by S1 nuclease.
4. As the subsequent extraction of the digested RNA with chloroform may cause some sample loss while recovering the upper aqueous layer, addition of certain volume of H_2O to the digested RNA will reduce the sample loss.
5. Extraction of the digested RNA with equal volume of phenol/chloroform/isopentanol may cause the relatively low recovery of nucleosides. Extraction with equal volume of chloroform has better nucleosides recovery than using phenol/chloroform/isopentanol.
6. The enzymatic digestion will introduce salts into the samples, which may contaminate the mass spectrometer. So, during LC-ESI-MS/MS analysis of the nucleosides, a valve-switch program should be adopted to cut the first several minutes of the elution to waste since salts normally are eluted out in the dead time.
7. The retention of nucleosides on reversed-phase column of C18 is relatively weak, so methanol instead of acetonitrile is preferred to be used as mobile phase since acetonitrile is a stronger eluent than methanol in reversed phase chromatography.
8. Addition of formic acid into mobile phase will decrease the formation of metal ion adducts, especially for Na^+ adducts, which therefore can increase the detection sensitivities of nucleosides.
9. Generally, it is not necessary to optimize the MRM parameters every time for the analysis unless the mass spectrometer needs calibration or the detection sensitivities of nucleosides obviously change.

Acknowledgments

The author thanks the financial support from the National Natural Science Foundation of China (21522507, 21672166).

References

1. He C (2010) Grand challenge commentary: RNA epigenetics? Nat Chem Biol 6:863–865
2. Machnicka MA, Milanowska K, Osman Oglou O, Purta E, Kurkowska M, Olchowik A, Januszewski W, Kalinowski S, Dunin-Horkawicz S, Rother KM, Helm M, Bujnicki JM, Grosjean H (2013) MODOMICS: a database of RNA modification pathways – 2013 update. Nucleic Acids Res 41:D262–D267
3. Chandola U, Das R, Panda B (2015) Role of the N6-methyladenosine RNA mark in gene regulation and its implications on development and disease. Brief Funct Genomics 14:169–179
4. Hastings MH (2013) m(6)A mRNA methylation: a new circadian pacesetter. Cell 155: 740–741
5. Wang X, Zhao BS, Roundtree IA, Lu Z, Han D, Ma H, Weng X, Chen K, Shi H, He C (2015) N(6)-methyladenosine Modulates Messenger RNA Translation Efficiency. Cell 161:1388–1399
6. Jia G, Fu Y, He C (2013) Reversible RNA adenosine methylation in biological regulation. Trends Genet 29:108–115
7. Niu Y, Zhao X, Wu YS, Li MM, Wang XJ, Yang YG (2013) N6-methyl-adenosine (m6A) in RNA: an old modification with a novel epigenetic function. Genomics Proteomics Bioinformatics 11:8–17
8. Fustin JM, Doi M, Yamaguchi Y, Hida H, Nishimura S, Yoshida M, Isagawa T, Morioka MS, Kakeya H, Manabe I, Okamura H (2013) RNA-methylation-dependent RNA processing controls the speed of the circadian clock. Cell 155:793–806
9. Wang X, Lu Z, Gomez A, Hon GC, Yue Y, Han D, Fu Y, Parisien M, Dai Q, Jia G, Ren B, Pan T, He C (2014) N6-methyladenosine-dependent regulation of messenger RNA stability. Nature 505:117–120
10. Schwartz S, Agarwala SD, Mumbach MR, Jovanovic M, Mertins P, Shishkin A, Tabach Y, Mikkelsen TS, Satija R, Ruvkun G, Carr SA, Lander ES, Fink GR, Regev A (2013) High-resolution mapping reveals a conserved, widespread, dynamic mRNA methylation program in yeast meiosis. Cell 155:1409–1421
11. Meyer KD, Saletore Y, Zumbo P, Elemento O, Mason CE, Jaffrey SR (2012) Comprehensive analysis of mRNA methylation reveals enrichment in 3' UTRs and near stop codons. Cell 149:1635–1646
12. Dominissini D, Moshitch-Moshkovitz S, Schwartz S, Salmon-Divon M, Ungar L, Osenberg S, Cesarkas K, Jacob-Hirsch J, Amariglio N, Kupiec M, Sorek R, Rechavi G (2012) Topology of the human and mouse m6A RNA methylomes revealed by m6A-seq. Nature 485:201–206
13. Liu J, Yue Y, Han D, Wang X, Fu Y, Zhang L, Jia G, Yu M, Lu Z, Deng X, Dai Q, Chen W, He C (2014) A METTL3-METTL14 complex mediates mammalian nuclear RNA N(6)-adenosine methylation. Nat Chem Biol 10:93–95
14. Jia G, Fu Y, Zhao X, Dai Q, Zheng G, Yang Y, Yi C, Lindahl T, Pan T, Yang YG, He C (2011) N6-methyladenosine in nuclear RNA is a major substrate of the obesity-associated FTO. Nat Chem Biol 7:885–887
15. Zheng G, Dahl JA, Niu Y, Fedorcsak P, Huang CM, Li CJ, Vagbo CB, Shi Y, Wang WL, Song SH, Lu Z, Bosmans RP, Dai Q, Hao YJ, Yang X, Zhao WM, Tong WM, Wang XJ, Bogdan F, Furu K, Fu Y, Jia G, Zhao X, Liu J, Krokan HE, Klungland A, Yang YG, He C (2013) ALKBH5 is a mammalian RNA demethylase that impacts RNA metabolism and mouse fertility. Mol Cell 49:18–29
16. Huang W, Qi CB, Lv SW, Xie M, Feng YQ, Huang WH, Yuan BF (2016) Determination of DNA and RNA methylation in circulating tumor cells by mass spectrometry. Anal Chem. 88:1378–1384.
17. Zhang HY, Xiong J, Qi BL, Feng YQ, Yuan BF (2015) The existence of 5-hydroxymethylcytosine and 5-formylcytosine in both DNA and RNA in mammals. Chem Commun (Camb) 52:737–740
18. Xiong J, Jiang HP, Peng CY, Deng QY, Lan MD, Zeng H, Zheng F, Feng YQ, Yuan BF (2015) DNA hydroxymethylation age of human blood determined by capillary hydrophilic-interaction liquid chromatography/mass spectrometry. Clin Epigenetics 7:72

19. Tang Y, Zheng SJ, Qi CB, Feng YQ, Yuan BF (2015) Sensitive and simultaneous determination of 5-methylcytosine and its oxidation products in genomic DNA by chemical derivatization coupled with liquid chromatography-tandem mass spectrometry analysis. Anal Chem 87:3445–3452
20. Shen F, Huang W, Huang JT, Xiong J, Yang Y, Wu K, Jia GF, Chen J, Feng YQ, Yuan BF, Liu SM (2015) Decreased N(6)-methyladenosine in peripheral blood RNA from diabetic patients is associated with FTO expression rather than ALKBH5. J Clin Endocrinol Metab 100: E148–E154
21. Jiang HP, Qi CB, Chu JM, Yuan BF, Feng YQ (2015) Profiling of cis-Diol-containing nucleosides and ribosylated metabolites by boronate-affinity organic-silica hybrid monolithic capillary liquid chromatography/mass spectrometry. Sci Rep 5:7785
22. Chu JM, Qi CB, Huang YQ, Jiang HP, Hao YH, Yuan BF, Feng YQ (2015) Metal oxide-based selective enrichment combined with stable isotope labeling-mass spectrometry analysis for profiling of ribose conjugates. Anal Chem 87:7364–7372
23. Yuan BF, Feng YQ (2014) Recent advances in the analysis of 5-methylcytosine and its oxidation products. Trend Anal Chem 54:24–35
24. Yuan BF (2014) 5-Methylcytosine and its derivatives. Adv Clin Chem 67:151–187
25. Tang Y, Xiong J, Jiang HP, Zheng SJ, Feng YQ, Yuan BF (2014) Determination of oxidation products of 5-methylcytosine in plants by chemical derivatization coupled with liquid chromatography/tandem mass spectrometry analysis. Anal Chem 86: 7764–7772
26. Wang ST, Huang W, Lu W, Yuan BF, Feng YQ (2013) TiO2-based solid phase extraction strategy for highly effective elimination of normal ribonucleosides before detection of 2'-deoxynucleosides/low-abundance 2'-O-modified ribonucleosides. Anal Chem 85:10512–10518
27. Tang Y, Chu JM, Huang W, Xiong J, Xing XW, Zhou X, Feng YQ, Yuan BF (2013) Hydrophilic material for the selective enrichment of 5-hydroxymethylcytosine and its liquid chromatography-tandem mass spectrometry detection. Anal Chem 85:6129–6135
28. Chen ML, Shen F, Huang W, Qi JH, Wang Y, Feng YQ, Liu SM, Yuan BF (2013) Quantification of 5-methylcytosine and 5-hydroxymethylcytosine in genomic DNA from hepatocellular carcinoma tissues by capillary hydrophilic-interaction liquid chromatography/quadrupole TOF mass spectrometry. Clin Chem 59:824–832
29. Chen ML, Liu YL, Xing XW, Zhou X, Feng YQ, Yuan BF (2013) Preparation of a hyper-cross-linked polymer monolithic column and its application to the sensitive determination of genomic DNA methylation. ChemEur J 19:1035–1041

Part II

Analysis of N6-Methyladenosine

Chapter 4

Genome-Wide Location Analyses of N^6-Methyladenosine Modifications (m^6A-Seq)

Benoit Molinie and Cosmas C. Giallourakis

Abstract

N^6-methyladenosine–sequencing (m^6A-seq) is a critical tool to obtain an unbiased genome-wide picture of m^6A sites of modification at high resolution. It allows the study of the impact of various perturbations on m^6A modification distribution and the study of m^6A functions. Herein, we describe the m^6A-seq protocol, which entails RNA immunoprecipitation (RIP) performed on fragmented poly(A) RNA utilizing anti-m^6A antibodies. The captured/enriched m^6A positive RNA fragments are subsequently sequenced by RNA-seq in parallel with background control non-immunoprecipitated input RNA fragments. Analyses reveal peaks of m^6A enrichment containing sites of modifications analogous to chromatin modification immunoprecipitation experiments.

Key words m^6A-Seq, Epitranscriptome, N^6-Methyladenosine, Genome-wide, METTL3, METTL14

1 Introduction

The emergence of observations that N^6-methyl-adenosine (m^6A) is a widespread reversible RNA chemical modification with proteins acting as "writers," "readers," and "erasers" of m^6A has led to a new field coined "epitranscriptomics." Although m^6A modifications have been recognized from yeast to humans since the 1970s, many aspects of m^6A modification, including function(s), are only beginning to be understood [1]. For example, the m^6A biogenesis machinery appears to be critical for developmental cell fate decisions from yeast to humans [2]. The development of m^6A "location analyses" called m^6A-seq has been a critical step in the field. In this protocol, RNA immunoprecipitation (RIP) is performed on fragmented poly(A) RNA utilizing anti-m^6A antibodies followed by RNA-seq of the captured/enriched m^6A positive RNA fragments with non-immunoprecipitated RNA fragments serving as input or background control. Utilizing m^6A-seq, studies have revealed that m^6A modification(s) sites occur on thousands of mRNAs and hundreds of noncoding RNAs in mouse and human cells [3, 4]. Topologically along the

Alexandra Lusser (ed.), *RNA Methylation: Methods and Protocols*, Methods in Molecular Biology, vol. 1562,
DOI 10.1007/978-1-4939-6807-7_4, © Springer Science+Business Media LLC 2017

transcriptome, m^{6}A modifications exhibit enrichment near 3′ end of both coding (often near stop codon or 3′UTR) and noncoding RNAs as well as long internal exons. Subsets of m^{6}A sites on RNAs appear to exhibit tissue and stimuli specificity, suggesting a regulated and dynamic m^{6}A epitranscriptome [5]. Indeed, it has been shown, for example, that under conditions of stress such as ultraviolet radiation and heatshock m^{6}A sites accumulate in the 5′UTR of genes to promote 5′ cap independent translation [6]. The "writing" of m^{6}A RNA modification is accomplished via an m^{6}A methyltransferase complex, including two known catalytic mammalian components encoded by METTL3 and METTL14 [7]. Additional critical components required for m^{6}A methyltranferase complex activity include WTAP and VIRILIZER although their exact functions and mechanism of action are unclear [8, 9]. m^{6}A modification(s) are reversible, based on the discovery that the protein encoded by fat mass and obesity gene, FTO, and a related protein ALKBH5, operate as m^{6}A demethylases or "erasers" [10, 11]. At the molecular level, m^{6}A has been implicated in many aspects of RNA metabolism through the binding of so-called YTH domain reader proteins including splicing, translational efficiency, cap-independent translation, RNA export, and RNA structure [12].

m^{6}A modifications location analyses by m^{6}A-seq are a critical assay to probe the functions of m^{6}A in normal physiology and in pathophysiological conditions. The basic schema of m^{6}A-seq is shown in Fig. 1 and allows for 50 nt resolution of sites of modifications. In short, poly(A) RNA is fragmented and subjected to anti-m^{6}A RNA immunoprecipitation (RIP) utilizing commercially

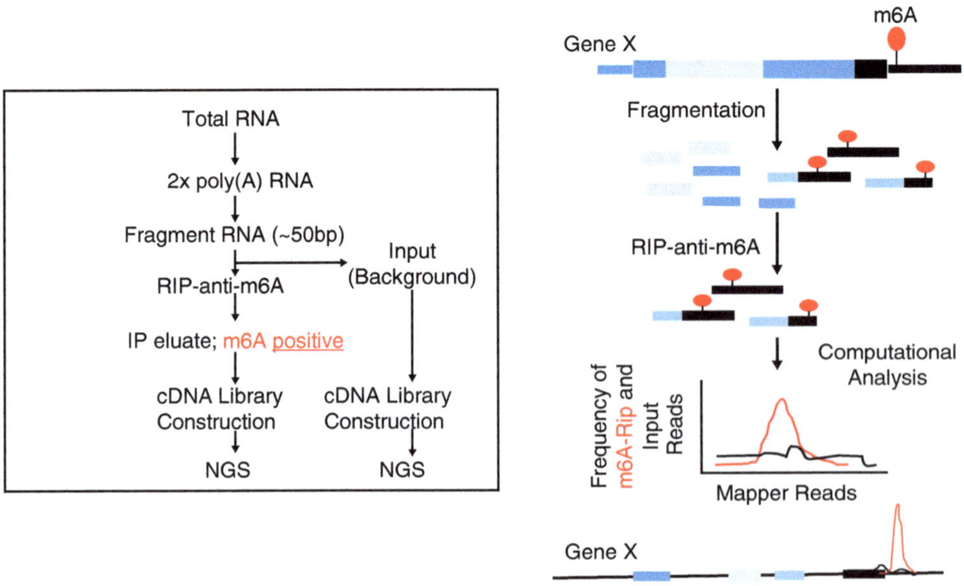

Fig. 1 Work flow and schematic diagram of m^{6}A-seq protocol

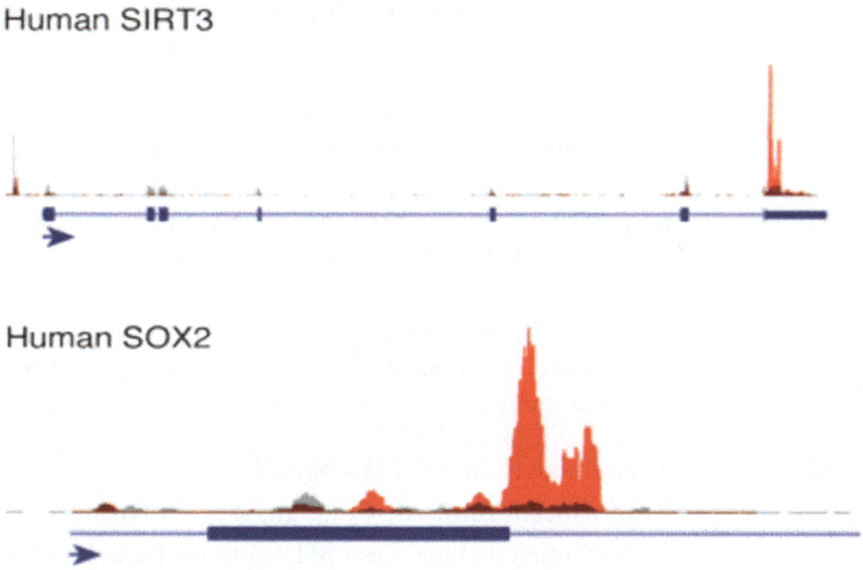

Fig. 2 m6A-seq example data tracks. UCSC genome browser tracks of m6A-seq position analyses for two genes of our previously published H1-ESC data (*red* for m6A-RIP and *gray* for input)

available anti-m6A antibodies. The m6A positive enriched fragments are subsequently sequenced and compared to the distribution of input/non-immunoprecipitated fragmented RNA as a background control. Peaks of m6A enrichment over background/input are then computationally identified utilizing either custom computer scripts or programs such as HOMER [13]. An example of the results of the m6A-seq utilizing sequence alignments to the UCSC genome browser is shown in Fig. 2.

2 Materials

Prepare all solutions using ultrapure autoclaved distilled water and analytical grade reagents that are RNase and DNase free. Use RNase/DNase-free consumables and glassware. All pipetting is performed using sterile RNase/DNase-free low retention filtered tips. Prior to be used, every stock solution, buffer solution, antibody batch, Dynabeads, and chemicals are independently tested on total RNA extract. The RNA is then run on a BioAnalyzer to control for potential degradation due to RNase contamination.

2.1 Buffers

1. Fragmentation buffer: 10 mM $ZnCl_2$, 10 mM Tris–HCl pH 7.0.
2. Stop Buffer: 0.5 M EDTA.
3. m6A Binding Buffer: 50 mM Tris–HCl pH 7.4, 150 mM $NaCl_2$, 1 % NP-40, 2 mM EDTA; Add RNAse inhibitor at the manufacturer's recommended concentration.

4. Low Salt Buffer: 0.2×SSPE, 0.001 M EDTA, 0.05 % Tween-20. Add RNAse inhibitor at the manufacturer's recommended concentration.
5. High Salt Buffer: 0.2×SSPE, 0.001 M EDTA, 0.05 % Tween-20,137.5 mM NaCl; Add RNAse inhibitor at the manufacturer's recommended concentration.
6. TET: 10 mM Tris–HCl pH 8.0, 1 mM EDTA pH 8.0, 0.05 % Tween-20; Add RNAse inhibitor at the manufacturer's recommended concentration.
7. Elution Buffer: 0.02 M DTT, 0.150 M NaCl, 0.05 M Tris–HCl pH 7.5, 0.001 M EDTA, 0.10 % SDS; Add RNAse inhibitor at the manufacturer's recommended concentration.

2.2 M6A RIP-Seq

1. Antibody: Anti-m^{6}A (*see* **Note 1**).
2. Dynabeads Antibody Coupling Kit: (*see* **Note 2**).
3. Acid-phenol:chloroform pH 4.5 (with IAA, 25:24:1).
4. Chloroform.
5. Absolute Ethanol (200 proof).
6. SUPERase-in RNAse inhibitor (Ambion).
7. Ultra Pure Glycogen: 20 mg/mL.
8. Magnets: (*see* **Note 3**).
9. mRNA Purification Kit: Ambion Dynabeads mRNA Purification Kit polyA isolation kit.
10. Kapa library quantification kit: (*see* **Note 4**).

3 Methods

3.1 M6A Rip

Carry out all procedures at room temperature unless otherwise specified. All procedures should be done in a clean RNAse-free environment.

1. Each biological replicate for m^{6}A-seq starts by using 400 μg of total RNA yielding approximately 10 μg of double poly(A) selected RNA (*see* **Note 5**).
2. Resuspend each sample obtained from **step 1** (10 μg of poly(A) RNA) in 50 μL of UltraPure H_2O.
3. Add 250 μL of fragmentation buffer to the 50 μL of isolated 2× poly(A) RNA to a final volume of 300 μL.

3.2 Fragmentation Step of the RIP

1. Proceed to the fragmentation of the 300 μL Poly(A) RNA solution at 94 °C for exactly 5 min using a thermocycler (*see* **Notes 6** and 7).
2. Stop the fragmentation reaction by adding 50 μL of Stop Buffer to a final volume of 350 μL and immediately put on ice.

3.3 RNA Binding to m^6A-Dynabeads

1. Add 150 μL of pre-equilibrated m^6A-Dynabeads (*see* **Notes 2** and **8**) to the 350 μL of fragmented RNA from **step2** of Subheading 3.2 to a final volume of 500 μL.
2. Allow the fragmented RNA to bind to the m^6A-Dynabeads at room temperature while rotating (tail-over-head) at 7 rotations per minute for 1 h.
3. Place the tubes containing the samples on a magnet allowing the bead complexes to cluster until the solution becomes clear (*see* **Note 3**).
4. Discard the 500 μL liquid phase or supernatant as this fraction represents the m^6A negative fragments not captured by the anti-m^6A antibody.

3.4 Washing of m6A-Dynabeads

The m^6A positive fragments that are retained on the surface of the m^6A-coupled Dynabeads are then subjected to a series of wash steps.

1. Resuspend m^6A-Dynabeads-RNA complexes in 500 μL of m^6A Binding Buffer, incubate for 3 min at room temperature, and remove clear supernatant after placing the beads in the magnet.
2. Repeat **step 1** with 500 μL of Low Salt Buffer.
3. Repeat **step 1** with 500 μL of High Salt Buffer. Do not exceed 3-min incubation time for this step to prevent release of the RNA from the beads.
4. Repeat **step 1** twice with 500 μL of TET buffer.

3.5 Elution of m6A-Positive RNA

1. Add 125 μL of 42 °C preheated Elution Buffer to the m^6A-Dynabead complexes from Subheading 3.4, **step 4** and incubate at 42 °C for 5 min.
2. At the end of the 5 min, vortex the beads gently and place them on the magnet.
3. Collect the liquid phase and transfer to a fresh tube, kept on ice, as it represents the eluate fraction containing the m^6A "enriched RNA."
4. Add an additional 125 μL of preheated Elution Buffer to the beads and process as described in **steps 1–3** above for three additional times, for a total of four elutions.
5. Collect the liquid phase obtained at each elution step and pool with the previous ones. Keep sample on ice while working on the next elution. After the fourth round of elution, the final total eluate volume of the m^6A positive RNA fraction is 500 μL.

3.6 Extraction and Clean-up Step of the RIP

1. Extract the 500 μL of m^6A positive RNA collected in the previous step by adding 500 μL of acid phenol-chloroform.
2. Centrifuge at 4 °C at 10,000 × *g* for 7.5 min.

3. Carefully collect the upper phase making sure not to touch the inter-phase and transfer to a fresh 1.5 ml tube.
4. Add 500 μL of previously tested RNAse-free chloroform to the fresh tube, vortex briefly, and centrifuge at 4 °C at 10,000 × *g* for 7.5 min.
5. Transfer the upper phase to a fresh 1.5 mL tube and proceed to RNA precipitation (*see* **Note 9**) overnight at −20 °C.
6. Centrifuge the sample at 4 °C for 20 min at 16,000 × *g*.
7. Wash the pellet twice in 70 % ethanol by centrifuging for 10 min at 4 °C at 16,000 × *g*.
8. Dry the pellet at room temperature for 10 min prior to resuspend it in the desired volume (typically 5–6 μL) of Ultra-Pure $H_2$0 (*see* **Note 10**).

3.7 Library Construction

We have generally utilized 100 ng of RNA (100 ng of input and 100 ng of post m^{6}A-IP positive fraction) for library construction utilizing the Illumina TrueSeq Stranded mRNA Sample Preparation Guide.

1. Add 13 μL of Fragment, Prime, Finish Mix to the 5 μL m^{6}A positive fragmented RNA obtained in **step 8** (to a final volume of 18 μL).
2. Skip the fragmentation step in the Illumina protocol given the RNA has already been fragmented, and proceed immediately to the synthesis of the First Strand cDNA.
3. Follow the Illumina protocol to the end.
4. Verify the fragment sizes of each individual library on an Agilent BioAnalyzer 2100 or equivalent using High Sensitivity DNA chip.
5. Quantify the library by qPCR or using the Kapa library quantification kit according to the manufacturer's instructions (*see* **Note 4**).
6. Submit the libraries for high-throughput sequencing (*see* **Notes 11–13**).

4 Notes

1. This protocol has been tested using Anti-M6A (N^6-methyladenosine) antibody from Synaptic Systems (Cat. No 202,003).
2. This protocol was tested using Dynabeads Antibody Coupling Kit following exactly the manufacturer's protocol. The anti-m6A antibody was coupled to the Dynabeads at a ratio of 5 μg of anti-m6A antibody per 1 mg of Dynabeads, as suggested by

the manufacturer (coupling range of 5–10 μg of antiboby per mg of Dynabeads). Based on the number of samples in the experiment, the amount of Dynabeads and antibody to be coupled has to be adjusted accordingly following the manufacturer's recommendations.

3. We often use either 16 or 96-well formatted magnets:
   - 16×1.5 mL tubes rack: Invitrogen, DYNAL Invitrogen based separations.
   - 96 Wells plate: Ambion, Magnetic Stand-96.
4. We used KAPABIOSYSTEMS Cat.NO KK4824.
5. To isolate 10 μg of double poly(A) RNA, we often start with approximately 400 μg of total RNA.
6. The conditions of fragmentation detailed in this protocol allow for approximately 50 bp fragments on multiple types of poly(A) RNA examined.
7. Individual users might need to optimize the fragmentation conditions based on their samples.
8. Preparation of equilibrated coupled m^6A-Dynabeads. 50 μL of coupled m^6A-Dynabeads is utilized per sample. The 50 μL m^6A-Dynabeads are equilibrated by resuspending them in 500 μL of m^6A Binding Buffer for 5 minutes at room temperature and then placed on the magnet. The supernatant is discarded and this step is repeated a second time. The 50 μL of equilibrated m^6A-Dynabeads is then resuspended in 150 μL of Binding Buffer and used in **step 1** of Subheading 3.3.
9. NaCl/ethanol precipitation is carried out overnight at −20 °C in the presence of 1 μL Ultra Pure Glycogen. To this end, 1/10 volume of 3 M NaCl and 2.5× volumes of absolute ethanol are added to the sample.
10. Following pellet resuspension you can choose to use 1 μL of your resuspended RNA to check the quantity and size of your m^6A immunoprecipated RNA by Nanodrop photometer and bioanalyzer respectively, or proceed to library construction directly.
11. The basic schema of m^6A-seq is shown in Fig. 1 and allows for 50–200 bp resolution of sites of modifications depending on fragmentation condition, depth of sequencing, and whether paired end reads or single end reads are used. The current protocol has been optimized so that the m^6A peaks are identified to approximately 50 nt resolution based on average length of fragmented RNA and paired end sequencing.
12. To date, the produced antibodies that recognize the m^6A modification also appear to bind another structurally related modification N6,2′-O-dimethyladenosine called m^6Am. This

m^{6}Am modification can be distinguished from m^{6}A modifications as it is always in the 5′UTR and if present always on the first nucleotide of a transcript if it starts with adenosine. Utilizing single site resolution CLIP based m^{6}A-seq, it was estimated that approximately 4–8 % of sites identified by m^{6}A-seq may actually be m6Am *(14)*. As transcripts often have heterogeneous start sites, the peak intensity of potential m6Am sites identified by m^{6}A-seq are often less then peak intensities of m^{6}A found in long internal exons or in the 3′end of genes.

13. In terms of identifying the site(s) of modifications, the consensus m^{6}A motif is RR(m^{6}A)CH. We then search for this motif surrounding the point of maximal peak enrichment and this is inferred to be the site of modification. However, it is possible that there is more then one consensus site near the peak and further experiments such as SCARLET or CLIP based m^{6}A-seq would be needed to determine if there are multiple m^{6}A sites harbored within the m^{6}A peak [14–16].

Acknowledgments

This work was supported by MGH Start-up funds to Cosmas Giallourakis. We thank Yi Xing, Ph.D. and Jinkai Wang, Ph.D. at UCLA who have been our computational biologist collaborators on m^{6}A related projects.

References

1. Meyer KD, Jaffrey SR (2014) The dynamic epitranscriptome: N^6-methyladenosine and gene expression control. Nat Rev Mol Cell Biol 15(5):313–326. doi:10.1038/nrm3785
2. Batista PJ, Molinie B, Wang J, Qu K, Zhang J, Li L, Bouley DM, Lujan E, Haddad B, Daneshvar K, Carter AC, Flynn RA, Zhou C, Lim KS, Dedon P, Wernig M, Mullen AC, Xing Y, Giallourakis CC, Chang HY (2014) m(6)A RNA modification controls cell fate transition in mammalian embryonic stem cells. Cell Stem Cell 15(6):707–719. doi:10.1016/j.stem.2014.09.019
3. Dominissini D, Moshitch-Moshkovitz S, Schwartz S, Salmon-Divon M, Ungar L, Osenberg S, Cesarkas K, Jacob-Hirsch J, Amariglio N, Kupiec M, Sorek R, Rechavi G (2012) Topology of the human and mouse m6A RNA methylomes revealed by m6A-seq. Nature 485(7397):201–206. doi:10.1038/nature11112
4. Meyer KD, Saletore Y, Zumbo P, Elemento O, Mason CE, Jaffrey SR (2012) Comprehensive analysis of mRNA methylation reveals enrichment in 3' UTRs and near stop codons. Cell 149(7):1635–1646. doi:10.1016/j.cell.2012.05.003
5. Schwartz S, Agarwala SD, Mumbach MR, Jovanovic M, Mertins P, Shishkin A, Tabach Y, Mikkelsen TS, Satija R, Ruvkun G, Carr SA, Lander ES, Fink GR, Regev A (2013) High-resolution mapping reveals a conserved, widespread, dynamic mRNA methylation program in yeast meiosis. Cell 155(6):1409–1421. doi:10.1016/j.cell.2013.10.047
6. Meyer KD, Patil DP, Zhou J, Zinoviev A, Skabkin MA, Elemento O, Pestova TV, Qian SB, Jaffrey SR (2015) 5' UTR m(6)A promotes cap-independent translation. Cell 163(4):999–1010. doi:10.1016/j.cell.2015.10.012
7. Liu J, Yue Y, Han D, Wang X, Fu Y, Zhang L, Jia G, Yu M, Lu Z, Deng X, Dai Q, Chen W, He C (2014) A METTL3-METTL14 complex mediates mammalian nuclear RNA N6-adenosine methylation. Nat Chem Biol 10(2):93–95. doi:10.1038/nchembio.1432
8. Schwartz S, Mumbach MR, Jovanovic M, Wang T, Maciag K, Bushkin GG, Mertins P,

Ter-Ovanesyan D, Habib N, Cacchiarelli D, Sanjana NE, Freinkman E, Pacold ME, Satija R, Mikkelsen TS, Hacohen N, Zhang F, Carr SA, Lander ES, Regev A (2014) Perturbation of m6A writers reveals two distinct classes of mRNA methylation at internal and 5' sites. Cell Rep 8(1):284–296. doi:10.1016/j.celrep.2014.05.048

9. Ping XL, Sun BF, Wang L, Xiao W, Yang X, Wang WJ, Adhikari S, Shi Y, Lv Y, Chen YS, Zhao X, Li A, Yang Y, Dahal U, Lou XM, Liu X, Huang J, Yuan WP, Zhu XF, Cheng T, Zhao YL, Wang X, Rendtlew Danielsen JM, Liu F, Yang YG (2014) Mammalian WTAP is a regulatory subunit of the RNA N^6-methyladenosine methyltransferase. Cell Res 24(2):177–189. doi:10.1038/cr.2014.3
10. Zheng G, Dahl JA, Niu Y, Fedorcsak P, Huang CM, Li CJ, Vagbo CB, Shi Y, Wang WL, Song SH, Lu Z, Bosmans RP, Dai Q, Hao YJ, Yang X, Zhao WM, Tong WM, Wang XJ, Bogdan F, Furu K, Fu Y, Jia G, Zhao X, Liu J, Krokan HE, Klungland A, Yang YG, He C (2013) ALKBH5 is a mammalian RNA demethylase that impacts RNA metabolism and mouse fertility. Mol Cell 49(1):18–29. doi:10.1016/j.molcel.2012.10.015
11. Jia G, Fu Y, Zhao X, Dai Q, Zheng G, Yang Y, Yi C, Lindahl T, Pan T, Yang YG, He C (2011) N^6-methyladenosine in nuclear RNA is a major substrate of the obesity-associated FTO. Nat Chem Biol 7(12):885–887. doi:10.1038/nchembio.687
12. Roundtree IA, He C (2016) RNA epigenetics-chemical messages for posttranscriptional gene regulation. Curr Opin Chem Biol 30:46–51. doi:10.1016/j.cbpa.2015.10.024
13. Heinz S, Benner C, Spann N, Bertolino E, Lin YC, Laslo P, Cheng JX, Murre C, Singh H, Glass CK (2010) Simple combinations of lineage-determining transcription factors prime cis-regulatory elements required for macrophage and B cell identities. Mol Cell 38(4):576–589. doi:10.1016/j.molcel.2010.05.004
14. Linder B, Grozhik AV, Olarerin-George AO, Meydan C, Mason CE, Jaffrey SR (2015) Single-nucleotide-resolution mapping of m6A and m6Am throughout the transcriptome. Nat Methods 12(8):767–772. doi:10.1038/nmeth.3453
15. Ke S, Alemu EA, Mertens C, Gantman EC, Fak JJ, Mele A, Haripal B, Zucker-Scharff I, Moore MJ, Park CY, Vagbo CB, Kussnierczyk A, Klungland A, Darnell JE Jr, Darnell RB (2015) A majority of m6A residues are in the last exons, allowing the potential for 3' UTR regulation. Genes Dev 29(19):2037–2053. doi:10.1101/gad.269415.115
16. Liu N, Parisien M, Dai Q, Zheng G, He C, Pan T (2013) Probing N^6-methyladenosine RNA modification status at single nucleotide resolution in mRNA and long noncoding RNA. RNA 19(12):1848–1856. doi:10.1261/rna.041178.113

Chapter 5

Mapping m6A at Individual-Nucleotide Resolution Using Crosslinking and Immunoprecipitation (miCLIP)

Anya V. Grozhik*, Bastian Linder*, Anthony O. Olarerin-George, and Samie R. Jaffrey

Abstract

N^6-methyladenosine (m6A) is the most abundant modified base in eukaryotic mRNA and has been linked to diverse effects on mRNA fate. Current m6A mapping approaches localize m6A residues to 100–200 nt-long regions of transcripts. The precise position of m6A in mRNAs cannot be identified on a transcriptome-wide level because there are no chemical methods to distinguish between m6A and adenosine. Here, we describe a method for using anti-m6A antibodies to induce specific mutational signatures at m6A residues after ultraviolet light-induced antibody-RNA crosslinking and reverse transcription. Then, we describe how to use these mutational signatures to map m6A residues at nucleotide resolution. Taken together, our protocol allows for high-throughput detection of individual m6A residues throughout the transcriptome.

Key words RNA, N^6-Methyladenosine, Crosslinking, High-throughput sequencing

1 Introduction

N^6-methyladenosine (m6A) is the most prevalent modified base in mRNA [1–5] and is found in several thousand transcripts, typically near the stop codon, but also in the coding sequence, 3′UTR, and 5′UTR of mRNAs [1–3].

The current m6A mapping approach, methyl-RNA immunoprecipitation and sequencing (MeRIP-Seq, also called m6A-Seq) [2, 3], has allowed researchers to map regions of RNA methylation. MeRIP-Seq involves immunoprecipitation of ~100 nt-long RNA fragments using m6A-specific antibodies, followed by high-throughput sequencing of the immunoprecipitated fragments. m6A-containing fragments then generate overlapping sequencing reads that produce a "peak" whose summit reflects an underlying m6A residue [2]. However, the current mapping approach does not identify specific m6A residues.

* These authors contributed equally to this work.

Alexandra Lusser (ed.), *RNA Methylation: Methods and Protocols*, Methods in Molecular Biology, vol. 1562, DOI 10.1007/978-1-4939-6807-7_5, © Springer Science+Business Media LLC 2017

Identifying m6A residues is challenging. Adenosine methylation is predominantly restricted to adenosines in a DRA*CHsequence context (D = A, G or U; R = purine; A* = methylatable A; H = A, C or U) [6]; however, not all DRACH motifs are methylated in vivo [1]. Exact positions of m6A residues can be bioinformatically predicted from MeRIP-Seq peaks by searching for the presence of a subset of DRACH motifs near the point of highest read coverage [7]. However, this approach is complicated because m6A often appears in clusters, which can result in large peaks spanning several m6A residues [2]. Additionally, multiple DRACH motifs can be present underneath a peak, making it difficult to predict the specific methylated adenosine.

Likewise, there is no chemical method that results in selective modification and detection of m6A residues. The nearly identical chemical properties of A and m6A have prevented the development of a chemical method to distinguish these nucleotides. Additionally, unlike other base modifications, m6A does not introduce errors during reverse transcription that would allow direct mapping of its position [8]. Thus, a major goal is to develop a method that provides a specific chemical signature that indicates the precise location of m6A residues in the transcriptome.

Here, we demonstrate how to use anti-m6A antibodies to induce specific mutational signatures that enable precise identification of m6A residues in RNA (Fig. 1). In this approach, anti-m6A antibodies are crosslinked to RNA using UV light to create antibody-RNA crosslinks. Reverse transcription of crosslinked RNA then results in a highly specific pattern of mutations or truncations in the cDNA. These mutational signatures are then computationally identified to reveal precise positions of m6A residues. Using these signatures we map m6A residues throughout the transcriptome at single-nucleotide resolution.

2 Materials

All buffers and solutions should be made in ultrapure, nuclease-, and nucleic-acid free water, and purified using vacuum filtration units (*see* below).

2.1 Buffers and Solutions

1. Binding/low-salt buffer: 50 mM Tris–HCl pH 7.4, 150 mM NaCl, 0.5 % Nonidet P-40 (NP-40).
2. High-salt buffer: 50 mM Tris–HCl pH 7.4, 1 M NaCl, 1 % NP-40, 0.1 % sodium dodecyl sulfate (SDS).
3. PNK wash buffer: 20 mM Tris–HCl pH 7.4, 10 mM $MgCl_2$, 0.2 % Tween-20.
4. 5X PNK pH 6.5 buffer: 350 mM Tris–HCl pH 6.5, 50 mM $MgCl_2$, 5 mM dithiothreitol (DTT).

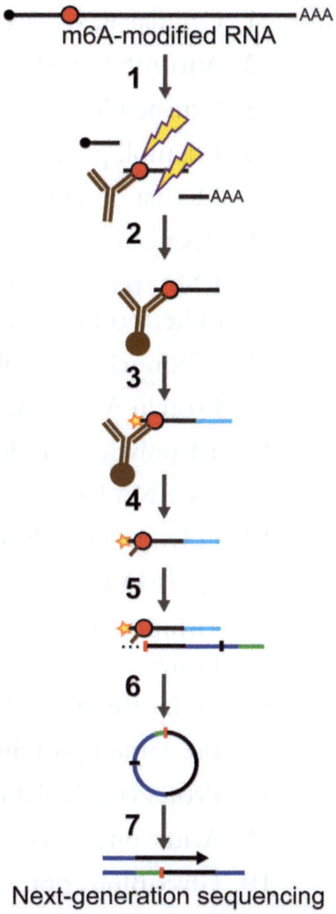

Fig. 1 Schematic of the miCLIP protocol. Cellular RNA containing m6A (*red circles*) is fragmented, incubated with an anti-m6A antibody and UV-crosslinked (1). Then, antibody-RNA complexes are recovered by protein A/G-affinity purification (2). Next, a 3′-adapter is ligated and the 5′-end is radiolabeled (3). RNA-protein complexes are then purified (4). RNA fragments are then reverse transcribed, generating mutations or truncations in the resulting cDNA (5). The cDNA is then circularized (6), relinearized and amplified by PCR (7)

5. Proteinase K (PK) buffer: 100 mM Tris–HCl pH 7.4, 50 mM NaCl, 10 mM ethylenediaminetetraacetic acid (EDTA).
6. Proteinase K urea (PKU) buffer: 100 mM Tris–HCl pH 7.4, 50 mM NaCl, 10 mM EDTA, 7 M urea.
7. RNase-free sodium acetate:3 M, pH 5.5.
8. RNase-free TE buffer: 10 mM Tris–HCl pH 7.5, 1 mM EDTA pH 8.0.
9. NuPage MES SDS Running Buffer: 20× (Novex), diluted to 1× in water.
10. Bis–Tris Transfer Buffer: 5× (Invitrogen), diluted to 1× in water.
11. TBE Running Buffer: 5× (Invitrogen), diluted to 1× in water.

2.2 Reagents

1. Cellular RNA (*see* **Note 1**).
2. Anti-m6A antibodies: 1 mg/mL (*see* **Note 2**).
3. Crushed ice.
4. Ethanol, pure.
5. Methanol, pure.
6. RNase-free DNase I: 1 U/μL.
7. RNA fragmentation reagent, including stop solution: 10× (Thermo Fisher).
8. Ribonuclease inhibitor: 40 U/μL.
9. Protein A/G magnetic beads.
10. T4 polynucleotide kinase supplied with buffer: 1000 U/μL.
11. T4 RNA ligase I supplied with buffer: 1000 U/μL.
12. Polyethylene glycol (PEG) 400.
13. [γ-^{32}P]ATP: 3000 Ci/mmol (Perkin Elmer) (*see* **Note 3**).
14. Lithium dodecyl sulfate (LDS) sample buffer: 4× (Thermo Fisher).
15. Dithiothreitol:1 M in water.
16. Prestained protein standard.
17. Proteinase K: 10 mg/mL.
18. Acidic phenol:chloroform:isoamyl alcohol, pH 6.5 (*see* **Note 4**).
19. GlycoBlue coprecipitant:15 mg/mL (Thermo Fisher).
20. Superscript III kit, supplied with buffer and 10 mM dNTPs (Thermo Fisher).
21. Deoxynucleotide triphosphates:10 mM in water.
22. Low molecular weight DNA ladder: 25 to 800 nucleotides.
23. TBE sample buffer: 5× (Thermo Fisher).
24. TBE-urea sample buffer: 2× (Thermo Fisher).
25. SYBR gold: 10,000× (*see* **Note 5**).
26. CircLigase II kit, supplied with buffer and 50 mM $MnCl_2$ (Epicentre).
27. FastDigest BamHI (800 reactions), supplied with buffer (Thermo Fisher).
28. Accuprime Supermix I (Thermo Fisher).
29. Agencourt Ampure XP magnetic beads (Beckman Coulter).

2.3 Oligonucleotides

All oligonucleotides are from Integrated DNA Technologies (IDT) with standard purification unless otherwise indicated.

1. L3 linker (HPLC-purified): /5rApp/AGATCGGAAGAGCG GTTCAG/3ddC/, 20 μM stock.

2. Cut oligo (PAGE-purified): GTTCAGGATCCACGACGCTCTTC/3ddC/, 10 μM stock.
3. P5 Solexa PCR primer: AATGATACGGCGACCACCGAGATCTACACTCTTTCCCTACACGACGCTCTTCCGATCT, 10 μM stock.
4. P3 Solexa PCR primer: CAAGCAGAAGACGGCATACGAGATCGGTCTCGGCATTCCTGCTGAACCGCTCTTCCGATCT, 10 μM stock.
5. Reverse transcription primers: 0.5 μM stocks of each.

   RT1: /5phos/NnaaccNNNAGATCGGAAGAGCGTCGTGgatcCTGAACCGC

   RT2: /5phos/NNacaaNNNAGATCGGAAGAGCGTCGTGgatcCTGAACCGC

   RT3: /5phos/NNattgNNNAGATCGGAAGAGCGTCGTGgatcCTGAACCGC

   RT4: /5phos/NNaggtNNNAGATCGGAAGAGCGTCGTGgatcCTGAACCGC

   RT5: /5phos/NNcgccNNNAGATCGGAAGAGCGTCGTGgatcCTGAACCGC

   RT6: /5phos/NNccggNNNAGATCGGAAGAGCGTCGTGgatcCTGAACCGC

   RT7: /5phos/NNctaaNNNAGATCGGAAGAGCGTCGTGgatcCTGAACCGC

   RT8: /5phos/NNcattNNNAGATCGGAAGAGCGTCGTGgatcCTGAACCGC

   RT9: /5phos/NNgccaNNNAGATCGGAAGAGCGTCGTGgatcCTGAACCGC

   RT10: /5phos/NNgaccNNNAGATCGGAAGAGCGTCGTGgatcCTGAACCGC

   RT11: /5phos/NNggttNNNAGATCGGAAGAGCGTCGTGgatcCTGAACCGC

   RT12: /5phos/NNgtggNNNAGATCGGAAGAGCGTCGTGgatcCTGAACCGC

   RT13: /5phos/NNtccgNNNAGATCGGAAGAGCGTCGTGgatcCTGAACCGC

   RT14: /5phos/NNtgccNNNAGATCGGAAGAGCGTCGTGgatcCTGAACCGC

   RT15: /5phos/NNtattNNNAGATCGGAAGAGCGTCGTGgatcCTGAACCGC

   RT16: /5phos/NNttaaNNNAGATCGGAAGAGCGTCGTGgatcCTGAACCGC

2.4 Supplies

1. Vacuum filter units for purifying buffers and solutions.
2. Tissue culture clusters, 12 wells per cluster.
3. Plastic wrap.
4. Microcentrifuge tubes of various sizes: 1.5, 0.5, 0.2 mL.
5. Pipettors and filtered tips in various sizes: 1000, 200, 20, 10 μL.
6. 4–12 % bis–tris protein gels (Thermo Fisher).
7. Nitrocellulose membrane, 0.45 μm pore size.
8. Carestream BioMax MS films, 8 × 10 in. (Kodak).
9. Carestream BioMax intensifying screen, 8 × 10 in. (Kodak).
10. Luminescent stickers.
11. Phase-lock gel heavy tubes (5 PRIME).
12. 6 % TBE-urea gels (Thermo Fisher).
13. 19G syringe needles or gel breaker tubes.
14. CoStar Spin-X columns, 0.22 μm pore size (Corning).
15. 6 % TBE gels (Thermo Fisher).

2.5 Equipment

1. Cold room and freezers (−20 and −80 °C).
2. End-over-end mixer for rotating microcentrifuge tubes.
3. Vortexer.
4. Magnetic rack for microcentrifuge tubes.
5. Shallow tray (~5 cm high) for crosslinking samples.
6. Stratalinker 2400 (Stratagene).
7. Electrophoretic gel chamber.
8. Electrophoretic transfer unit.
9. Geiger counter.
10. Autoradiography cassette.
11. Temperature-adjustable microcentrifuge mixer.
12. Refrigerated microcentrifuge capable of spins >/=20,000 RCF.
13. UV transilluminator or imager for visualizing SYBR gold-stained gels.
14. Thermocycler.
15. Access to a facility capable of Illumina high-throughput sequencing.

2.6 Software

1. Flexbar2.5 (available at: https://github.com/seqan/flexbar.git).
2. pyCRAC1.1.3 (available at: https://bitbucket.org/sgrann/pycrac).

3. Novoalign3.02.12 (available at: http://www.novocraft.com/products/novoalign).
4. CIMS and CITS packages (available at: http://zhanglab.c2b2.columbia.edu/index.php/CIMS_Documentation).
5. Bedtools2.25.0 (available at: https://github.com/arq5x/bedtools2.git).
6. Samtools1.2 (available at: http://samtools.sourceforge.net/).
7. Perl, python, and basic unix command line tools.
8. File of the reference genome in .fasta format (various reference genomes are available at: http://hgdownload.cse.ucsc.edu/downloads.html).
9. Bedtools-compatible genome index file in .fasta.fai format (produced with samtools index command).

3 Methods

For any steps requiring the use or addition of water, use ultrapure, nuclease-, and nucleic-acid free water.

3.1 Fragment RNA (See Note 6)

1. Start with 5–20 μg of DNase-treated, polyA+–selected (optional) RNA suspended in 20 μL of water in a sterile, nuclease-free microcentrifuge tube.
2. Add 2 μL of RNA fragmentation reagent to the RNA, and mix gently with a pipettor. Incubate the mixture in a heat block at 75 °C for exactly 12 min.
3. Immediately after fragmentation has completed, place the tube on ice, add 2.2 μL of the stop solution to the mixture, and mix gently with the pipettor (*see* **Note 7**).

3.2 Crosslink RNA to Anti-m6A Antibody

1. Bring up the volume of the sample to 500 μL with an appropriate volume of the binding/low-salt buffer and mix gently with the pipettor.
2. Add 10 μL of anti-m6A antibody and 2 μL of ribonuclease inhibitor to the sample and mix gently with the pipettor (*see* **Note 8**).
3. Rotate the sample tube at 4 °C (e.g., in a cold room) for 1.5 to 2 h.
4. Prior to crosslinking, prechill a sterile 12-well tissue culture plate on ice (*see* **Note 9**).
5. After RNA-antibody binding has completed, transfer the sample to a single well of the 12-well tissue culture plate. Place the plate on crushed ice in a shallow (~5 cm) tray.

6. Remove the lid from the plate, and crosslink twice in a Stratalinker 2400 using 254 nm light and 150 mJ/cm^2. Between each irradiation, gently agitate the tray to mix the sample.
7. Transfer the crosslinked sample to a new microcentrifuge tube, and place on ice (*see* **Note 10**).

3.3 Immunoprecipitate the RNA-Antibody Complexes

1. Resuspend protein A/G magnetic beads by gentle vortexing. Aliquot 50 μL of the magnetic beads to a new microcentrifuge tube. Wash the beads twice in 500 μL of binding/low-salt buffer. Resuspend the washed beads in 100 μL of binding/low-salt buffer.
2. Add the bead slurry to the crosslinked sample and mix gently with the pipettor.
3. Rotate the sample tube in a cold room for 1.5 h to overnight (*see* **Note 11**).
4. Wash the magnetic beads twice in 900 μL of high-salt buffer (*see* **Note 12**).
5. Wash the magnetic beads twice in 900 μL of binding/low-salt buffer.
6. Wash the magnetic beads twice in 500 μL of PNK wash buffer.

3.4 Dephosphorylate 3′ Ends of the RNA Fragments

1. Resuspend the magnetic beads in 20 μL of dephosphorylation master mix:

   15 μL water.

   4 μL 5× PNK pH 6.5 buffer.

   0.5 μL T4 polynucleotide kinase.

   0.5 μL ribonuclease inhibitor.
2. Incubate the sample in a thermal mixer at 37 °C for 20 min, shaking constantly at 1100 RPM.
3. Wash the beads once in 500 μL of PNK wash buffer.
4. Wash the beads once in 500 μL of high-salt buffer.
5. Wash the beads twice in 500 μL of PNK wash buffer.

3.5 Ligate Linker to 3′ Ends of RNA Fragments

1. Resuspend the magnetic beads in 20 μL of ligation master mix:

   11 μL water.

   2 μL 10× T4 RNA ligase buffer.

   1 μL T4 RNA ligase I.

   0.5 μL ribonuclease inhibitor.

   1.5 μL L3 linker.

   4 μL PEG400.

2. Incubate the sample in a thermal mixer at 16 °C overnight, shaking intermittently at 1100 RPM (30 s shaking followed by 4 min and 30 s without shaking).
3. Wash the beads once in 500 μL of PNK wash buffer.
4. Wash the beads twice in 500 μL of high-salt buffer. For each wash, rotate the sample tube at 4 °C for 5 min.
5. Wash the beads twice in 500 μL of PNK wash buffer.

3.6 Radiolabel the 5′ Ends of RNA Fragments (See Note 13)

1. Resuspend the magnetic beads in 40 μL of radiolabeling master mix:

   33 μL water.

   4 μL 10× T4 polynucleotide kinase buffer.

   1 μL[γ-32P]ATP.

   2 μL T4 polynucleotide kinase.
2. Incubate the sample in a thermal mixer at 37 °C for 20 min, shaking constantly at 1100 RPM.
3. Wash beads twice in 500 μL of binding/low-salt buffer.
4. Wash beads twice in 500 μL of PNK wash buffer.

3.7 Purify RNA-Antibody Complexes Using SDS-PAGE and Membrane Transfer

1. Resuspend the magnetic beads in 20 μL of sample buffer:

   14 μL water.

   5 μL 4× LDS sample buffer.

   1 μL 1 M DTT.
2. Incubate the sample in a thermal mixer at 70 °C for 15 min, shaking constantly at 1100 RPM.
3. Place the sample tube in a magnetic rack and allow the magnetic beads to pellet completely for 2 min (*see* **Note 14**).
4. Load the supernatant on a 4–12 % bis–tris protein gel according to the manufacturer's instructions using MES buffer. In a separate lane, load 10 μL of a prestained protein standard.
5. Run the gel at 4–8 °C in a cold room at 200 V for 35 min (*see* **Note 15**).
6. After the gel has finished running, remove the wells and the running front of the gel and discard in radioactive waste.
7. Set up transfer of the RNA-antibody complexes to a nitrocellulose membrane in an electrophoretic transfer unit according to the manufacturer's instructions.
8. Transfer the RNA–protein complexes to the membrane in a cold room at 400 mA for 30 min (*see* **Note 16**).
9. Wrap the membrane in plastic wrap and expose to a Kodak BioMax MS film in an autoradiography cassette with an intensifying screen at −80 °C (*see* **Notes 17** and **18**).

3.8 Elute RNA-Antibody Complexes

1. Align the developed film to the membrane and mark regions of the membrane containing the RNA-antibody complexes (Fig. 2).
2. Using a sterile razor blade or small surgical scissors, cut the marked region of the membrane into small fragments (~1 × 2 mm) and place the fragments into a new microcentrifuge tube.
3. Quick-spin the membrane fragments and add 200 μL of PK buffer to the sample tube. The membrane pieces must be submerged.
4. Add 10 μL proteinase K to the sample and mix gently with the pipettor.
5. Incubate the sample in a thermal mixer at 37 °C for 25 min, shaking constantly at 1100 RPM.
6. Add 200 μL PKU buffer to the sample and mix gently with the pipettor.
7. Incubate the sample in a thermal mixer at 37 °C for an additional 25 min, shaking constantly at 1100 RPM.
8. Pre-spin a phase-lock gel heavy tube for 2 min at 20,000 RCF in a microcentrifuge.

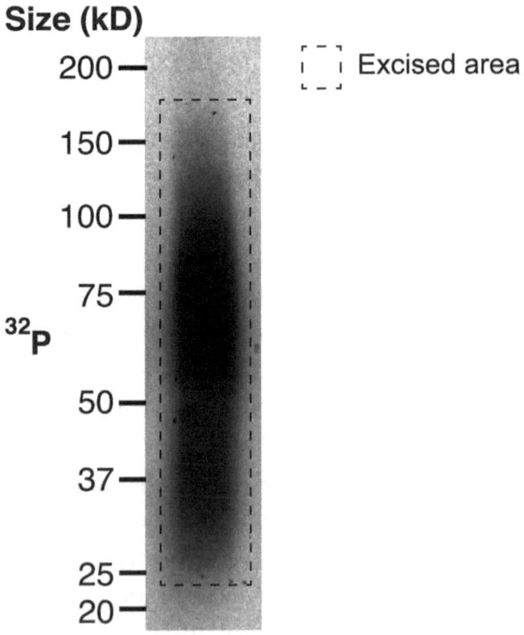

Fig. 2 Sample miCLIP autoradiogram. Here, poly(A)$^+$ RNA was crosslinked to an anti-m6A antibody. The crosslinked RNA-protein complexes were then reduced and purified with SDS-PAGE and nitrocellulose membrane transfer. Crosslinked RNA fragments are found as smears extending from the light and heavy chains of the reduced antibody, at 25 kD and 50 kD, respectively

9. Quick-spin the membrane fragments and transfer the supernatant to the phase-lock gel heavy tube. Add 400 μL of acidic phenol:chloroform:isoamyl alcohol to the tube.
10. Incubate the sample in a thermal mixer at 30 °C for 5 min, shaking constantly at 1100 RPM.
11. Centrifuge the sample tube at 20,000 RCF for 5 min.
12. Transfer the supernatant to a new microcentrifuge tube, and properly discard the phase-lock gel heavy tube (*see* **Note 19**).
13. Add the following to the supernatant in order, mixing gently and thoroughly at each step:

    1 μL GlycoBlue coprecipitant.

    40 μL 3 M sodium acetate, pH 5.5.

    1000 μL 100 % ethanol.

14. Precipitate the RNA fragments at −20 °C overnight.

3.9 Reverse Transcribe RNA Fragments

1. Spin down precipitated RNA fragments in a refrigerated centrifuge at 20,000 ×g for 20 min. Discard the supernatant.
2. Add 800 μL of 75 % ethanol to the sample tube. Spin the sample tube in a refrigerated centrifuge at 20,000 ×g for 5 min. Discard the supernatant.
3. Invert the sample tube onto a clean tissue paper and allow the RNA pellet to air dry for 5 to 10 min.
4. Resuspend the RNA pellet in 7 μL of denaturing master mix:

    5 μL water.

    1 μL RT primer (*see* **Note 20**).

    1 μL dNTPs.

5. Transfer the resuspended RNA to a new 0.2 mL PCR tube. Incubate the sample in a thermocycler using the following settings:

    70 °C for 5 min.

    Pause at 25 °C.

6. Add reverse transcription master mix to the sample

    7 μL water.

    4 μL 5× First strand buffer.

    1 μL DTT.

    0.5 μL ribonuclease inhibitor.

    0.5 μL SuperScript III.

7. Incubate the sample in the thermocycler using the following settings

25 °C for 5 min.

42 °C for 20 min.

50 °C for 40 min.

80 °C for 5 min.

Pause at 4 °C.

8. After reverse transcription has completed, transfer the sample to a microcentrifuge tube.
9. Add the following to the sample in order, mixing gently and thoroughly at each step:

   350 μL TE buffer.

   1 μL GlycoBlue coprecipitant.

   40 μL 3 M sodium acetate, pH 5.5.

   1000 μL 100 % ethanol.
10. Precipitate the cDNA at −20 °C overnight or at −80 °C for 2 h.

3.10 Purify the cDNA Fragments

1. Spin down precipitated cDNA in a refrigerated centrifuge at 20,000 ×*g* for 20 min. Discard the supernatant.
2. Add 800 μL of 75 % ethanol to the sample tube. Spin the sample tube in a refrigerated centrifuge at 20,000 ×*g* for 5 min. Discard the supernatant.
3. Invert the sample tube onto a clean tissue paper and allow the cDNA pellet to air dry for 5–10 min.
4. Resuspend the cDNA pellet in 6 μL of water.
5. Add 6 μL of 2× TBE-urea sample buffer to the cDNA and mix gently with the pipettor.
6. Incubate the sample at 65 °C for 5 min to denature the cDNA. Immediately place the tube on ice when finished.
7. Load the sample on a 6 % TBE-urea gel according to the manufacturer's instructions. In a separate lane, load 0.5 μL of a low molecular weight DNA ladder.
8. Run the gel at 200 V for 35 min, or until the dark blue dye of the sample buffer has migrated approximately three-quarters of the way through the gel.
9. Using a sterile razor blade, remove the lane of the gel carrying the ladder and stain this portion of the gel using a 1:10,000 dilution of SYBR gold in gel running buffer for 5–10 min.
10. Place the gel and the stained ladder lane on clean plastic wrap on a UV transilluminator. Using the stained ladder and dye markers as a guide, cut three bands with a sterile razor blade: 120–200 nt (high), 85–120 nt (medium), and 70–85 nt (low) (*see* **Note 21**).

11. Transfer each of the gel bands into a gel breaker tube placed into a new microcentrifuge tube (*see* **Note 22**).
12. Centrifuge each sample at 20,000 ×*g* for 2 min to crush the gel slices. Discard the gel breaker tubes.
13. Add 400 μL of 1× TE buffer to the crushed gel. Freeze the sample in a −80 °C freezer for 5 min. Then, incubate the sample in a thermal mixer at 4 °C overnight, shaking constantly at 1100 RPM.
14. The following morning, transfer the sample to a CoStar SpinX column and centrifuge at 20,000 ×*g* for 2 min. Transfer the eluate to a phase-lock gel heavy tube.
15. Add 400 μL of acidic phenol:chloroform:isoamyl alcohol to the tube.
16. Incubate the samples in a thermal mixer at 30 °C for 5 min, shaking constantly at 1100 RPM.
17. Centrifuge the sample tube at 20,000 RCF for 5 min.
18. Transfer the supernatant to a new microcentrifuge tube, and properly discard the phase-lock gel heavy tube.
19. Add the following to the sample in order, mixing gently and thoroughly at each step:

    1 μL GlycoBlue coprecipitant.

    40 μL 3 M sodium acetate, pH 5.5.

    1000 μL100 % ethanol.
20. Precipitate the cDNA at −20 °C overnight or at −80 °C for 2 h.

3.11 Circularize and Cut the cDNA Fragments

1. Spin down precipitated cDNA in a refrigerated centrifuge at 20,000 ×*g* for 20 min. Discard the supernatant.
2. Add 800 μL of 75 % ethanol to the sample tube. Spin the sample tube in a refrigerated centrifuge at 20,000 ×*g* for 5 min. Discard the supernatant.
3. Invert the sample tube onto a clean tissue paper and allow the cDNA pellet to air dry for 5–10 min.
4. Resuspend the cDNA pellet in 8 μL of circularization master mix:

    6.5 μL water.

    0.8 μL CircLigase buffer.

    0.4 μL $MnCl_2$.

    0.3 μL CircLigase II.
5. Transfer the sample to a 0.2 mL PCR tube and incubate in a thermocycler using the following settings

60 °C for 60 min.

Pause at 4 °C.

6. Add 30 μL of oligo annealing master mix to the sample and mix gently

   26 μL water.

   3 μL FastDigest buffer.

   1 μL Cut oligo.

7. Anneal the cut oligo to the circularized cDNA using the following settings

   95 °C for 2 min.

   Decrease temperature by 1 °C every 20 sec until 25 °C is reached.

   Pause at 25 °C.

8. Add 2 μL FastDigest BamHI to the sample and linearize the cDNA using the following settings

   37 °C for 30 min.

   80 °C for 5 min.

9. Transfer the cDNA to a new microcentrifuge tube. Add the following in order, mixing gently at each step:

   350 μL TE.

   1 μL GlycoBlue coprecipitant.

   40 μL 3 M sodium acetate, pH 5.5.

   1000 μL 100 % ethanol.

10. Precipitate the cDNA at −20 °C overnight or at −80 °C for 2 h.

3.12 PCR-Amplify the cDNA Fragments

1. Spin down precipitated cDNA in a refrigerated centrifuge. Centrifuge at 20,000 ×*g* for 20 min. Discard the supernatant.
2. Add 800 μL of 75 % ethanol to the sample tube. Spin the sample tube in a refrigerated centrifuge at 20,000 ×*g* for 5 min. Discard the supernatant.
3. Invert the sample tube onto a clean tissue paper and allow the cDNA pellet to air dry for 5–10 min.
4. Resuspend the cDNA pellet in 21 μL of water.
5. Prepare three test PCR reactions for each sample, consisting of:

   3.75 μL water.

   1 μL cDNA.

   0.25 μL P5 and P3 Solexa PCR primer mix.

   5 μL Accuprime I supermix.

6. Incubate the reactions in a thermocycler using the following settings:

   94 °C for 2 min.

   Then, 15, 20, or 25 cycles of:

   94 °C for 15 s.

   65 °C for 30 s.

   68 °C for 30 s.

   Pause at 4 °C.

7. After the PCR reactions have completed, place the reactions on ice and add 2.5 μL of 5× TBE sample buffer to each.
8. Load the samples on a 6 % TBE gel according to the manufacturer's instructions. In a separate lane, load 0.5 μL of a low molecular weight DNA ladder.
9. To visualize the PCR products, stain the gel using a 1:10,000 dilution of SYBR gold in gel running buffer for 5–10 min. Visualize the stained gel on a UV transilluminator or imager (Fig. 3).

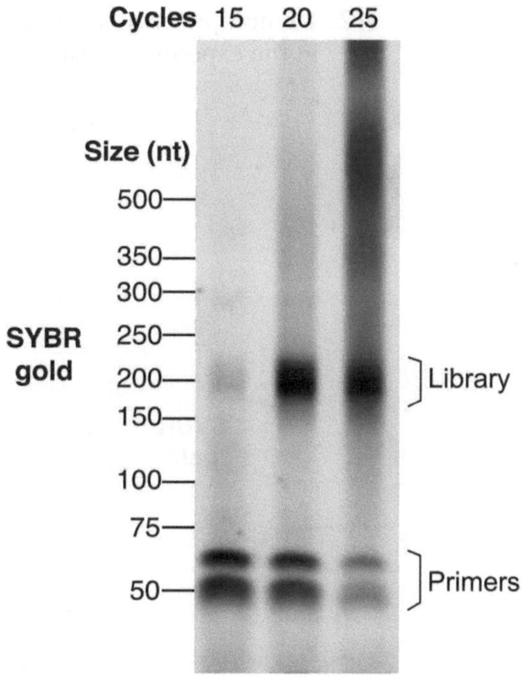

Fig. 3 Test PCR reactions. This step helps determine the number of cycles used in the preparative PCR reaction. Here, a miCLIP library of medium cDNA size was amplified using PCR for 15, 20, or 25 cycles, and visualized with gel electrophoresis and SYBR gold. While 15 cycles lead to under-amplification, 20 or 25 cycles lead to overamplified products

10. Determine the appropriate number of PCR cycles for the preparative PCR reaction (*see* **Note 23**).
11. Assemble the preparative PCR reaction, consisting of:

    11 μL water.

    8 μL cDNA.

    1 μL of P5 and P3 Solexa primer mix.

    20 μL Accuprime I Supermix.
12. Amplify the cDNA in a thermocycler for the appropriate number of cycles using the settings described in **step 6**, Subheading 3.12. (Fig. 4).
13. Purify the PCR products using Agencourt Ampure XP magnetic beads using the manufacturer's instructions. Elute the purified PCR products in 30–50 μL of water (*see* **Note 24**).
14. Submit half of the purified library for high-throughput sequencing (*see* **Note 25**).

3.13 Preprocess Sequencing Reads and Align Them to the Reference Genome

1. Trim 3′ adaptor sequences using flexbar [9] (*see* **Note 26**):

    flexbar -r forward_reads.fastq -p reverse_reads.fastq -f i1.8 -a flexbar_adapters.fasta --pre-trim-phred 30 -s -t exp_prefix.
2. Demultiplex reads using the barcodes in the RT primers used in the experiment using the command below. We routinely use

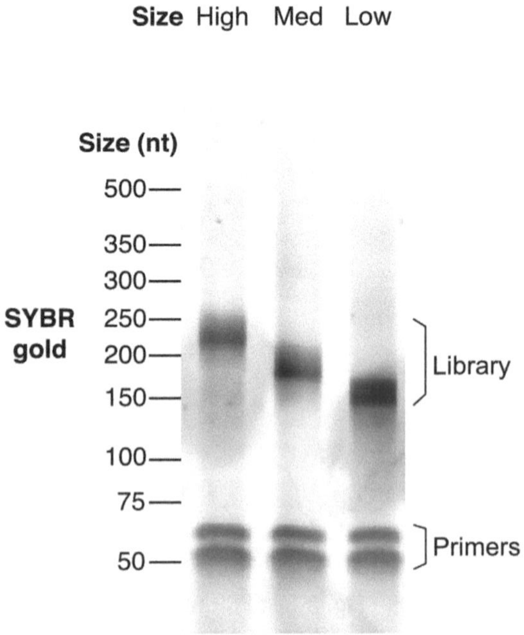

Fig. 4 Preparative PCR reactions. Here, miCLIP libraries of high, medium, and low cDNA sizes were PCR-amplified for the number of cycles determined to be appropriate in the test PCR reactions. The intensity of the SYBR gold signal here represents an appropriate amount of amplification prior to sequencing of libraries

the pyCRAC suite [11] because it preserves the random portion of the barcode when demultiplexing paired-end reads. This step will also move the first three nucleotides of the random barcode into the header of each read (*see* **Note 27**).

pyBarcodeFilter.py –f exp_prefix_1.fastq –r exp_prefix_2.fastq –b pyCRAC_barcodes.txt.

3. Move the remainder of the random barcode—the two nucleotides following the experimental barcode—to the read headers (*see* **Note 28**):

   awk -F "##" '{sub(/.../,"##"$2, $2); getline($3); $4 = substr($3,1,2); $5 = substr($3,3); print $1 $2 $4"\n"$5}' exp_prefix_1_NNNGGTT_cond1.fastq >exp_prefix_1_NNNGGTTNN_cond1.fastq.

4. Collapse PCR duplicates based on the read sequence using pyCRAC[10]:

   pyFastqDuplicateRemover.py -f exp_prefix_1_NNNGGTTNN_cond1.fastq -r exp_prefix_2_NNNGGTT_cond1.fastq -o dedup_reads.

5. Transform the header in the forward read file to be compatible with downstream CIMS analysis:

   awk-F'[_/]''/^>/{print$1"_"$2"_"$3"/"$4"#"$3"#"$2; getline($9); print $9}' dedup_reads_1.fasta > dedup_reads_1.cims.fasta.

6. Use the header information of the forward reads to create a CIMS-compatible header for the reverse reads, and reverse-complement these reads using the command below. This strategy reduces the noise of mutations introduced by sequencing errors. We have developed a perl script for this purpose that can be found at https://github.com/jaffreylab/miCLIP_MiMB_2016 (*see* **Note 29**).

   perl match_barcodes_and_reverse_complement.pl dedup_reads_1.cims.fasta dedup_reads_2.fasta.

7. Concatenate the forward reads and modified reverse reads using the command below. Note that orphan reads generated in **step 1,** Subheading 3.13 can be concatenated into this file.

   cat dedup_reads_1.cims.fasta dedup_reads_2.cims.rc.fasta >reads.cims.cat.fasta.

8. Use Novoalign to align the concatenated reads to an index of the reference genome (*see* **Note 30**):

   novoalign -t 85 -d genome_index.nix -f reads.cims.cat.fasta -F FA -l 16 -s 1 -o Native -r None -a >reads.cims.cat.novoalign.

3.14 Call m6A Sites with the CIMS Pipeline

1. Parse novoalign output to generate a bed file of read coordinates and a separate file of mutation coordinates:

   perl novoalign2bed.pl -v --mismatch-file reads.cims.mutation.txt reads.cims.cat.novoalignreads.cims.tag.bed.

2. Collapse PCR duplicates based on read coordinates and barcode identities (*see* **Note 31**):

   perl tag2collapse.pl -v --random-barcode -EM 30 --seq-error-model alignment --weight-in-name --keep-max-score --keep-tag-name reads.cims.tag.bed reads.cims.tag.uniq.bed.

3. Cluster overlapping CLIP tags:

   perl tag2cluster.pl -v -s -maxgap "-1" reads.cims.tag.uniq.bedreads.cims.tag.uniq.cluster.bed.

4. Use the CIMS algorithm to call mutation sites:

   python joinWrapper.py reads.cims.mutation.txt reads.cims.tag.uniq.bed 44 N reads.cims.tag.uniq.mutation.txt.

5. Filter for C→T transitions and generate a .bed file of their coordinates:

   awk '{if($6=="+" && $8=="C" && $9==">" && $10=="T" || $6=="-" && $8=="G" && $9==">" && $10=="A") {print $0}}' reads.cims.tag.uniq.mutation.txt | cut -f 1–6> reads.cims.tag.uniq. C2T.bed.

6. Evaluate the reproducibility of mutations by permutation:

   perl CIMS.pl -v -n 5 -p -c. /$1_cache_C2T --keep-cache reads.cims.tag.uniq.bed reads.cims.tag.uniq. C2T.bed reads.cims.tag.uniq. C2T.CIMS.txt.

7. Create a sorted bed file of C→T transitions (*see* **Note 32**):

   awk -v prefix="rep1" '{if($9<=1){print $1"\t"$2"\t"$3"\t"prefix"_"$4"\t"$5"\t"$6}}' reads.cims.tag.uniq. C2T.CIMS.txt | sort -k 9,9n -k 8,8nr -k 7,7n > reads.cims.tag.uniq. C2T.CIMS.p1.bed.

8. Call and annotate m6A sites using the script below, available at https://github.com/jaffreylab/miCLIP_MiMB_2016 (*see* **Note 33**). This generates a .bed file centered on coordinates of putative m6A residues (i.e., at position −1 of the C→T transition), and extracts their sequence environment. To identify putative m6A residues, this script filters C→T transitions based on the number of transitions ($m \geq 2$) and the transition frequency ($1\,\% \geq m/k \leq 50\,\%$) at any given coordinate. Additionally, the script generates statistics describing the ratio of called adenosines, and those called within the m6A consensus motif relative to the total number of filtered transitions. This information allows for a quick assessment of the success of the experiment.

   kmer.annotate.cims.shreads.cims.tag.uniq. C2T.CIMS.p1.bed genome.fasta

3.15 Calling m6A Sites with CITS

We recommend using only forward reads for calling truncations. If only CITS analysis is desired, preprocessing of reverse reads in **step 5** and **6**, Subheading 3.13 can be omitted. After alignment of the forward reads, proceed to **steps 2–4**, Subheading 3.14, to generate files necessary for CITS analysis. Finally, proceed to CITS analysis as described below:

1. Identify coordinates of truncation sites:

   perl bedExt.pl -n up -l "-1" -r "-1" -v reads.cims.tag.uniq.bed reads.cims.tag.uniq.trunc.bed.

2. Shuffle truncation events to evaluate their significance using the command below. The cluster file used here is derived from **step 3**, Subheading 3.14 (*see* **Note 34**).

   perl tag2peak.pl -c./reads.cims_cache_trunc -ss -v -gap 25 -p 0.05 reads.cims.tag.uniq.cluster.bed reads.cims.tag.uniq.trunc.bed reads.cims.tag.uniq.CITS.p005.bed.

3. Annotate m6A sites with their sequence environment and generate a summary of called DRACH sites using the script below, available in the supplementary information:

   kmer_annotate_cits.sh reads.cims.tag.uniq.CITS.p005.

4 Notes

1. We usually prepare sample RNA by harvesting total RNA from cultured cells using TRIzol Reagent (Thermo Fisher) according to the manufacturer's instructions. Then the RNA is treated with DNase I (*see* Subheading 2) to remove DNA contamination. Finally, if desired, we enrich for messenger RNA by selecting polyadenylated RNAs using oligo d(T) magnetic beads. We recommend that the total amount of RNA should be 10 μg or higher for each sample subjected to miCLIP.
2. We have successfully used rabbit polyclonal antibodies from Synaptic Systems and Abcam. Other anti-m6A antibodies may also be tested in miCLIP.
3. ^{32}P is a high-energy beta emitter that poses a dose hazard. Perform all work using this isotope behind Plexiglas shielding of 3/8-inch thickness or greater, and wear protective equipment, including gloves, wrist guards, a lab coat, and eye protection. Additionally, radioisotopes must be used and disposed of in accordance with federal and institutional regulations and guidelines. Consult your institution's radiation safety specialist for institution-specific radiation safety requirements.
4. Phenol is corrosive to the skin and mucous membranes, and should only be handled in a chemical hood with proper protective equipment, including gloves, a lab coat, and eye protection.
5. SYBR Gold binds DNA and is therefore a potential mutagenic agent. Wear protective equipment, including gloves, a lab coat, and eye protection when handling SYBR Gold.
6. This step will generate RNA fragments of 30–130 nt in length which are compatible with downstream radiolabeling and linker ligation procedures. In order to preserve the greatest

number of fragments for sequencing, we avoid generating too many fragments under 30 nt or over 130 nt in length. Fragments that are too short or too long are not appropriate for downstream sequencing and will be lost during size selection of cDNA.

7. If desired, a small portion of the fragmented RNA (2 μL) can be saved at this point. This is important if RNA-seq analysis of the same sample is desired. These RNA fragments are compatible with a variety of small RNA library preparation protocols and commercial kits (such as those from Illumina or NEB).
8. For samples with lower quantities of input RNA (5–10 μg), it may be possible to generate a miCLIP library. To do so, we recommend using a greater amount of antibody at this step to help capture the greatest possible number of methylated RNA fragments. For example, for 5 μg of input RNA, we recommend using 25 μL of antibody.
9. We recommend using the 12-well tissue culture cluster because the diameter of each well allows for the sample to have optimum depth and surface area for efficient crosslinking, and allows for crosslinking of multiple samples at once. Alternatively, a similarly sized sterile round petri dish may be used.
10. At this point, crosslinked samples may be frozen at −80 °C until ready for immunoprecipitation. The RNA-antibody complexes will remain stable for months. However, we do not recommend thawing and refreezing the samples.
11. After immunoprecipitation has completed, place the sample tube in a magnetic rack to pellet the magnetic beads, and save the supernatant (nonbound fraction) in a new microcentrifuge tube if desired. This is useful if downstream analysis of the nonbound fraction is desired.
12. For steps involving the washing of magnetic beads, remove the tube from the magnet, resuspend the beads in the new wash buffer, and mix gently by pipetting. Then, return the tube to the magnet, allow the beads to pellet, and remove the wash buffer.
13. The following steps involve use of a radioisotope. All steps involving use of radioactive materials require use of protective equipment, including gloves, wrist guards, a lab coat, and eye protection, and must be carried out behind Plexiglas shielding of 3/8-inch thickness or greater. Radioactive materials must be used and disposed of in strict accordance with federal and institutional regulations and guidelines.
14. At this point, the supernatant can be transferred to a new microcentrifuge tube and frozen at −80 °C until ready for

SDS-PAGE. However, we recommend that SDS-PAGE be carried out within several days to prevent loss of radioactive signal.

15. The gel may be run for longer times to better resolve the antibody (150 kD) and its respective reduced fragments (50 and 25 kD). We recommend that the prestained protein standard be used as a guide for determining whether the gel should be run for longer than 35 min.
16. After transfer has completed, verify the presence of radiolabeled RNA–protein complexes on the membrane and gel using a Geiger counter.
17. We recommend that the gel be sealed in plastic wrapping (such as a Kapak SealPAK pouch, Ampac Flexibles 400–24) and exposed to the film along with the membrane to validate that transfer was complete. We usually find that some radioactive signal remains in the gel from loading of residual magnetic beads into the wells of the gel. However, the resolved RNA-antibody complexes are always nearly completely transferred to the membrane.
18. Exposure times will vary depending on the amount of input RNA, antibody binding and immunoprecipitation efficiency, and activity of the [γ-^{32}P]ATP at the time of labeling. We recommend developing the film after 30–60 min to determine whether the membrane should be reexposed to a new film for a longer period (up to 2–3 days).
19. Avoid touching or disrupting the gel interphase at this step and subsequent steps involving the use of the phase-lock gel tubes.
20. Use an RT primer with a unique experimental barcode for each sample.
21. We recommend placing a clean paper under the portion of the plastic wrap underlying lanes of the gel that contain the cDNA products. This way, the UV illumination will allow visualization of the ladder without illuminating the cDNA products, potentially introducing mutations.

    Avoid cutting gel slices that are higher than 200 nt or lower than 70 nt to prevent inclusion of cDNA inserts that are too long or unligated L3 linker sequence. We strongly recommend that for each sample (i.e., each individual gel lane), the three gel slices (high, medium, and low) be collected separately.
22. As an alternative to using gel breaker tubes, you may use 0.5 mL tubes by piercing their bottom with a 19–20 G syringe needle. Because other CLIP-based protocols do not use gel elution kits or dissolving agents for cDNA extraction, we do not recommend using such kits for this protocol.

23. We recommend determining the optimum number of PCR cycles for each sample using the outcome of the three test PCR reactions. In general, the optimum number of PCR cycles is the number required to faintly see the PCR products on the gel. However, because the amount of cDNA template used for the preparative PCR is greater than the amount used for the test PCR, we recommend that the number of cycles used for the preparative PCR be one less than the optimum number determined in the test PCR.
24. The unique barcodes used during reverse transcription allow for pooling of samples without loss of sample identity. Thus, given that the proper number of preparative PCR cycles was chosen individually for each sample, we pool completed preparative PCR reactions from multiple samples prior to purification. To elute the purified PCR products, we use 30 μL to 50 μL of water depending on the number of samples pooled for the purification.
25. We recommend sequencing miCLIP libraries in paired-end mode with a read length of 50 base pairs. For libraries of up to 6 unique samples, we find that sequencing the samples on a single lane of a high-throughput flowcell is sufficient for adequate sequencing coverage. However, for a greater number of unique samples, or if higher sequencing depth is desired, we recommend sequencing the libraries on up to three lanes.
26. If this step results in a substantial number of orphan reads, which are missing a mate due to quality filters but are written to a unique file, we recommend running subsequent steps for the orphan reads as well as paired reads. The orphan reads can be concatenated into the processed read file before alignment during subsequent steps as described.
27. The "barcodes.txt" file should contain one line for each barcode. The barcode is written in the format NNNXXXX, where NNN is the first part of the iCLIP random barcode and XXXX is the experiment barcode (e.g., barcode 1 is NNNGGTT—note that the experimental portion of the barcode is the reverse complement of the RT primer). Like flexbar, the pyBarcodesFilter.py script will also generate files of orphan reads. If this constitutes a substantial amount of reads, we recommend processing them in parallel with paired reads and including them in the downstream analysis by concatenating them at a downstream step as described.
28. The "getline" function used here requires that this command is run with the "mawk" implementation of awk on some systems. If this is not the standard implementation used by your system (e.g., on Debian Wheezy, which uses gawk), you should explicitly write "mawk" instead of "awk" in this command.

29. This script reverse-complements the reverse reads and their barcodes to use them as a sequencing replicate of the forward reads, thereby filtering for sequencing errors. Note that if your experiment uses an even number of random barcode nucleotides (e.g., when not using standard iCLIP barcodes), reads with palindromic barcodes will be flagged as duplicates and removed in step two of the CIMS analysis.
30. The option "-c 36" starts Novoalign with 36 processing threads. Note that multithreading is only available for a licensed version of Novoalign; otherwise, only a single thread can be used. When multithreading, more or less processing threads may be used depending on the computational power of your machine.
31. At this step, it is essential that the random barcode in the header of the reverse reads was reverse complemented (**step 5**, Subheading 3.13). Otherwise, read pairs will be collapsed as PCR duplicates.
32. The "if($9<=1)" statement defines a threshold for the false-discovery-rate determined by the CIMS analysis. Here, we allow an FDR of 1 because we apply downstream filtering to acquired C→T transitions. Also, the "prefix=" portion of this command allows for adding an experimental identifier to the fourth column of the resulting .bed file. If sites from multiple experiments are to be concatenated later, we recommend using a unique identifier for each experiment.
33. This script makes extensive use of the bedtools suite [11]. The second argument of the command refers to a .fasta file of the reference genome that has an accompanying index file in the same directory (e.g., genome.fasta and an accompanying genome.fasta.fai). Note that this script uses genomic coordinates to extract the sequence context of m6A residues. Thus, some m6A residues may be misannotated if located at an exon boundary.
34. We use an initial p-value cutoff of 0.05 for filtering truncation sites. Depending on the enrichment of DRACH sites that is evaluated in the next step, the stringency of this filter can be adjusted accordingly.

Acknowledgments

We thank members of the Jaffrey lab for their helpful comments and support. This work was supported by NIH grants NIDA DA037150 (S.R.J.), T32 CA062948 (A.O.-G.), and a German Research Foundation (DFG) fellowship (B.L.).

References

1. Linder B, Grozhik A, Olarerin-George A et al (2015) Single-nucleotide-resolution mapping of m6A and m6Am throughout the transcriptome. Nat Methods 12:767–772
2. Meyer K, Saletore Y, Zumbo P et al (2012) Comprehensive analysis of mRNA methylation reveals enrichment in 3′UTRs and near stop codons. Cell 149:1635–1646
3. Dominissini D, Moshitch-Moshkovitz S, Schwartz S et al (2012) Topology of the human and mouse m6A RNA methylomes revealed by m6A-seq. Nature 485:201–206
4. Perry R, Kelley D, Friderici K et al (1975) The methylated constituents of L cell messenger RNA: evidence for an unusual cluster at the 5′ terminus. Cell 4:387–394
5. Desrosiers R, Friderici K, Rottman F (1974) Identification of methylated nucleosides in messenger RNA from Novikoff hepatoma cells. Proc Natl Acad Sci U S A 71:3971–3975
6. Schibler U, Kelley D, Perry R (1977) Comparison of methylated sequences in messenger RNA and heterogeneous nuclear RNA from mouse L cells. J Mol Biol 115:695–714
7. Schwartz S, Mumbach M, Jovanovic M et al (2014) Perturbation of m6A writers reveals two distinct classes of mRNA methylation at internal and 5′ sites. Cell Rep 8:294–296
8. Sugimoto Y, Konig J, Hussain S et al (2012) Analysis of CLIP and iCLIP methods for nucleotide-resolution studies of protein-RNA interactions. Genome Biol 13:R67
9. Dodt M, Roehr J, Ahmed R et al (2012) FLEXBAR—flexible barcode and adapter processing for next-generation sequencing platforms. Biology 1:895–905
10. Webb S, Hector R, Kudla G et al (2014) PAR-CLIP data indicate that Nrd1-Nab3-dependent transcription termination regulates expression of hundreds of protein coding genes in yeast. Genome Biol 15:R8
11. Quinlan A, Hall I (2010) BEDTools: a flexible suite of utilities for comparing genomic features. Bioinformatics 26:841–842

Chapter 6

Detection and Quantification of N^6-Methyladenosine in Messenger RNA by TLC

Zsuzsanna Bodi and Rupert G. Fray

Abstract

The base-modified nucleotide, N^6-methyladenosine, is a relatively abundant modification found in the mRNA of most higher eukaryotes. Methylation levels can change dependent upon environmental conditions, cell differentiation state, or following knockdown of members of the methylase complex, and it is often useful to directly measure and compare N^6-methyladenosine levels between samples. Two dimensional chromatography of radiolabeled nucleotides, following specific nuclease treatments, provides a robust, sensitive, and reproducible assay for this modification.

Key words N^6-Methyladenosine, m^6A, Thin Layer Chromatography, TLC, mRNA Methylation, Epitranscriptome

1 Introduction

In most eukaryotes, transcripts that are synthesized by RNA polymerase II are modified at their 5′end by the addition of a 7-methyl guanosine (m^7G) cap and at their 3′ end by the addition of a poly(A) tail. Additional modifications can also be made to nucleotides of the primary transcript itself, and the most common of these is the post- or co-transcriptional methylation of adenosine to form *N6*-methyladenosine (m^6A). The presence of m^6A has recently been shown to play important roles in stem cell maintenance [1], mRNA turnover [1, 2], and translation efficiency, including cap-independent translation [3, 4]. Early direct biochemical detection [5, 6] and in vitro methyltransferase assays [7], as well as more recent sequencing of anti-m^6A immunoprecipitated RNA fragments [8], revealed a consensus site for methylation of RGACU (where R=purine, and the central A is the target of methylation). While direct mapping of m^6A sites within specific transcripts is important, it is often valuable to measure global changes in the level of m^6A in the mRNA population as a whole. Several methods have been described for this including HPLC of digested

Alexandra Lusser (ed.), *RNA Methylation: Methods and Protocols*, Methods in Molecular Biology, vol. 1562,
DOI 10.1007/978-1-4939-6807-7_6, © Springer Science+Business Media LLC 2017

nucleosides [2], estimation by dot blot using an anti-m^6A antibody [9], and detection following labeling by thin layer chromatography [10]. There are issues to be considered when selecting any of these methods; most eukaryotes contain modified nucleotides (including m^6A) within ribosomal RNA (rRNA), this is particularly relevant for HPLC approaches, as even small amounts of rRNA will result in a m^6A signal. The use of dot blots is not quantitative and also suffers from the fact that all of the anti-m^6A antibodies that we have tested show some level of avidity for poly(A) (and even polypurine) tracts and thus differences in signal could be reporting differences in poly(A) tail length (or abundance) rather than differences in methylation levels. Again, any contamination with rRNA could also lead to an overestimation of m^6A abundance. The TLC method described here is more robust as it specifically detects nucleotides in a GA context. This means that (for most eukaryotes) it will not detect m^6A derived from rRNA, as this methylation is carried out by different enzymes and is not deposited in a GA context. Furthermore, it gives a value (normally expressed as the percentage of As in a GA context that are methylated) that allows direct comparisons between samples. It should be remembered that only those methyladenosines in a GA context will be reported, the small number of m^6As that are preceded by other nucleotides will not be assayed by this method. However, the fact that only nucleotides in a GN context are detected increases the sensitivity of detection as the m^6A signal is not diluted by adenosines from the poly(A) tail or in other irrelevant contexts.

2 Materials

2.1 mRNA Digestion and Labeling

1. Poly(A) RNA (following two rounds of purification using oligo(dT)) (*see* **Note 1**) at a concentration of 10 ng/μL or greater in water or TE buffer.
2. T1 nuclease:1000 U/μL.
3. T4 polynucleotide kinase (PNK): 10 U/μL.
4. 10× PNK buffer A.
5. γ^{32}P-ATP: 6000 Ci/mmol.
6. m^6ATP nucleotide standard, used for in vitro transcription of control template: 100 mM.
7. In vitro transcription system (such as Promega RiboprobeSystem-T7, Promega Corp., USA) where each of the nucleotide triphosphates is supplied separately.

2.2 Precipitation

1. Absolute ethanol.
2. Sodium acetate, pH 5.2: 3 M; autoclave to sterilize.

2.3 P1 Nuclease Digestion

1. P1 nuclease reconstitution buffer: 20 mM sodium acetate (pH 5.4), 5 mM $ZnCl_2$, 50 mM sodium chloride, and 40 % glycerol (added after autoclaving).
2. Reconstituted P1 nuclease: P1 nuclease is supplied as a lyophilized powder at ≥200 units/mg protein. Reconstitute with P1 reconstitution buffer to a concentration of 0.1 mg/mL. The reconstituted enzyme can be stored at −20 °C for more than 6 months.
3. 10× Nuclease P1 buffer: 200 mM sodium acetate (pH 5.3), 50 mM $ZnCl_2$, 500 mM sodium chloride.

2.4 Running TLCs

1. TLC Cellulose F glass plates 20 × 20 cm (plastic backed plates from the same manufacturer can also be used) (*see* **Note 2**).
2. First dimension running buffer: isobutyric acid,0.5 M NH_4OH (5:3, v/v) (*see* **Note 3**).
3. Second dimension running buffer: isopropanol, concentrated HCl (specific gravity 1.18), H_2O (70:15:15, v/v/v).
4. Chromatographic tanks capable of taking 20 × 20 cm plates.

2.5 Scanning and Analysis

1. Thin polythene sleeves or bags to accommodate dried 20 × 20 cm glass TLC plates.
2. Phosphorimaging screen such as Kodak Imaging Screen-K.
3. Bio-Rad Molecular Imager FX Phospho scanner.
4. Bio-Rad "Quantity One" analysis software version 4.6.3.

3 Methods

When estimation of m^6A levels in a poly(A) mRNA sample is performed it is advisable to carry out a parallel analysis on the total RNA sample from which it is derived. As rRNAs of most species do not contain m^6A after Gs, the m^6A signal should be barely or non-detectable before poly(A) enrichment. The increase in m^6A in the mRNA sample should be obvious (Fig. 1). The total RNA sample will, however, contain other modified nucleotides (especially pseudouridine and 2′ ribose methylated nucleotides) and these should largely be absent from the mRNA sample (Fig. 1). Standards should also be run to establish the mobilities of labeled adenosine and N^6-methyladenosine 5′-monophosphates. The preparation of m^6A and A containing synthetic transcripts is described below. We have also used a similar approach for the generation of other standards for 2′ ribose methylated nucleotides.

3.1 Preparation of A and m^6A Containing RNA Standards

1. Prepare an unmodified adenosine standard RNA by in vitro transcription reactions using a Promega Riboprobe kit with the supplied control template, following the manufacturer's instructions.

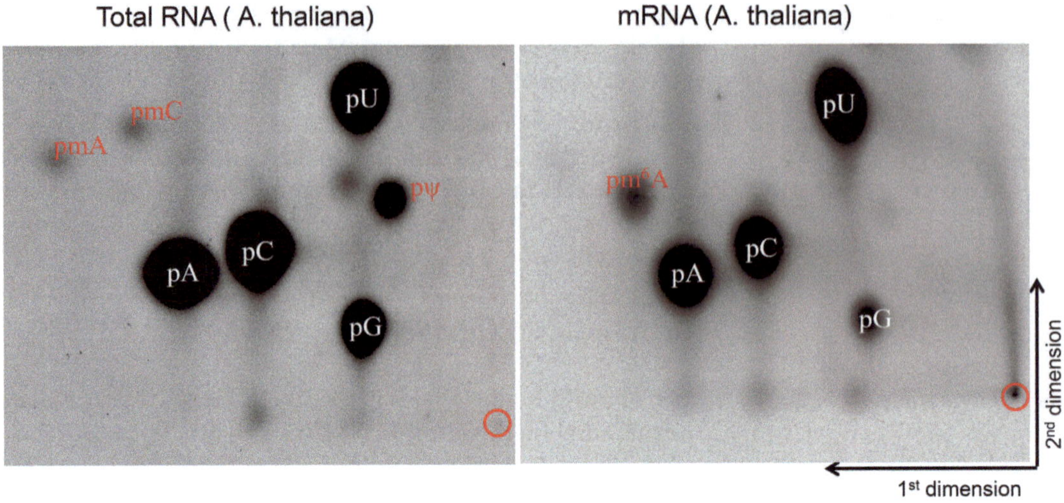

Fig. 1 Comparison of separation patterns of total RNA and poly(A) RNA samples. The total RNA samples were extracted from *A. thaliana* and used directly for TLC analysis and for 2 round poly(A) purification followed by TLC analysis. White letters are the labeled 5′ nucleotide spots, where pA is adenosine 5′ monophosphate, pC is cytosine 5′ monophosphate, pU is uridine 5′ monophosphate, and pG is guanosine 5′ monophosphate. pGs are underrepresented in all samples as T1 cuts after every G. Thus, Gs that are available for labeling are guanosine 3′ monophosphates that are not as efficiently labeled by T4 PNK as di or polynucleotides. The red circles are the sites of sample loading. The red letters are modified nucleotides where pmA is 2′-O-methyl-adenosine 5′ monophosphate, pm^{6}A is N^6-methyladenosine 5′ monophosphate, pmC is 2′-O-methyl-cytosine 5′ monophosphate, and pψ is pseudouridine

2. Prepare a standard RNA in which all As are replaced with m^{6}A by in vitro transcription of the same template following the supplier's instructions, but replacing the ATP supplied with the kit with an equivalent amount of m^{6}ATP.

3.2 T1 Digestion and 5′ End Labeling

1. In a 1.5 mL microcentrifuge tube, mix 50 ng of poly(A) RNA (*see* **Note 1**) with 1 μL 10× PNK buffer A and 1 μL of Ribonuclease T1 (*see* **Note 4**) and make up to 10 μL with sterile deionized water. Set up three similar control reactions in fresh microcentrifuge tubes using 50 ng of the total RNA or the in vitro transcribed A- and m^{6}A-containing standard transcripts instead of the poly(A) RNA (*see* **Note 5**). Incubate for 20 min at 37 °C.
2. Start labeling reactions by the sequential addition of 1 μL 10× PNK buffer A, 7 μL sterile deionized water, 1 μL γ^{32}P-ATP, and 1 μL of T4 polynucleotide kinase to each tube. Incubate for an additional hour at 37 °C.
3. Precipitate 5′ labeled fragments by the addition of 2 μL sterile 3 M sodium acetate pH 5.2 and 80 μL of absolute ethanol to each tube. Leave overnight (16 h) at −20 °C.

3.3 Digestion to Nucleotide 5′ Monophosphates

1. Pellet the labeled nucleotide fragments in a microcentrifuge at 12,000 rpm for 20 min.
2. Carefully remove the supernatant (*see* **Note 6**).
3. Dry the pellets by placing the open microcentrifuge tubes in a 37 °C heating block for 5 min (*see* **Note 7**).
4. Make up a master mix consisting of 5 μL of 10× P1 buffer, 5 μL of P1 nuclease stock and 40 μL of sterile deionized water. Use 10 μL of this solution to carefully resuspend each of the dried, labeled pellets.
5. Incubate the P1 digestions for 2 h at 37 °C.

3.4 Thin Layer Chromatography

1. Pour 200–300 mL of first dimension running buffer into the chromatography tank (*see* **Note 8**) and replace the lid, applying a smear of silicon vacuum grease around the edges to ensure an airtight seal. The tank should be left at room temperature for at least half an hour to allow the vapor phase to become saturated (*see* **Note 9**).
2. Gently mark a TLC plate with a soft lead pencil in the left bottom corner, 3 cm in from each edge. Add an identifying mark at the top left corner. Using a micropipette, load 1 μL of the digest to the marked position on the plate to create a spot of 3 mm or less in size (*see* **Note 10**). Repeat this process with the total RNA and the synthetic RNA standards using a fresh TLC plate for each sample.
3. Place the TLC plates in the tank and replace the lid and leave until the solvent front is 2 cm away from the top of the plate. At 20 °C, this normally takes 15–16 h for glass-backed plates. (*see* **Note 11**).
4. Remove the plate and air dry in a fume hood for at least 5 h.
5. Prepare the second dimension running buffer in a fresh tank in the same way as for Subheading 3.4, **step 1** above.
6. Rotate the TLC plates 90° to the direction of the first dimension (counter clockwise) and place it into the second dimension chromatography tank (the original lower left corner should now be at the lower right). Allow separation until the solvent front reaches 3 cm from the top of the TLC plate. This usually takes 19 h using glass-backed place, or 12 h with plastic-backed plates.
7. Remove the plate and dry for 2–5 h. Seal it in a plastic sleeve, and place in an X ray cassette with a storage phosphor screen for detection and quantification. (The storage phosphor screen should be fully erased before use by exposure to white light (light box) for 15–20 min).

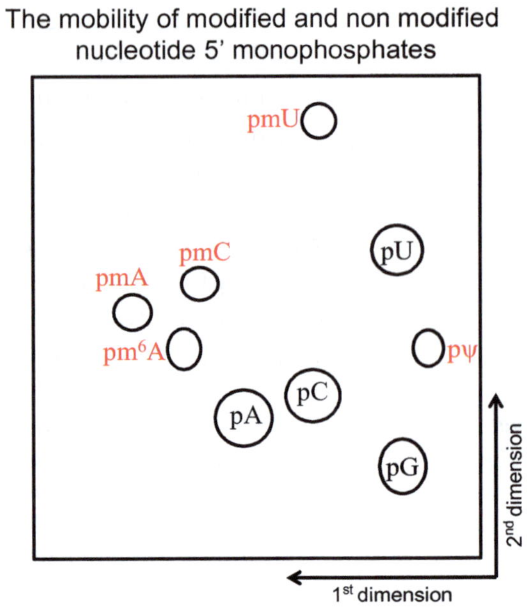

Fig. 2 The schematic migration pattern of modified and nonmodified nucleotides. The expected position of the different nucleotide monophosphate and modified nucleotide monophosphate from RNA samples after T1digestion followed by T4 PNK radiolabeling and P1 digestion *black letters* are the nonmodified nucleotides, where pA is adenosine 5′ monophosphate, pC is cytosine 5′ monophosphate, pU is uridine 5′ monophosphate, and pG is guanosine 5′monophosphate. All of these nucleotides are expected to be present in both total RNA and poly(A) RNA samples. The *red letters* are modified nucleotides where pm^{6}A is N^6-methyladenosine 5′ monophosphate, pmA is 2′-O-methyl-adenosine 5′monophosphate, pmC is 2′-O-methyl-cytosine 5′ monophosphate, pmU is 2′-O-methyl uridine 5′monophosphate, and pψ is pseudouridine

3.5 Detection and Quantification of m^{6}A Nucleotides

1. Remove phosphor screen from cassette, and quickly place into the FX scanner (Bio-Rad Molecular Imager FX Phospho scanner) (*see* **Note 12**).
2. Start scanning using the QuantityOne software default settings for ^{32}P (*see* **Note 13**).(*see* Figs. 1 and 2 for explanation of expected spots).
3. Transform the saved image using QuantityOne image analysis tools. Select the pA, pm^{6}A spots, and define general background. Carry out quantification of signal intensities.
4. Using the signal intensities calculate the pm^{6}A to pA ratio, and express it as % (*see* **Note 14**).

4 Notes

1. The purity of the poly-adenylated RNA is critical for producing good quality TLCs. At least two rounds of purification on oligo(dT) cellulose or beads should be used, and the input and

poly(A) samples should be checked by Agilent Bioanalyzer to confirm the initial RNA quality and the removal of all rRNA species following poly(A) purification (Fig.1). A number of good protocols exist for poly(A) purification [11] and several commercial kits are available. One of the main reasons for carry-over of non-polyadenylated RNAs is due to the nonspecific binding of such RNAs to the mRNA being purified rather than to the oligo(dT) itself [12]. Protocols or kits that use isostabilizing salts such as tetramethyl ammonium chloride strengthen A:T bonding and thus give better quality poly(A) RNA, as nonspecific binding of rRNA to the target RNAs can be reduced [12]. For this reason, we routinely use the Ambion PolyA purist kit. An aliquot of the starting total RNA should also be retained for analysis.

2. Do not use aluminum-backed plates as these are not compatible with the acid conditions of buffer 2. Plastic-backed plates can be cut into smaller sizes for quick checking of samples.
3. Prepare the 0.5 M ammonium hydroxide solution fresh from concentrated stock using deionized sterile water.
4. T1 nuclease only cuts after Gs, leaving polynucleotides that end with a 3′ phosphate G and which start with a 5′ OH nucleotide. Thus, any nucleotide following a G can be labeled using γ^{32}P-ATP and polynucleotide kinase, which adds the γ phosphate from ATP to the 5′OH of the polynucleotide.
5. When processing large numbers of samples, the starting RNA can often be omitted, as can the synthetic standards once the mobility patterns have been established.
6. Care should be taken as both supernatant and pellet will be radioactive. It is often useful to transfer the supernatant to a fresh microcentrifuge tube so that a rough estimation of the relative incorporation can be made (Geiger counter). The pellet will not normally be visible, but will normally contain >50 % of the radioactive counts. The radioactive supernatant can then be disposed of according to local radioisotope usage rules.
7. It is not necessary to wash the pellet, but it is important to ensure that all the ethanol is removed. Trace amounts of ethanol can partially inhibit the P1 nuclease and can lead to additional spots on the TLC (Fig. 3).
8. The volume of the buffer used can be varied depending upon tank dimensions. The volume used should give a depth of 1.5–2 cm in the bottom of the tank, it is important that when the TLC plate is placed in the tank, the buffer is below the point of loading on the plate.
9. Fluctuations in temperature can result in variations in separation rate.

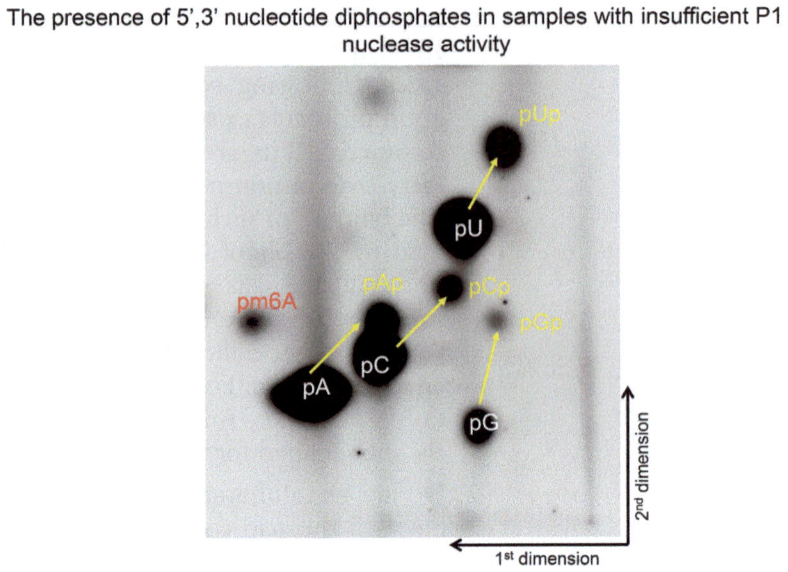

Fig. 3 Insufficient P1 nuclease activity can give rise to additional spots due to the presence of 5′3′ nucleotide diphosphates. The extra spots on the TLC plates are due to the presence of 5′3′ dinucleotides, where pAp is 5′3′ adenosine diphosphate, pCp is 5′3′ cytosine diphosphate, pUp is 5′3′ uridine diphosphate, and pGp is 5′3′ guanosine diphosphate. The new spots show a shift to the direction of the right top corner of the TLC plate due to slower migration in the first dimension and faster migration in the second. The sample was an *A. thaliana* 2 round oligo(dT) purified mRNA sample. This error usually occurs when the P1 nuclease digestion is insufficient, due to traces of ethanol in the digest, old P1 enzyme, or short digestion time combined with high concentration RNA sample. It can also be an indication of low level RNase contamination

10. The sample can be loaded in two or three aliquots with drying in between using a hair dryer on cold or low heat setting.
11. Plastic-backed plates will run significantly faster, typically 8 h for a full 20 × 20 cm plate and less than this if the plate is cut down.
12. Exposure time to the TLC plates is dependent on the radiation intensity of the nucleotide spots on the TLC. Typically, this would be from 4 h (>100 cpm) to overnight (<100 cpm). It is important to keep the exposure of phosphor screens to light to a minimum during removing from cassette before FX scanning, to prevent signal weakening.
13. When saving the image before the start of scanning have the saturated pixels warning on. The m^6A spots are unlikely to become saturated but in long exposures of hot samples, the unmodified nucleotide spots may become saturated. If this is the case, the phosphor screens have to be erased, and reexposed for a shorter time. This is very important as quantification of

an image with saturated A spot would lead to an overestimation of the m^{6}A content of the sample.

14. The final step of quantification is calculating the ratio of pm^{6}A to pA and this is expressed as %. This ratio gives a % of GAs that are methylated. There are literature reports where the ratio of pm^{6}A to pC has been used as a typical value for adenosine methylation. However, this is potentially misleading and is largely meaningless. As the ratio of pCs to pAs is not necessarily fixed between samples, the changes of m^{6}A values, when pm^{6}A to pC ratio is used, may only reflect the changes in pA to pC ratios.

References

1. Wang Y, Li Y, Toth JI, Petroski MD, Zhang Z, Zhao JC (2014) *N*^{6}-methyladenosine modification destabilizes developmental regulators in embryonic stem cells. Nat Cell Biol 16:191–198. doi:10.1038/ncb2902
2. Wang X, Lu Z, Gomez A, Hon GC, Yue Y et al (2014) *N*^{6}-methyladenosine-dependent regulation of messenger RNA stability. Nature 505:117–120 /nature12730
3. Bodi Z, Bottley A, Archer N, May ST, Fray RG (2015) Yeast m^{6}A methylated mRNAs are enriched on translating ribosomes during meiosis, and under rapamycin treatment. PLoS One 10(7):e0132090. doi:10.1371/journal.pone.0132090
4. Meyer KD, Patil DP, Zhou J, Zinoviev A, Skabkin M, Elemento O et al (2015) 5′UTR m^{6}A promotes cap-independent translation. Cell 163:999–1010
5. Horowitz S, Horowitz A, Nilsen TW, Munns TW, Rottman FM (1984) Mapping of *N*^{6}-methyladenosine residues in bovine prolactin mRNA. Proc Natl Acad Sci U S A 81: 5667–5671
6. Kane SE, Beemon K (1985) Precise localization of m^{6}A in Rous sarcoma virus RNA reveals clustering of methylation sites: Implications for RNA processing. Mol Cell Biol 5:2298–2306
7. Harper JE, Miceli SM, Roberts RJ, Manley JL (1990) Sequence specificity of the human mRNA *N*^{6}-adenosine methylase in vitro. Nucleic Acids Res 18:5735–5741
8. Schwartz S, Mumbach MR, Jovanovic M, Wang T, Maciag K et al (2014) Perturbation of m^{6}A writers reveals two distinct classes of mRNA methylation at internal and 5′ sites. Cell Rep 8:284–296. doi:10.1016/j.celrep.2014.05.048
9. Meyer KD, Saletore Y, Zumbo P, Elemento O, Mason CE, Jaffrey SR (2012) Comprehensive analysis of mRNA methylation reveals enrichment in 3′ UTRs and near stop codons. Cell 149:1635–1646
10. Zhong S, Hongying L, Bodi Z, Button JD, Vespa L, Herzog M, Fray RG (2008) *MTA* is an Arabidopsis messenger RNA adenosine methylase and interacts with a homolog of a sex-specific splicing factor. Plant Cell 20:1278–1288
11. Sambrook J, Russel DW (2001) Molecular cloning : a laboratory manual. Cold Spring Harbor Laboratory Press, Cold Spring Harbor, NY, pp. 7.13–7.17
12. Conrad RC, AMBION, INC. High efficiency mRNA isolation. United States Patent Application Publication, Pub. No.: US 2004/0230048 A1 Conrad Pub. Date: 18 Nov 2004.

Part III

Analysis of 5-Methylcytosine

Chapter 7

Illustrating the Epitranscriptome at Nucleotide Resolution Using Methylation-iCLIP (miCLIP)

Harry George, Jernej Ule, and Shobbir Hussain

Abstract

Next-generation sequencing technologies have enabled the transcriptome to be profiled at a previously unprecedented speed and depth. This yielded insights into fundamental transcriptomic processes such as gene transcription, RNA processing, and mRNA splicing. Immunoprecipitation-based transcriptomic methods such as individual nucleotide resolution crosslinking immunoprecipitation (iCLIP) have also allowed high-resolution analysis of the RNA interactions of a protein of interest, thus revealing new regulatory mechanisms. We and others have recently modified this method to profile RNA methylation, and we refer to this customized technique as methylation-iCLIP (miCLIP). Variants of miCLIP have been used to map the methyl-5-cytosine (m5C) or methyl-6-adenosine (m6A) modification at nucleotide resolution in the human transcriptome. Here we describe the m5C-miCLIP protocol, discuss how it yields the nucleotide-resolution RNA modification maps, and comment on how these have contributed to the new field of molecular genetics research coined "epitranscriptomics."

Key words Epitranscriptome, Epitranscriptomics, RNA methylation, Methylation-iCLIP, miCLIP, NSun2

1 Introduction

1.1 RNA Methylation and the Development of Transcriptomic Techniques to Detect Their Occurrence at Nucleotide Resolution

Covalent modifications to cellular RNA molecules can occur both on the sugar-phosphate RNA backbone and on nucleobases, and can include those that occur at internal positions as well as those present on 5′ or 3′ termini. A few of these such as the m7G cap that is ubiquitously present on the 5′ terminus of cellular mRNAs have been well characterized. However, it is known that well over a hundred types of modified RNA nucleobases can exist, though the characterization of these has been hampered largely due to a lack of suitable techniques to detect their sequence-specific occurrence. For example, it had for decades been suspected that internal methyl-6-adenosine (m6A) and methyl-5-cytosine (m5C) base methylations may be a common occurrence on mRNAs in mammalian cells [1], but it was only much more recently that the development of refined transcriptomic approaches allowed these

Alexandra Lusser (ed.), *RNA Methylation: Methods and Protocols*, Methods in Molecular Biology, vol. 1562,
DOI 10.1007/978-1-4939-6807-7_7, © Springer Science+Business Media LLC 2017

modifications to be comprehensively mapped [2–6] thus greatly aiding efforts to investigate their mechanistic roles [6–10].

Previous important studies were also critical for setting the stage for the recent flurry of activity in the field. It had been shown that a multiprotein complex, which includes the METTL3 and METTL14 subunits, catalyzes the formation of m6A in RNA [11] and that an additional enzyme encoded by the FTO gene can remove these methylation marks [12]. The latter finding was particularly exciting as it suggested that these RNA methylation events could be dynamically controlled during the execution of important biological processes. Two seminal studies then appeared simultaneously in the literature describing the use of essentially the same immunoprecipitation-based method to map m6A modifications in the transcriptome [2, 3]. The approach consisted of fragmenting cellular RNA into small pieces and then using an antibody specific for the methylated base to immunoprecipitate m6A-containing fragments. Libraries were then prepared from the purified RNAs for high-throughput sequencing on a Next-Generation-Sequencing (NGS) platform. Although the m6A sites could in reality be located anywhere within the sequencing reads, computational algorithms were used to predict the likely methylation site. Importantly, it was also found by both studies that most of the RNA fragments mapped to within close vicinity of the stop codon of mRNAs, although the mechanistic significance of this finding remains to be determined.

An additional study published in 2012 also reported transcriptomic identification of m5C using bisulfite sequencing [4], which had been used previously to detect DNA m5C modifications. The authors optimized the protocol so that RNA could seemingly withstand the harsh chemical conditions of the bisulfite treatment. The analysis revealed widespread occurrence of the modification in several RNA biotypes, including mRNAs as previously predicted by Dubin and Taylor (1975)[1].The integrity of the RNA libraries following bisulfite treatment however could not be fully assessed and the method also does not enrich for modified RNA sites, as in the m6A-immunoproecptitaion protocol, meaning that methylation-site detection and calling was likely suboptimal. Independent approaches were also clearly necessary and were provided by two studies again appearing fortuitously in quick succession in the literature that described distinct but related cross-linking immunoprecipitation (CLIP)-based techniques for detecting m5CRNA modifications [5, 6].

Initially described for the use of detecting protein–RNA interactions, efficient CLIP-based techniques employed UV covalent crosslinking to enable the RNA interaction partners of a protein of interest to be identified with high stringency and at a remarkably high signal-to-noise ratio [13, 14]. The methylation of m5C RNA is catalyzed via a covalent catalytic intermediate between the RNA methylase enzyme and the RNA substrate: if these catalytic intermediates could be frozen or "trapped," they could be immunoprecipitated to detect the RNA methylation sites. Khoddami and Cairns (2013) described their 5-azacytidine-mediated RNA immunoprecipitation (Aza-IP) approach where they employed the cytidine nucleoside analogue5-azacytidine [5], which becomes incorporated into DNA and RNA during replication and transcription respectively. m5C DNA and RNA methylases will methylate cytosine residues as well as 5-azacytosine residues via a covalent catalytic intermediate, but when they methylate the latter, the resolution of the covalent adduct cannot take place due to the altered chemistry of the nucleobases. Thus cells treated with 5-Azacytidine enabled CLIP-based experiments to be performed where an antibody against an epitope-tagged m5C RNA methylase was used to immunoprecipitate the frozen catalytic intermediates, enabling identification of the methylation substrates. In CLIP experiments, during the cDNA synthesis step of the library preparation, mutations in the sequence are incorporated as the reverse transcriptase traverses the crosslink site; these crosslink-induced mutation sites (CIMS) are estimated to occur at a frequency of ~8–20 % of reverse transcription events, and their presence can be used to pinpoint the crosslink/interaction site [15]. Similarly, in the Aza-IP libraries, it was observed that the sequence reads often also contained CIMS at the crosslink/methylation site and hence these could be used to infer the methylation site at nucleotide resolution [5].

Aza-IP involves altering the physiology of cells by requiring a 12 h pre-incubation with the cytotoxic agent 5-azacytidine, which is known to lead to hypomethylation of DNA even at low doses, and its effect of the nucleoside analogue following incorporation into different RNA species is not fully characterized but may interfere with RNA translation and metabolism. Further, more than 80 % of reverse transcription events are estimated to stall at the protein-RNA crosslink site during cDNA synthesis, rather than reading through the crosslink site [16]; in CLIP-based techniques, these truncated cDNAs are lost during NGS library preparation as they lack the terminal adapter sequence required for PCR amplification. This is overcome by the individual-nucleotide resolution crosslinking immunoprecipitation (iCLIP) method, which was developed to enable amplification of such truncated cDNAs [17]. In iCLIP, the last nucleotide of the truncated cDNAs can be used to identify the crosslink site with increased efficiency [16, 17]. The methylation-iCLIP (miCLIP) technique developed in Hussain et al. (2013a) describes a customized version of iCLIP [6] that takes advantage of a mutation within the key cysteine residue of m5C RNA methylases that is required for resolution of catalytic intermediates [18–20] (Fig. 1). Substitution of this catalytic cysteine to alanine led to effective trapping of catalytic intermediates, which enabled to crosslink the methylases to its endogenous RNA targets without the need of photocrosslinking. These crosslinked

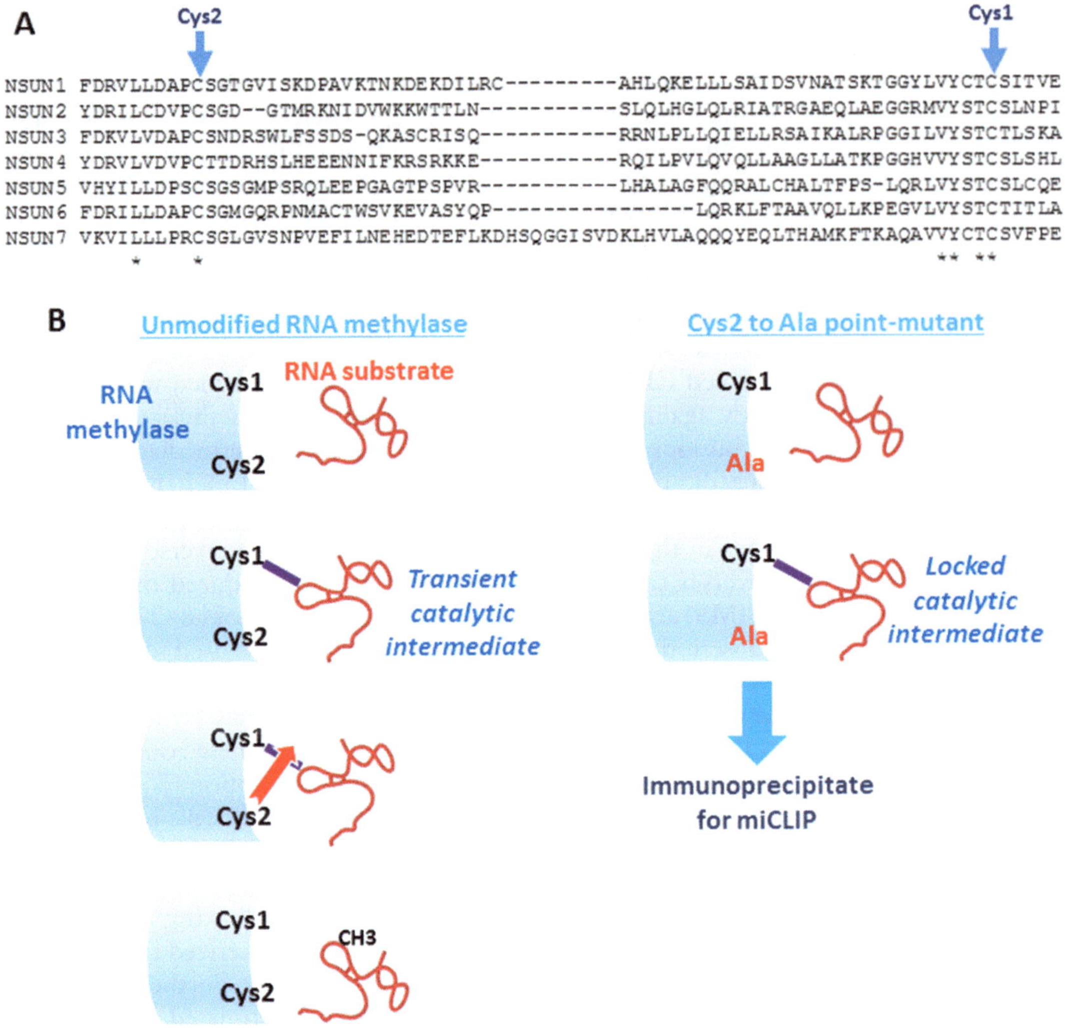

Fig. 1 (**a**) Sequence alignment of a region of the catalytic domain of the seven known members of the human family of m5C RNA methylases. Two key conserved cysteine residues (Cys1 & Cys2) play critical roles in the methylation process. Identical residues in the alignment are indicated by asterisk. (**b**) The conserved cysteine residue Cys1 forms a transient covalent bond (*purple line*) with the cytosine being methylated during the methylation process. A second cysteine Cys2 is then critical for the resolution of this covalent bond to complete the catalytic process (*left*). In methylation-iCLIP, Cys2 is mutated to alanine (*right*) which results in the capture of the catalytic intermediate where the methylase is covalently bound to its substrate. This complex can then be immunoprecipitated using an antibody against the methylase or an epitope tag, and the trapped RNAs are then amenable to purification under highly stringent conditions. Libraries prepared from recovered RNA substrates are then subjected to high-throughput sequencing

epitope-tagged enzyme-substrate covalent adducts are then immunoprecipitated, the cDNA library is prepared using the iCLIP protocol, and the positions of cDNA truncations identified the m5C methylation sites with nucleotide-resolution.

The miCLIP study demonstrated that identified m5C substrates were likely of biological relevance. For example, miCLIP for the NSun2 m5C RNA methylase detected methylation sites at various nucleotide positions within the Vault ncRNAs; these were then confirmed by performing bisulfite sequencing of Vault ncRNA extracted from cells obtained from human patients with NSun2-deficiency syndromes [6]. The study also showed that the methylation events mechanistically impacted on the processing of Vault hairpin RNAs into small RNAs with microRNA-like function. Importantly, there are six other members of the NSun family of m5C RNA methylases which should also be amenable to miCLIP (Fig. 1).

Recently, a further iCLIP-based technique was developed to identify m6A sites at nucleotide resolution in transcriptomes [21]. The technique involves incubation of a m6A-antibody with purified cellular RNA followed by in vitro UV-crosslinking. These crosslinked antibody-RNA complexes are then immunoprecipitated, the cDNA library is prepared using the iCLIP protocol, and the positions of cDNA truncations identified the m6A methylation sites with nucleotide-resolution. The procedure was named m6A-iCLIP and abbreviated to miCLIP. To allow a broader use of this term, we suggest that miCLIP is described as methylation-iCLIP, which can then be applied both for cases where protein-RNA complexes are crosslinked within intact cells [6] or purified RNA [21] to identify sites of RNA methylation.

1.2 An Overview of m5C-iCLIP

The aforementioned work [2–6, 21] as well as important studies that have also described novel transcriptomic techniques to detect other types of RNA modifications [21–23] have set the stage for "epitranscriptomics." This area of research concerns the sequence-context specific characterization and investigation of RNA modifications in complex transcriptomes [24–26]. Methylation-iCLIP will continue to play an important role in this fledgling field, and in this chapter we describe the m5C-iCLIP protocol. In classic iCLIP experiments, UV crosslinking is required to introduce covalent crosslinks between proteins and interacting RNAs; this step is omitted from the miCLIP protocol. Instead, expression of a point-mutant form of an RNA methylase of interest is required which will lead to RNA substrate trapping; we have previously and are currently performing this using point mutant forms of the NSun family of RNA m5C methylases. Experiments can be performed by transfecting suitable expression vectors into cells which also offers the advantage that an antibody against an epitope tag can be used for immunoprecipitation. Genome editing methods, however, may also potentially be used to achieve endogenous expression of the point mutant. Following expression, cells are lysed and the enzyme-substrate covalent adducts are recovered by immunoprecipitation; the use of a good, highly specific antibody is the key to the experiment and we generally recommend using an antibody against an epitope tag. A multistep library preparation procedure is then employed which is designed to enable methylation sites to be determined at nucleotide resolution following sequencing of prepared libraries on an NGS platform. Bioinformatics analysis of recovered data can be performed by using the online PIPE-CLIP tool [27]. As in iCLIP, the primers used for cDNA synthesis during library prep contain a randomized barcode (often referred to as unique molecular identifiers or UMIs), which allow individual cDNAs to be counted during analyses, thus yielding a reliable quantitative estimate of crosslinking efficiencies at each nucleotide.

2 Materials

2.1 Cell Culture and Transfection

1. HEK293 cells.
2. Growth medium: Dulbecco's Modified Eagles Medium (DMEM), 10 % Fetal Bovine Serum, Penicillin-Streptomycin.
3. Trypsin-EDTA: 1× solution.
4. Phosphate Buffered Saline (PBS): 137 mM NaCl, 2.7 mM KCl, 10 mM Na_2HPO4, 1.8 mM KH_2PO_4.
5. Lipofectamine 2000.

2.2 Immunoprecipitation and Linker Ligation

1. Protein G or A magnetic Dynabeads.
2. RNase I: 100 U/μL.
3. High RNase: 1:50 dilution of RNaseI.
4. Low RNase: 1:1000 dilution of RNaseI.
5. Turbo DNase: 2 U/μL.
6. T4 PNK (with 3′ phosphatase activity) plus 10× PNK buffer.
7. RNasin: 20–40 U/μL.
8. T4 RNA Ligase I: 30 U/μL.
9. PEG400.

2.3 SDS PAGE Analysis and Nitrocellulose Transfer

1. [γ-^{32}P]-ATP; 3000 Ci/mmol: 10 mM.
2. NuPAGE loading buffer.
3. 4–12 % NuPAGE Bis-Tris gel.
4. NuPAGE MOPS SDS running buffer.
5. Prestained protein marker.
6. Nitrocellulose membrane.
7. Autoradiography cassette.
8. X-ray film.

2.4 Library Preparation

1. Proteinase K: 20 mg/mL in water.
2. Phenol-chloroform (24:23 v/v).
3. Phase-lock gel heavy tubes.
4. Glycoblue: 15 mg/mL in water.
5. Sodium acetate solution: 3 M in water, pH 5.5.
6. Superscript III reverse transcriptase and 5× RT buffer (Life Technologies).
7. dNTP mix: 10 mM dATP, 10 mM dTTP, 10 mM dCTP, 10 mM dGTP.
8. DTT solution: 0.1 M in water.
9. RNase I: 100 U/μL
10. 2× TBE-Urea loading buffer (Thermofisher Scientific).
11. 6 % TBE-Urea polyacrylamide gel.
12. Costar SpinX columns.
13. Glass pre-filters.
14. CircLigase II enzyme and 10× CircLigase II buffer (Epicentre).
15. FastDigest BamHI enzyme and 10× FastDigest Buffer.
16. P5/P3 solexa primer mix: 10 μM each primer.
17. Accuprime Supermix (Life Technologies).
18. 6 % TBE polyacrylamide gel.
19. 10,000×SYBR green I (Thermofisher Scientific).

2.5 Buffers

1. Lysis buffer: 50 mM Tris–HCl, pH 7.4, 100 mM NaCl, 1 % Igepal CA-630, 0.1 % SDS.
2. High-salt wash buffer: 50 mM Tris–HCl, pH 7.4, 1 M NaCl, 1 mM EDTA, 1 % Igepal CA-630, 0.1 % SDS.
3. PNK buffer: 20 mM Tris–HCl, pH 7.4, 10 mM $MgCl_2$, 0.2 % Tween-20.
4. 5× PNK pH 6.5 buffer: 350 mM Tris–HCl, pH 6.5, 50 mM $MgCl_2$, 25 mM dithiothreitol (DTT).
5. 4× ligation buffer: 200 mM Tris–HCl, pH 7.8, 40 mM $MgCl_2$, 40 mM dithiothreitol (DTT).
6. PK buffer: 100 mM Tris–HCl, pH 7.4, 50 mM NaCl, 10 mM EDTA.
7. PK buffer-7 M Urea buffer: 100 mM Tris–HCl, pH 7.4, 50 mM NaCl, 10 mM EDTA, 7 M urea.
8. TE Buffer: 10 mM Tris, 1 mM EDTA, pH 8.

2.6 Oligonucleotides

1. Pre-adenylated linker (L3-App): 20 μM.
   rApp/AGATCGGAAGAGCGGTTCAG/ddC; rApp = adenylation, ddC = dideoxycytidine (*see* **Note 1**).

2. RT primers (*see* **Note 2**):10 μM each.
   Rt1clip/5Phos/NNAACCNNNAGATCGGAAGAGCGTC GTGgatcCTGAACCGC.
   Rt2clip/5Phos/NNACAANNNAGATCGGAAGAGCGTC GTGgatcCTGAACCGC.
   Rt3clip/5Phos/NNATTGNNNAGATCGGAAGAGCGTC GTGgatcCTGAACCGC.
   Rt4clip/5Phos/NNAGGTNNNAGATCGGAAGAGCGTC GTGgatcCTGAACCGC.
   Rt6clip/5Phos/NNCCGGNNNAGATCGGAAGAGCGTC GTGgatcCTGAACCGC.
   Rt7clip/5Phos/NNCTAANNNAGATCGGAAGAGCGTC GTGgatcCTGAACCGC.
   Rt8clip/5Phos/NNCATTNNNAGATCGGAAGAGCGTC GTGgatcCTGAACCGC.
   Rt9clip/5Phos/NNGCCANNNAGATCGGAAGAGCGTC GTGgatcCTGAACCGC.
   Rt11clip/5Phos/NNGGTTNNNAGATCGGAAGAGCGTC GTGgatcCTGAACCGC.
   Rt12clip/5Phos/NNGTGGNNNAGATCGGAAGAGCGTC GTGgatcCTGAACCGC.
   Rt13clip/5Phos/NNTCCGNNNAGATCGGAAGAGCGTC GTGgatcCTGAACCGC.
   Rt14clip/5Phos/NNTGCCNNNAGATCGGAAGAGCGTC GTGgatcCTGAACCGC.
   Rt15clip/5Phos/NNTATTNNNAGATCGGAAGAGCGTC GTGgatcCTGAACCGC.
   Rt16clip/5Phos/NNTTAANNNAGATCGGAAGAGCGTC GTGgatcCTGAACCGC.
   N = random nucleotides.
3. BamHI-cut-oligo (GTTCAGGATCCACGACGCTCTTCaaaa): 10 μM.
4. Illumina PCR primers:10 μM each.
   P5Solexa: AATGATACGGCGACCACCGAGATCTACACT CTTTCCCTACACGACGCTCTTCCGATCT.
   P3Solexa: CAAGCAGAAGACGGCATACGAGATCGGTC TCGGCATTCCTGCTGAACCGCTCTTCCGATCT.

3 Methods

3.1 Cell Culture and Transfection

1. In a 10 cm dish, grow HEK293 cells in growth medium in a humidified incubator with 5 % CO_2 until they reach 50–70 % confluence.
2. Use Lipofectamine 2000 (Life Technologies) to transfect cells with mammalian expression vector encoding the RNA methylase of interest harboring the substrate trapping point mutation (*see* **Note 3**) and incubate cells for 24 h.

3. Following incubation, wash cells twice in PBS. Disperse cells by incubating cells for 1 min in Trypsin-EDTA and then add 30 mL of growth medium.
4. Collect cells by centrifugation and then wash twice more in PBS via centrifugation and resuspension of cell pellet.
5. Transfer cells in 1 mL PBS to microcentrifuge tubes, and subject to centrifugation to pellet cells for a final time. Cell pellets may be used immediately or frozen at −80 °C for later use.

3.2 Immunoprecipitation

All the subsequent steps in the protocol should be performed on wet ice unless otherwise stated.

3.2.1 Dynabead Preparation

1. Transfer 100 μL of resuspended magnetic dynabeads (*see* **Note 4**) to two 1.5 mL microtubes and label one "High RNase" and the other "Low RNase". Also repeat this for a no-antibody control sample.
2. Wash the beads twice with 1 mL of lysis buffer.
3. After the last wash, resuspend the beads in 500 μL of lysis buffer and then add 5–10 μg of antibody.
4. Rotate tubes for 30 min at room temperature.
5. Wash beads 5 times with 1 mL lysis buffer and leave in last wash until ready to proceed to Subheading 3.2.3.

3.2.2 Lysis and Partial RNA Digestion

1. Add 1 mL of cold lysis buffer to cell pellet, and then pipette to resuspend cell pellet. Incubate on ice for 30 min and syringe lysate through a 19-gauge needle every 10 min during the incubation.
2. Add 10 μL of high or low RNase (*see* **Note 5**) to the corresponding tube and 2 μL Turbo DNase to the cell lysate and then place the samples into a thermomixer set at 37 °C for 3 min, shaking at 1100 rpm. Following the 3 min incubation immediately transfer the microtube onto wet ice and leave for at least 3 min.
3. Centrifuge for 20 min at 20,000 × *g* at 4 °C and then transfer the supernatant to a 1.5 mL microtube. The remaining pellet can be discarded.

3.2.3 Immunoprecipitation

1. Add the clarified lysate to the beads from Subheading 3.2, **step 5** and then rotate for 2–3 h at 4 °C.
2. Use a magnetic rack to collect the dynabeads, discard the supernatant, and then wash the beads twice with 1 mL high salt wash buffer.
3. Wash three more times with 1 mL PNK buffer for each wash and discard the supernatant from the final wash.

3.3 RNA 3′ Dephosphorylation

1. Resuspend beads to homogeneity in 20 μL 3′ dephosphorylation mix containing 4 μL5× PNK pH 6.5 buffer, 0.5 μL PNK (with 3′ phosphatase activity), 0.5 μL RNasin, 15 μL water.
2. Incubate for 20 min at 37 °C in a thermomixer at 1100 rpm.
3. Remove from thermomixer and add 1 mL of PNK buffer to wash beads. Perform four further washes in 1 mL PNK buffer.

3.4 Preadenylated L3 Adapter Ligation

1. Remove the supernatant and resuspend the beads in 20 μL adapter ligation mix containing 8 μL water, 5 μL 4× ligation buffer, 1 μL T4 RNA ligase, 0.5 μL RNasin, 1.5 μL preadenylated linker L3-App, 4 μL PEG400.
2. Incubate overnight at 16 °C in a thermomixer at 1100 rpm.
3. Remove from thermomixer and add 1 mL of PNK buffer to wash beads. Perform four further washes in 1 mL PNK buffer, and then leave in 1 mL of PNK buffer.

3.5 RNA 5′ End Labeling

1. Transfer 200 μL of the resuspended beads (the remaining 800 μL will be used in **step 4**) to a new RNAse-free microtube and remove the supernatant.
2. Resuspend beads in 10 μL of radioactive PNK mix containing 0.5 μL PNK, 1 μL ^{32}P-γ-ATP, 1 μL 10× PNK buffer, 7.5 μL water.
3. Incubate for 5 min at 37 °C and then discard the supernatant as radioactive waste.
4. Resuspend beads in 20 μL 1× NuPAGE loading buffer and add this to the remaining cold beads from Subheading 3.4, **step 3**, following removal of the supernatant from them.
5. Incubate for 5 min at 80 °C.
6. Place back on the magnet and collect the supernatant which will be loaded directly onto the gel in the next step.

3.6 SDS–PAGE and Nitrocellulose Transfer

1. Load the samples on a 4–12 % NuPAGE Bis-Tris gel. Also load 5 μL of a prestained protein size marker. Use 0.5 L 1×NuPAGE MOPS SDS running buffer.
2. Run the gel at 180 V for 50 min and once complete, cut off the foot of the gel containing the dye front, and discard it as solid radioactive waste.
3. Transfer the protein–RNA complexes from the gel to a nitrocellulose membrane at 30 V for 1 h.
4. Once the transfer is complete, wrap the nitrocellulose membrane in saran wrap and fix it into an autoradiography cassette that contains a fluorescent marker.
5. Expose the membrane to an X-ray film and then transfer the cassette into a −80 °C freezer. Exposures can be performed for 30 min, 1 h, and overnight.

3.7 Library Preparation

3.7.1 RNA Isolation

1. Use the fixed fluorescent marker in the autoradiography cassette to realign the processed film and then mark the area of the nitrocellulose membrane that contains the protein-RNA complex smear in the low RNase condition (*see* **Note 6**). Using a sharp blade, carefully cut out this region of nitrocellulose, dissect further into small pieces, and then transfer them to a RNase-free 1.5 mL microtube.
2. Add 10 μL proteinase K in 200 μL PK buffer to the nitrocellulose fragments so that they are all submerged. Transfer tubes to a thermomixer and incubate for 20 min shaking at 1100 rpm at 37 °C.
3. Add 200 μL of PK buffer-7 M urea and incubate for further 20 min at 37 °C and 1100 rpm.
4. Collect the solution and add it to a 2 mL Phase Lock Gel Heavy tube. Add 400 μL phenol/chloroform and then incubate in a thermomixer for 5 min at 30 °C shaking at 1100 rpm.
5. Separate the phases by spinning for 5 min at 20,000 × *g* and room temperature.
6. Transfer the aqueous layer into a nonstick RNase-free 1.5 mL microfuge tube and add 1 μL glycoblue and 40 μL sodium acetate solution. Mix and then add 1 mL 100 % ethanol. Mix again and then place at −20 °C overnight.
7. Centrifuge at 20,000 × *g* at 4 °C for 20 min and then remove the supernatant. Wash the pellet with 1 mL 80 % ethanol and spin again for 5 min. Resuspend the pellet thoroughly in 5 μL H_2O and transfer to an RNase-free PCR tube.

3.7.2 Reverse Transcription

1. Add 1 μL of RT primer (*see* **Note 7**) and 1 μL dNTP mix to the resuspended pellet and then incubate in a thermocycler set to the following programme: 70 °C for 5 min, then hold at 25 °C until the RT mix is added next.
2. Add 13 μL of the RT mix containing 7 μL H_2O, 4 μL 5× RT buffer, 1 μL DTT, 0.5 μL RNasin, 0.5 μL Superscript III reverse transcriptase, transfer tube to a thermal cycler and perform the reverse transcription with the following program: 25 °C for 5 min, 42 °C for 20 min, 50 °C for 40 min, 80 °C for 5 min, then hold at 4 °C.
3. Once the RT program is complete, add 350 μL TE buffer, 1 μL GlycoBlue, and 40 μL sodium acetate pH 5.5. Mix, and then add 1 mL 100 % ethanol. Mix again and precipitate overnight at −20 °C.

3.7.3 Gel Purification (See Note 8)

1. Centrifuge for 20 min at 20,000 × *g* at 4 °C. Remove the supernatant and add 1 mL 80 % ethanol. Repeat centrifugation, remove supernatant, and then resuspend the pellet in 6 μL H_2O.

2. Add 6 μL 2×TBE-urea loading buffer to the samples and also to 6 μL DNA size marker. Heat samples at 70 °C for 2 min immediately before loading onto a 6 % TBE-urea gel (Life Technologies).
3. Run gel in 1× TBE buffer for 40 min at 180 V until the lower dye is close to the bottom.
4. Cut off the last lane containing the size marker and incubate it in 20 mL 1X TBE buffer with 2 μL SYBR green II for 10 min. Wash the gel twice with 1XTBE and visualize by UV transillumination. Print the result with 100 % scale, and use it as a mask to guide the excision of cDNA bands from the rest of the gel.
5. Together with the preadenylated L3 adapter sequence, the primer sequence accounts for 52 nt of the cDNA. Cut three bands, at 70–80 nt, 80–150 nt, and 150–200 nt and transfer them each to a nonstick RNase-free 1.5 mL microtube (*see* **Note 9**).
6. Add 400 μL TE buffer to the gel piece and crush it into tiny pieces with a 1 mL pipette tip.
7. Transfer to a thermomixer and incubate with shaking at 1100 rpm for 1 h at 37 °C. Then place on dry ice for 5 min, and then place back for 1 h at 1100 rpm at 37 °C in the thermomixer.
8. Transfer the liquid portion of the supernatant into a Costar SpinX column into which you have placed two 1 cm glass pre-filters.
9. Centrifuge at 20,000 × *g* for 1 min at room temperature and transfer the flow-through to a nonstick RNase-free 1.5 mL microtube.
10. Add 1 μL glycoblue and 40 μL sodium acetate solution and mix. Add 1 mL 100 % ethanol, mix again, and precipitate at −20 °C overnight.

3.7.4 cDNA Circularization (See Note 10)

1. Centrifuge for 20 min at 20,000 × *g* at 4 °C. Remove the supernatant and add 1 mL 80 % ethanol. Repeat centrifugation, remove supernatant, and then resuspend the pellet in 8 μL ligation mix containing 6.5 μL water, 0.8 μL 10× CircLigase Buffer II, 0.4 μL 50 mM $MnCl_2$, 0.3 μL CircLigase II. Transfer the mix to nuclease-free PCR tubes and incubate for 1 h at 60 °C.
2. Add 30 μL oligonucleotide annealing mix: 26 μL H_2O, 3 μL FastDigest Buffer, 1 μL BamHI Cut_oligo and anneal in a thermal cycler using the following program: 95 °C for 2 min; then successive cycles of 20 s, starting from 95 °C and decreasing the temperature by 1 °C each cycle down to 25 °C; then hold at 25 °C.

3. Add 2 μL FastDigest BamHI enzyme and incubate for 30 min at 37 °C, followed by an incubation at 80 °C for 5 min.
4. Add 350 μL TE, 1 μL GlycoBlue, 40 μL sodium acetate solution and mix. Then add 1 mL 100 % ethanol, mix again, and precipitate overnight at −20 °C.

3.8 PCR Amplification and Library Submission

1. Centrifuge for 20 min at 20,000 × *g* at 4 °C. Remove the supernatant and add 1 mL 80 % ethanol. Repeat centrifugation, remove supernatant, and then resuspend the pellet in 20 μL H_2O and transfer to PCR tube. 10 μL of this will be used in the PCR in the next step and the remainder can be stored at −20 °C for further future use if necessary.
2. Prepare the PCR mix as follows: 10 μL cDNA, 1 μL primer mix of P5Solexa and P3Solexa, 20 μL Accuprime Supermix, 9 μL H_2O.
3. Place tube in thermal cycler and run the following program: 94 °C for 2 min, followed by 25 cycles of: 94 °C for 15 s, 65 °C for 30 s, 68 °C for 30 s; complete with 68 °C for 3 min, then hold at 25 °C.
4. Once the program is complete, mix 8 μL PCR product with 2 μL 5× TBE loading buffer, load on a 6 % TBE polyacrylamide gel, and run at 180 V and 120 mA for 30 min. Stain and visualize with SYBR green I (*see* **Note 11**).
5. PCR amplified miCLIP libraries can then be quantified using a Qubit fluorometer (Life Technologies).
6. Dilute libraries to 10 nM, or to a sequencing facility-specified concentration, and sequence using 50 nt single or paired-end runs on an Illumina sequencer. We have routinely performed sequencing on the GAII and also the MiSeq platform to generate sufficient RNAseq reads.
7. Bioinformatics analysis may be performed using the online PIPE-CLIP tool [27].

4 Notes

1. Pre-adenylation (rApp) of the DNA linker is required for efficient ligation to the 3′-OH group of RNAs. The dideoxycytidine group (ddC) is added to the 3′ end of the adenylated linker to prevent circularization.
2. The 5′end of each primer contains a 9-nucleotide barcode consisting of 2 random nucleotides (N) followed by a 4-nucleotide fixed barcode and finished off with 3 more random nucleotides. The fixed barcode allows multiplexing of replicates or samples, which is important to reduce the costs of the downstream next-generation sequencing. The random

barcode allows for tracing and counting of individual cDNA synthesis events, thus avoiding the overrepresentation of abundant RNAs which would otherwise be encountered by analyzing exponentially amplified PCR product. The last 9 nucleotides of the RT primers have sequence complementarity to the final 9 nucleotides of the L3-App linker.

3. Enzymes that catalyze RNA modification via a two-step catalytic process as described in Fig. 1 will be amenable to methylation-iCLIP. Several enzymes have been described which are known to catalyze RNA modifications via a covalent intermediate [28]; however, the biochemical mechanisms via which these adducts are resolved are often poorly defined. It is possible that amino acid residues within the catalytic domain are critical for resolution, as is observed in the NSun family of RNA m5C methylases, but these will require identification for each particular enzyme before methylation-iCLIP can be performed.
4. Generally, protein G for mouse antibodies, and protein A for rabbit antibodies is used.
5. RNA digestion conditions will need to be preoptimized for each protein-RNA complex under study. We initially recommend using a 1:50 dilution of RNaseI for "high RNase" and a 1:1000 dilution for "low RNase." High RNase treatment should result in a sharp band around the expected molecular weight of your RNA methylase of interest during the SDS-PAGE analysis (Subheading 3.6), whereas the low RNase treatment should result in a smear above the expected molecular weight of the protein.
6. The high RNAse sample is only used in the protocol to check for the presence of the desired protein-RNA complex following immunoprecipitation, but is not used to prepare libraries as the RNAs would be too short in length to map uniquely to the genome. The RNAs present in the higher molecular weight smear in the low RNase condition are thus used for library preparation only.
7. A distinct RT primer (one of Rt1clip to Rt16clip) can be used for each of several replicates or samples if desired. The barcode on these RT primers (*see* **Note 2**) will then allow for multiplexing of samples, thus enabling the cost of next-generation sequencing to be managed.
8. Gel purification is necessary to remove residual RT primer and also for size fractionation of cDNAs (*see* **Note 9**).
9. The band cut at 70–80 nt will contain a cDNA insert size of 18–28 nt; this band usually consists of a high cDNA complexity, but due to the small size of inserts they do not always map to the genome. If the RNA methylase of interest is suspected to methylate microRNAs, then this band should be included for sequencing. The band cut at 80–150 nt will contain insert

sizes of 29–98 nt with a high cDNA complexity; these cDNAs will also map uniquely to the genome and are the best cDNAs. The band cut at 150 nt–200 nt will contain insert sizes of 98–148 nt but this fraction tends to consist of a limited cDNA complexity.

10. Most cDNA synthesis events will truncate at the crosslink site, thus allowing the methylation site to be subsequently determined. In order to achieve PCR amplification of these cDNAs however, ligation of a 5′ terminal adapter is required. This is achieved by performing a cDNA circularization followed by a BamHI digestion of a restriction site originally located within the 3′ terminal adapter (**step 3**).This results in a portion of the 3′ adapter now being ligated to the 5′ end, thus serving as a PCR priming site.
11. The size of the cDNA insert will be the size of PCR product minus the combined length of the P3/P5Solexa primers and the barcode (128 nt). Thus, for example, a band cut at 70–80 nt on the cDNA gel (with 18–28 nt cDNA+ 52 nt primer) will generate 146–156 nt PCR products.

Acknowledgments

We wish to acknowledge Dr. Julian Konig who codeveloped the original iCLIP protocol, and Dr. Yoichiro Sugimoto for helpful feedback and discussions during the development of methylation-iCLIP. Research in the SH laboratory is supported by a Seed Award in Science from the Wellcome Trust (WT108285MA), and a Responsive Mode Project Grant from the Biotechnology and Biosciences Research Council (BBSRC) UK (BB/N000749/1).

References

1. Dubin DT, Taylor RH (1975) The methylation state of poly A-containing messenger RNA from cultured hamster cells. Nucleic Acids Res 2:1653–1668
2. Dominissini D, Moshitch-Moshkovitz S, Schwartz S, Salmon-Divon M, Ungar L, Osenberg S et al (2012) Topology of the human and mouse m^{6}A RNA methylomes revealed by m^{6}A-seq. Nature 485:201–206
3. Meyer KD, Saletore Y, Zumbo P, Elemento O, Mason CE, Jaffrey SR (2012) Comprehensive analysis of mRNA methylation reveals enrichment in 3′ UTRs and near stop codons. Cell 149:1635–1646
4. Squires JE, Patel HR, Nousch M, Sibbritt T, Humphreys DT, Parker BJ et al (2012) Widespread occurrence of 5-methylcytosine in human coding and non-coding RNA. Nucleic Acids Res 40:5023–5033
5. Khoddami V, Cairns BR (2013) Identification of direct targets and modified bases of RNA cytosine methyltransferases. Nat Biotechnol 31:458–464
6. Hussain S, Sajini AA, Blanco S, Dietmann S, Lombard P, Sugimot Y et al (2013a) NSun2-mediated cytosine-5 methylation of vault non-coding RNA determines its processing into regulatory small RNAs. Cell Rep 4:255–261
7. Wang X, Lu Z, Gomez A, Hon GC, Yue Y, Han D et al (2014) N^6-methyladenosine-dependent regulation of messenger RNA stability. Nature 505:117–120
8. Wang X, Zhao BS, Roundtree IA, Lu Z, Han D, Ma H et al (2015) N(6)-methyladenosine modulates messenger RNA translation efficiency. Cell 161:1388–1399
9. Meyer KD, Patil DP, Zhou J, Zinoviev A, Skabkin MA, Elemento O, Pestova TV, Qian

SB, Jaffrey SR (2015) 5' UTR m(6)A promotes cap-independent translation. Cell 163:999–1010

10. Hussain S, Bashir ZI (2015) The epitranscriptome in modulating spatiotemporal RNA translation in neuronal post-synaptic function. Front Cell Neurosci 9:420
11. Bokar JA, Shambaugh ME, Polayes D, Matera AG, Rottman FM (1997) Purification and cDNA cloning of the AdoMet-binding subunit of the human mRNA (N6-adenosine)-methyltransferase. RNA 3:1233–1247
12. Jia G, Fu Y, Zhao X, Dai Q, Zheng G, Yang Y et al (2011) N^6-methyladenosine in nuclear RNA is a major substrate of the obesity-associated FTO. Nat Chem Biol 7:885–887
13. Ule J, Jensen KB, Ruggiu M, Mele A, Ule A, Darnell RB (2003) CLIP identifiesNova-regulated RNA networks in the brain. Science 302:1212–1215
14. Licatalosi DD, Mele A, Fak JJ, Ule J, Kayikci M, Chi SW, Clark TA, Schweitzer AC, Blume JE, Wang X, Darnell JC, Darnell RB (2008) HITS-CLIP yields genome-wide insights into brain alternative RNA processing. Nature 456:464–469
15. Zhang C, Darnell RB (2011) Mapping in vivo protein-RNA interactions at single-nucleotide resolution from HITS-CLIP data. Nat Biotechnol 29:607–614
16. Sugimoto Y, Konig J, Hussain S, Zupan B, Curk T, Frye M, Ule J (2012) Genome Biol 13:R67
17. König J, Zarnack K, Rot G, Curk T, Kayikci M, Zupan B, Turner DJ, Luscombe NM, Ule J (2010) iCLIP reveals the function of hnRNP particles in splicing at individual nucleotide resolution. Nat Struct Mol Biol 17:909–915
18. King MY, Redman KL (2002) RNA methyltransferases utilize two cysteine residues in the formation of 5-methylcytosine. Biochemistry 41:11218–11225
19. Redman KL (2006) Assembly of protein-RNA complexes using natural RNA and mutant forms of an RNA cytosine methyltransferase. Biomacromolecules 7:3321–3326
20. Hussain S, Benavente SB, Nascimento E, Dragoni I, Kurowski A, Gillich A, Humphreys P, Frye M (2009) The nucleolar RNA methyltransferase Misu (NSun2) is required for mitotic spindle stability. J Cell Biol 186:27–40
21. Linder B, Grozhik AV, Olarerin-George AO, Meydan C, Mason CE, Jaffrey SR (2015) Single-nucleotide-resolution mapping of m6A and m6Am throughout the transcriptome. Nat Methods 12:767–772
22. Carlile TM, Rojas-Duran MF, Zinshteyn B, Shin H, Bartoli KM, Gilbert WV (2014) Pseudouridine profiling reveals regulated mRNA pseudouridylation in yeast and human cells. Nature 515:143–146
23. Schwartz S, Bernstein DA, Mumbach MR, Jovanovic M, Herbst RH, León-Ricardo BX et al (2014) Transcriptome-wide mapping reveals widespread dynamic-regulated pseudouridylation of ncRNA and mRNA. Cell 159:148–162
24. Saletore Y, Meyer K, Korlach J, Vilfan ID, Jaffrey S, Mason CE (2012) The birth of the epitranscriptome: deciphering the function of RNA modifications. Genome Biol 13:175
25. Hussain S, Aleksic J, Blanco S, Dietmann S, Frye M (2013b) Characterizing 5-methylcytosine in the mammalian epitranscriptome. Genome Biol 14:215
26. Rinn JL, Ule J (2014) Oming in on RNA–protein interactions. Genome Biol 15:401
27. Chen B, Yun J, Kim MS, Mendell JT, Xie Y (2014) PIPE-CLIP: a comprehensive online tool for CLIP-seq data analysis. Genome Biol 15:R18
28. Guelorget A, Golinelli-Pimpaneau B (2011) Mechanism-based strategies for trapping and crystallizing complexes of RNA-modifying enzymes. Structure 19:282–291

Chapter 8

Detection of 5-Methylcytosine in Specific Poly(A) RNAs by Bisulfite Sequencing

Thomas Amort and Alexandra Lusser

Abstract

RNA bisulfite sequencing (RNA-BS-seq) represents a method for the detection of methylated cytosines in RNA. Developed originally for the analysis of DNA methylation, a modified version of this method can be used for the analysis of methylated cytosine in RNA. Treatment of nucleic acids with HSO_3^- ions under acidic conditions results in deamination of cytosine (C) to uracil, while 5-methylcytosine (m5C) or 5-hydroxymethylcytosine (hm5C) exhibit low reactivity in this reaction and remain unchanged. Subsequent PCR amplification and sequencing of specific targets allows for the assessment of the methylation status of single Cs in their native sequence context at nucleotide resolution. Here, we describe the application of this method for the analysis of cytosine methylation in low abundance poly(A)RNA using a combination of commercially available kits and standard lab methods to ensure reproducible results. Furthermore, useful information on optimizing the method, suitable controls for almost all steps, and general troubleshooting guides are provided.

Key words Bisulfite sequencing, RNA methylation, 5-Methylcytosine, 5-Hydroxymethylcytosine, mRNA modification

1 Introduction

RNAs are subject to over 100 posttranscriptional modifications of both bases and ribose moieties with roughly two thirds of those being methylations [1–3]. The majority of these modifications has been identified and studied in the two most abundant RNA species, rRNA and tRNA. In poly(A)RNA, methylation of the cap structure (*m7G(5′)ppp(5′)N*) is the best understood modification [4]. A recent surge in method development has led to the discovery of widespread modification of internal bases, such as methylation of N6 in adenosine (m6A), methylation of C5 in cytosine (m5C), or conversion of uridine to pseudouridine [5–10].

Bisulfite sequencing has been the method of choice to assess cytosine methylation in DNA since the early 1990s [11]; its application to RNA, however, has been hampered by the greater instability of RNA phosphodiesterbonds under conditions applied during

Alexandra Lusser (ed.), *RNA Methylation: Methods and Protocols*, Methods in Molecular Biology, vol. 1562,
DOI 10.1007/978-1-4939-6807-7_8, © Springer Science+Business Media LLC 2017

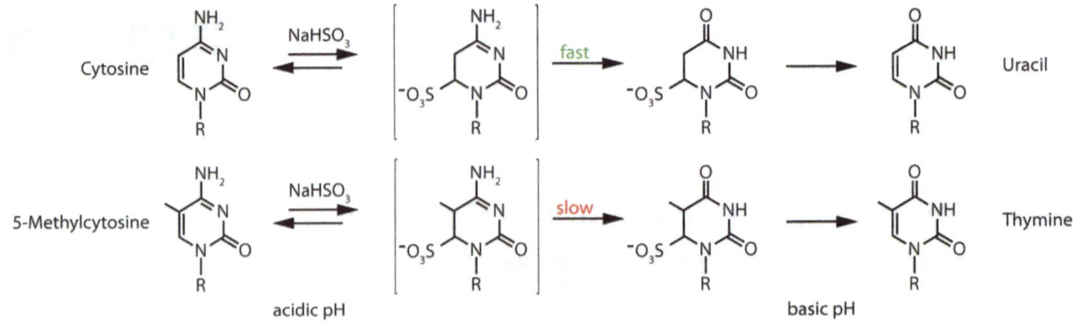

Fig. 1 Bisulfite-mediated conversion of cytosine or 5-methylcytosine to uracil or thymine

bisulfite treatment like high temperature and basic pH. Adaptation of bisulfite (BS) sequencing for RNA [12] allowed for the first time to study RNA methylcytosine at nucleotide resolution in a native sequence context. This method takes advantage of the high reactivity of C with HSO_3^- in acidic conditions causing cytosine to form a sulfonate, which results in deamination to uracil-sulfonate. Changing the pH from acidic to basic removes the bisulfite and gives uracil (U). The analogous conversion of m5C (or 5-hydroxymethylcytosine, hm5C) to thymine (T), however, is much slower, allowing for selective deamination of C, but not m5C or hm5C, by choosing suitable BS treatment conditions (Fig. 1) [13, 14].

We describe here a BS sequencing protocol for fast and cost-effective analysis of cytosine methylation in candidate RNAs [15]. Total RNA or, dependent on the research question, specific RNA types (rRNA, tRNA, mRNA, lncRNA) are isolated and subjected to BS-treatment. The resulting products are reverse transcribed, the RNAs of interest are amplified by PCR, subcloned into a suitable vector, and subsequently a certain number of clones is sequenced. Comparison of the obtained sequences with the original sequence allows for the identification of methylated Cs, since they will appear as C in the sequence, while all unmethylated Cs will appear as T.

Although the method is relatively straightforward, there are a number of potential caveats that need to be considered. First, since BS-induced deamination only works efficiently on single stranded nucleic acids, and many RNAs are known to possess extensive secondary structures, it is essential to ensure complete denaturation of the input material. Likewise, the RNA needs to be free of protein contamination, as RNA-protein complexes will also inhibit the reaction (many commercially available RNA isolation kits, however, give excellent results in this respect) and this will give rise to high numbers of false-positive methylation calls. Second, also the opposite outcome, i.e., loss of true methylation sites, can be a problem, because bisulfite will ultimately also convert m5C and hm5C to U, although this occurs at least two orders of magnitude more slowly [14, 16]. Thus, finding the balance between efficient conversion of unmethylated cytosines while avoiding overconversion and loss of methylation information can be challenging and requires some initial optimization of the method. Third, as RNA is highly unstable in high temperatures and at basic pH, BS treatment results in extensive degradation, which can be a problem for the identification of methylated Cs in rare RNAs. Therefore, it is important to ensure that the input RNA is of the highest quality. Finally, artifacts can be introduced by the PCR amplification step. On one hand, the loss of sequence complexity (due to C->U conversion) renders primer design more difficult and usually requires testing of several sets of primers. On the other hand, biased amplification of templates can occur early on in the PCR reaction. As a consequence, quantification of m5C will be impossible and even the detection of modified C can be prevented. The use of barcoded primers for reverse transcription can control this issue. Taken together, if some optimization is performed and appropriate controls are used, the "low-throughput" BS-sequencing analysis described below is a relatively fast, robust, and cost-effective method for the analysis of methylated cytosine at nucleotide resolution in different types of RNA.

2 Materials

Special care should be taken while handling RNA. Use RNase-free plastics, filter tips, and wear gloves at all time. Kit components should be stored as indicated by the provider. Pure and integer input RNA is crucial for robust and reproducible results. The source of input RNA is dictated by the experimental design. Take precautions to avoid RNA degradation and DNA contamination. The protocol below is described for human HEK293 cells but it should be compatible with a wide variety of cells as well as tissue types.

2.1 RNA Isolation

1. Nuclease-free water: Add 1 mL DEPC per 1 L of water and stir for 1 h at room temperature to completely dissolve DEPC. Autoclave two times for 20 min to remove DEPC completely.
2. TRIzol® (Sigma).
3. Chloroform.
4. 75 % ethanol.
5. Isopropanol.
6. DNase I: 2 U/μL.
7. DNase I Reaction Buffer: 10×.

2.2 Bisulfite Treatment of RNA

EZ RNA methylation kit (Zymo Research) containing RNA Conversion Reagent, RNA Binding Buffer, RNA Wash Buffer, RNA Desulphonation Buffer, DNase/RNase-Free Water, and Zymo-Spin IC Columns (*see* **Note 1**).

2.3 cDNA Synthesis

GoScript™ Reverse Transcription System (Promega):

GoScript™ Reverse Transcriptase.

GoScript™ 5× Reaction Buffer.

$MgCl_2$ solution: 25 mM.

PCR Nucleotide Mix: each 10 mM.

Random Primers: 0.5 μg/μL.

Nuclease-Free Water.

Recombinant RNasin® Ribonuclease Inhibitor.

2.4 Barcode Labeling and PCR

1. Forward primer: 10 μM in a.d.
2. Reverse primer: 10 μM in a.d.
3. Adapter barcode primer: 10 μM in a.d.
4. Adapter primer: 10 μM in a.d.
5. dNTP solution: 5 mM dATP, 5 mM dTTP, 5 mM dCTP, 5 mM dGTP in a.d.
6. EpiMark® Hot Start *Taq* Polymerase: 5 U/μL (New England BioLabs) (*see* **Note 2**).
7. EpiMark® Buffer: (5×) (New England BioLabs).

2.5 Agarose Gel Electrophoresis

1. Agarose gel: 2 % agarose in 1× TBE buffer with 0.5 μg/mL ethidium bromide.
2. 10× TBE buffer: Dissolve 108 g Tris, 55 g boric acid and 9.3 g Na_2EDTA, pH 8 in 1 L distilled water. Stir on a magnetic stirrer until completely dissolved and store at room temperature. Use 0.5× TBE buffer for gel electrophoresis supplemented with 0.5 μg/mL ethidium bromide.
3. Ethidium bromide stock solution: 10 mg/mL ethidium bromide in nuclease-free water; store at 4 °C in the dark.

2.6 Gel Purification

Commercial DNA gel purification system.

2.7 Amplicon Analysis

1. Cloning vector containing a single 3′ T-overhang at the cloning site, e.g., pGEM®-T (Promega): 50 ng/μL.
2. Competent *E. coli* DH5α cells: competence > 10^7 cfu/μg DNA.
3. LB medium: Dissolve 10 g tryptone, 5 g yeast extract and 10 g NaCl in 950 mL deionized water. Autoclave for 20 min. Store at room temperature.

4. LB plates: Prepare LB medium and add 15 % (w/v) agar before autoclaving. Allow cooling to approximately 55 °C and add ampicillin to 100 μg/mL. For blue-white selection add 0.5 mM IPTG and 80 μg/mL X-Gal.

2.8 Instrumentation

1. Thermal cycler.
2. UV/Vis spectrophotometer.
3. Vortex mixer.
4. Microcentrifuge.
5. Agarose gel electrophoresis chamber: clean thoroughly to avoid contamination with other nucleic acids.
6. Shaking incubator.
7. Heat block.
8. Fume hood.

3 Methods

All the pipetting steps can be carried out at room temperature unless otherwise stated.

3.1 RNA Isolation

Depending on the experimental question, different types of input RNA can be used, e.g., total RNA, fractionated RNA (i.e., poly(A) RNA, tRNA, rRNA), cytoplasmic or nuclear RNA (*see* also Chapter 9 in this volume).

Isolate the RNA with a procedure of your choice, e.g., TRIzol® or an RNA extraction kit following the manufacturer's suggestions (*see* **Note 3**). We describe below the purification of total RNA from human HEK293 cells using the TRIzol® method.

1. Harvest 0.5–1 × 10^7 cells (alternatively use 50–100 mg fresh or flash frozen tissue), add 1 mL TRIzol® to the cell pellet, and mix thoroughly by pipetting up and down. If working with tissue samples, use a small sterile pistil for homogenization in a microcentrifuge tube.
2. Incubate samples for 5 min at room temperature, then add 0.2 mL chloroform per mL TRIzol®, and shake vigorously on a vortex mixer for at least 15 s (*see* **Note 4**).
3. Incubate at room temperature for 15 min. Centrifuge at 12,000 × *g* for 15 min at 4 °C.
4. Transfer upper aqueous phase to a new 1.5 mL reaction tube and add 0.5 mL of isopropanol per mL of TRIzol® to precipitate RNA.
5. Incubate for 10 min at room temperature.

6. Centrifuge at 12,000 × *g* for 15 min at 4 °C, remove supernatant, and wash the pellet with 1 mL 75 % ethanol.
7. Centrifuge at 12,000 × *g* for 10 min (*see* **Note 5**).
8. Remove supernatant, air-dry the RNA pellet for 5 min, and dissolve in 100 μL nuclease-free water.

3.2 DNase I Digestion and RNA Purification

1. To remove residual DNA by DNase I digestion (*see* **Note 6**), add 11 μL of DNase I buffer and 2 μL of DNase I to the RNA sample, mix carefully, and incubate at 37 °C for 20 min (*see* **Note 7**).
2. Repurify the RNA using any commercial kit that uses spin column purification and elute with an appropriate volume of nuclease-free water (50–100 μL).
3. Determine concentration and purity of the RNA sample by measuring absorbance at 260 nm and 280 nm in a UV/Vis-spectrophotometer. Highly pure RNA should give an A_{260}/A_{280} reading of ~2 (*see* **Note 8**).

3.3 Bisulfite Treatment

The amount of RNA required for bisulfite sequencing depends on the experimental design. For instance, 0.5–2 μg of total RNA is needed for the analysis of highly abundant rRNA and tRNAs. For low abundance poly(A)RNAs, such as mRNA and lncRNA, 1–2 μg poly(A)-enriched RNA is used.

1. Prepare 0.5–2 μg DNA-free RNA diluted in 20 μL of nuclease-free water (*see* **Note 9**).
2. Mix by pipetting with 130 μL bisulfite solution (conversion reagent) provided in the EZ RNA methylation kit in a 200 μL thin-walled PCR reaction tube. Briefly spin sample to ensure that there are no droplets at the side or the cap of the tube (critical step) (*see* **Note 10**).
3. Incubate the tube in a thermal cycler using the following program:
   (a) 70 °C for 5 min (denaturation).
   (b) 64 °C for 45 min (conversion).
   (c) *Optional*: Repeat **steps (a)** and **(b)** once or twice to ensure complete denaturation of the RNA and optimal efficiency of the bisulfite reaction (*see* **Note 11**).
4. Place a spin column provided in the kit into a collection tube and add 250 μL of RNA Binding Buffer to the column.
5. Load the deaminated RNA sample from **step 3** into the spin column containing the RNA Binding Buffer and mix by pipetting up and down (*see* **Note 12**).
6. Add 400 μL of 100 % ethanol to the sample/Binding Buffer mixture in the column. Mix immediately by inverting the column several times.

7. Centrifuge at full speed for 30 s and discard the flow-through.
8. Add 200 μL RNA Wash Buffer to the column and centrifuge at full speed for 30 s.
9. Add 200 μL of RNA Desulphonation Buffer to the column and incubate at room temperature for 30 min. After the incubation, centrifuge at full speed for 30 s. Discard the flow-through (*see* **Note 13**).
10. Add 400 μL RNA Wash Buffer to the column and centrifuge at full speed for 30 s. Repeat the wash step and discard the flow-through.
11. Centrifuge the column in the emptied collection tube at full speed for at least 2 min to remove residual liquid. Transfer the column carefully from the collection tube into an RNase-free tube provided in the kit.
12. Elute RNA by directly pipetting 10 μL of nuclease-free water onto the column matrix. Let stand at room temperature for 1 min. Centrifuge at full speed for 1 min (*see* **Note 14**).

3.4 cDNA Synthesis

1. Use the eluted RNA (~10 μL) and mix with 1 μL random primers. If necessary, fill up with nuclease-free water to 11.5 μL (*see* **Note 15**).
2. Incubate at 65 °C for 5 min and subsequently place tube immediately on ice for 5 min.
3. In the meantime, prepare the following reaction mix on ice by mixing the following components in the indicated order: 4 μL GoScript™ 5× reaction buffer, 2 μL $MgCl_2$, 1 μL PCR nucleotide mix, 0.5 μL RNasin® Ribonuclease Inhibitor, and 1 μL GoScript™ Reverse Transcriptase.
4. Add the reverse transcription mix to the RNA/primer mix and mix by pipetting or flicking the tube.
5. Incubate in a thermal cycler using the following program:
   (a) 25 °C for 5 min.
   (b) 45 °C for 1 h.
   (c) 70 °C for 15 min.
   (d) cool to 4 °C.
6. The cDNA can be stored at −20 °C and should be used within several months.

3.5 Primer Design and PCR Amplification

It is highly recommended to establish the RNA-BS-seq protocol first with primers that will amplify regions in specific RNAs containing known methylated Cs. Analysis of such targets will serve to determine the reaction conditions that will ensure good (>99 %) overall C deamination but at the same time will conserve specific methylation sites. Suitable positive control targets are, for example, 28S rRNA with its confirmed m5C site at position 4447 (NR_003287) or tRNA_Asp^{GTC} (GD0002843), which has been shown to be methylated at positions C38, C47, and C48 (*see* **Note 16**).

3.5.1 Standard PCR Amplification of BS-Treated RNA Samples

1. When designing primers for your RNA of interest, take into account the specific properties of bisulfite-treated RNA, which are the virtual replacement of C with U and the short size that is due to degradation during bisulfite treatment.
2. Design primers with a length between 25 and 30 bases to ensure specificity. Assuming that the majority of Cs of a given RNA will not be methylated [7] in the forward primer all Cs are converted to T, while the reverse primer should contain A instead of G to guarantee binding of the primers to deaminated sequences (critical step) (*see* **Note 17**).
3. Take into account that the amplicon length should not exceed 250 bp to increase chances for amplification of degraded RNA fragments. Longer amplicons may be possible but intact templates will be underrepresented due to degradation of the RNA in the conversion step (*see* **Note 11**).
4. For PCR add the following components in the indicated order: 5 μL EpiMark® buffer, 1 μL dNTPs, 1 μL forward primer, 1 μL reverse primer, 1 μL cDNA, 15.5 μL nuclease-free water, and 0.5 μL EpiMark® *Taq* polymerase (*see* **Note 2**).
5. Incubate the reaction in a thermal cycler using the following program:
   (a) 95 °C for 30 s.
   (b) 95 °C for 30 s.
   (c) 50–60 °C for 20 s (optimal annealing temperature has to be determined by gradient PCR).
   (d) 68 °C for 30 s.
   (e) Repeat steps (b)–(d) 35–40 times.
   (f) 68 °C for 10 min.
   (g) Hold at 4 °C.
6. Check PCR products on a 2 % agarose gel. Load the entire PCR reaction (25 μL) mixed with loading dye along with a suitable DNA ladder on the gel, run for 30 min at 100 V, and check ethidium bromide-stained PCR products using a UV detection system (*see* **Note 18**).
7. Excise the PCR product of the correct size using a clean blade wearing eye protection (*see* **Note 19**).

3.5.2 Alternative Protocol for Second Strand cDNA Labeling to Introduce Barcodes (See Note 20)

1. Design a primer containing a 4 nt random sequence (barcode) preceded by a 18 nt adapter sequence and followed by a 25–30 nt long sequence specific for the RNA of interest following the guidelines in Subheading 3.5.1, **step 2** (Fig. 2).

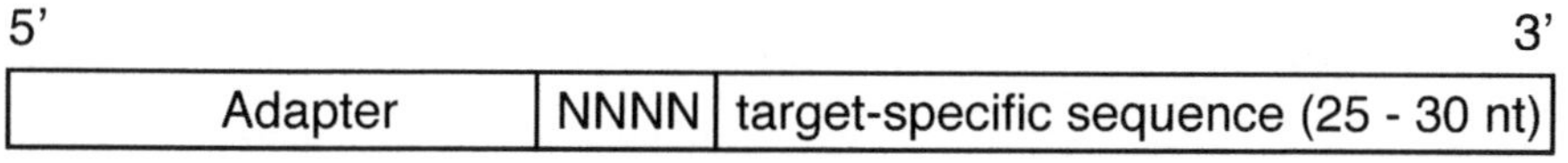

Fig. 2 Schematic representation of primer composition for barcode-labeling of BS-derived cDNA. NNNN, random nucleotides

2. Set up a reaction in a total volume of 25 μL by adding the following components in the indicated order: 5 μL EpiMark® buffer, 1 μL dNTPs, 1 μL adapter barcode primer, 10 μL cDNA (Subheading 3.4), 8.5 μL nuclease-free water, and 0.5 μL EpiMark® *Taq* polymerase.
3. Incubate the reaction in a thermal cycler for 5 min at 95 °C, 20 s at the annealing temperature of the sequence-specific parts of the primers, 30 s at 68 °C.
4. Purify the barcoded DNA using a commercial DNA purification kit according to the manufacturer's instructions (*see* **Note 21**).

1. Set up the PCR reaction using the purified barcoded DNA as described in Subheading 3.5.1, **steps 4–7**. Use 1 μL adapter primer as a forward primer (*see* **Note 22**).

3.6 Sequencing Analysis of Amplicons

1. Gel purify the PCR product of the expected size from the agarose gel using a commercial kit following the provided protocol.
2. Ligate into pGEM®-T vector or into another vector containing single 3′ T-overhangs as indicated by the provider taking advantage of the 3′ A-overhangs produced by *Taq* polymerase (*see* **Note 23**).
3. Incubate 3 h at room temperature or overnight at 4 °C.
4. Transform half of the ligation reaction into highly competent *E. coli* cells using standard protocols.
5. Plate cells onto LB agar plates containing ampicillin, IPTG, and X-Gal and incubate overnight at 37 °C.
6. Pick white colonies and incubate in LB/ampicillin medium overnight for subsequent plasmid minipreparation (*see* **Note 24**). Perform plasmid miniprep using standard protocols or commercial kits and verify correct insert by digestion with restriction enzymes suitable for the used plasmid.
7. Analyze plasmid inserts by Sanger sequencing using universal primers appropriate for the respective cloning vector.

3.7 Data Analysis and Interpretation

1. Retrieve the sequence of your candidate RNA from public databases and delimit to the region amplified during PCR; exclude primer-binding regions in the analysis.
2. Perform a complete in silico conversion of all cytosines to thymine (Fig. 3a). This can be done by using the "search and replace" function available in many nucleic acid-analyzing

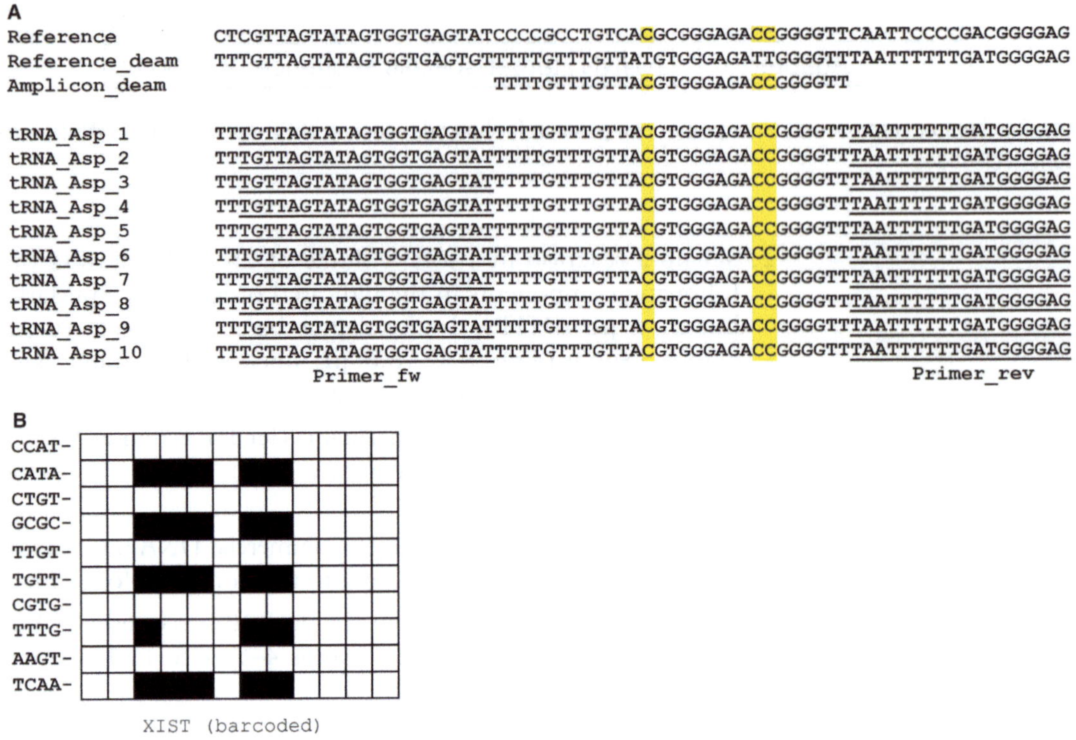

Fig. 3 Example results for a candidate RNA-BS analysis. (**a**) Alignment of sequences from ten clones of a tRNA_Asp^{GTC} amplicon to the reference sequence as well as the *in silico* deaminated reference sequence. PCR primers are underlined. (**b**) Graphical representation of sequenced clones of human XIST RNA repeat 8 after BS-analysis [15]. The presence of different barcodes (sequences at the left) demonstrates the origin of each clone from a different RNA molecule. Each row corresponds to one clone and each square designates a C. Black squares are non-deaminated (= methylated) Cs, white squares are deaminated Cs. Numbers designate the position of the methylated Cs in the transcript sequence

tools, e.g., MacVector, VectorNTI, or even using text-processing software like Microsoft word.

3. Trim the obtained sequencing reads of each clone to remove plasmid and primer sequences.
4. Align your sequencing reads to the converted reference sequence using software solutions suitable for this task or free available web-tools like Clustal Omega (http://www.ebi.ac.uk/Tools/msa/clustalo/).
5. Calculate overall deamination rates as the percentage of deaminated cytosines out of all sequenced cytosines of a particular amplicon.
6. Any unconverted C in the sequenced clones, each of which is assumed to be representative of a single RNA molecule, can represent true methylation or an artifact due to incomplete deamination. Although arbitrary, we consider each C that

remains unconverted in more than 20 % of the sequences as truly methylated. Calculate the methylation rate of each position passing this qualification as the percentage of non-deaminated cytosines at each position (*see* **Note 24**).

7. Figure 3 shows an example methylation analysis of tRNA_Asp^{GTC} as well as the lncRNA XIST in human HEK293 cells. In tRNA_Asp, C38, C47, and C48 show high methylation rates as reported previously [17, 18]. In XIST, a much less abundant lncRNA, methylation rates of 5 Cs in repeat 8 of the XIST A-region vary between 20 % and 50 % [15] (*see* **Note 25**).

4 Notes

1. We find that using commercial kits for RNA methylation analysis, such as from Zymo Research or Epigentek, facilitates the generation of reproducible results; Bisulfite treatment by adapting the protocols provided in DNA bisulfite kits is also possible [12], but requires careful optimization to prevent artifacts. However, it is still possible and sometimes necessary to modify several steps to improve performance of the kit as stated in the protocol main text, especially when using a new lot of the respective kit. To this end, use appropriate and proven target RNAs, such as tRNA_Asp^{GTC} or 28S rRNA (*see* Subheading 3.5).
2. Although in theory any polymerase should work (including proof reading enzymes), in our hands EpiMark® DNA polymerase consistently gave the best results. Since the amplification of deaminated templates, especially those of low abundance, is notoriously difficult, the use of specific polymerases may decide between obtaining a product or not.
3. When using TRIzol®, purified RNA should be additionally purified through spin columns (*see* Subheading 3.2), since RNA quality is critical for the outcome of the experiment. Alternatively, many commercially available kits for RNA isolation give excellent results.
4. Initial incubation for 5 min will ensure complete dissociation of RNA-protein complexes. This is important, since bisulfite treatment will not work if RNA is protected by bound proteins.
5. If you wish to pause your procedure, this is an optimal point, since precipitated RNA in ethanol is protected from degradation and can be easily stored at 4 °C for several weeks.
6. It is important to remove all contaminating DNA, since it is typically not possible to distinguish at later steps if identified potential m5C sites originate from DNA or RNA.

7. Take care with DNase I, since it is sensitive to physical denaturation. Do not vortex and just mix gently by inverting the tube or pipetting up and down.
8. Absence of DNA in the purified RNA sample can be assessed by PCR amplification using RNA without reverse transcription as a template.
9. Do not exceed the suggested RNA amount; too much input RNA will not be efficiently deaminated and will lead to artifacts and difficult to interpret data.
10. Complete mixing of the RNA sample with the conversion solution is crucial for efficient deamination. Droplets can harbor RNA molecules, which are protected from deamination and will lead to artifacts in the subsequent analysis.
11. Reaction and deamination of non-methylated Cs with the bisulfite reagent is highly dependent on the structure of the RNA. For the reaction to occur efficiently, the RNA needs to be in a single stranded form. Hence, to ensure the absence of any local secondary structures, denaturation and conversion steps can be repeated. Thereby, the repeated denaturation step not only increases the chances for melting of secondary structures, the deamination achieved during the first and/or second cycle in addition progressively weakens secondary structures. Caution is necessary, however, to not exaggerate denaturation, since this will exacerbate the already significant degradation of RNA that occurs in high pH coupled to high temperature.

    It should be noted that m5C or hm5C are not inert to reacting with bisulfite, and prolonged incubation with the reagent will ultimately lead to deamination of also the methylated Cs resulting in the loss of the signal upon sequencing. Should this occur, the bisulfite reagent can be diluted with a.d. (e.g., to 80 % or 50 % bisulfite solution) and/or the optional repetition of the reaction cycles can be omitted. In any case, it is necessary to test reactions with different conditions (e.g., different numbers of treatment cycles and/or different bisulfite concentrations) to achieve optimal results. To this end, RNA species for which m5C positions are well known, such as tRNA_Asp^{GTC} or 28S rRNA, can be used (*see* Subheading 3.5 and **Note 16**).
12. Make sure that the RNA is completely mixed with the Binding Buffer to avoid loss during centrifugation.
13. The desulfonation step is critical, as this step leads to the removal of the sulfonate moiety from cytosine by a pH change from acidic to basic.
14. Although it is possible to store bisulfite-converted samples at −20 °C, for best results we recommend immediately proceeding with cDNA synthesis.

15. Using random primers for reverse transcription allows for studying methylation in different candidate RNAs at the same time. Alternatively, specific primers can be used to reverse transcribe specific RNAs. If possible, such primers should be designed in a way that their binding site does not contain any Cs, because nonconverted Cs (due to methylation) would interfere with primer binding and cause biased reverse transcription. Assuming that the majority of the Cs in the RNA will be unmethylated, the primer sequence should reflect the C→U conversion occurring during BS-treatment. Carefully determine the suitability of such primers before starting a complete experiment to avoid loss of the RNA sample due to failure of reverse transcription caused by inadequate priming of the specific primer to the converted RNA.
16. 28S rRNA or tRNA_Asp^{GTC} are convenient targets because the methylated positions appear to be widely conserved. The following primers can be used to amplify a 201 bp product from human 28S rRNA (NR_003287): Forward primer 5′-GGGGTTTTAYGATTTTTTTGATTTTTTGGG, Reverse primer 5′-CCAACTCACRTTCCCTATTAATAAATAAAC; human tRNA_Asp^{GTC}(GD0002843) can be amplified with forward primer 5′- TAGTATAGTGGTGAGTATT and reverse primer 5′-CTCCCCATCAAAAAATCA. In our hands, the optimal annealing temperatures were 56 °C for the 28S rRNA fragment and 45 °C for tRNA_Asp^{GTC}. If no PCR product is obtained, insufficient deamination or excessive degradation of the RNA has occurred during bisulfite treatment. In that case, the treatment step requires optimization (*see* **Note 11**).
17. The use of a primer design tool (e.g., Primer3) is recommended. Primer design for deaminated samples can be challenging and often several primer pairs need to be tested to identify one that allows for specific amplification of the candidate RNA. If possible, primer pairs proven in the literature to be suitable for amplification should be used to assess positive controls, such as 28S rRNA or tRNA_Asp^{GTC} (*see* **Note 16**).
18. If no PCR product is detectable, although all recommendations were followed and positive control primers work on the prepared samples, design new primers.
19. If analyzing mRNA or lncRNA candidates, the amount of the starting material will typically have been very low. In addition, RNA loss will have occurred during the procedure and PCR amplification of deaminated sequences is generally poor. Therefore, expect in most cases very weak bands. However, even if barely visible with the naked eye, there is usually enough material for cloning.
20. PCR may sometimes lead to uneven amplification of the template cDNA molecules and therefore the final PCR product

may contain mostly copies of a few or even just one original cDNA (=RNA) molecule, which will confound determination of true methylation rates. The use of barcoded reverse transcription primers can help to alleviate this problem, but is usually very inefficient due to the long 5′ overhangs of the primers that are required to accommodate barcodes and adapter sequences. We found that labeling of random hexamer-generated cDNA with subsequent barcoding of the second strand works much more efficiently. The barcoded primer is subsequently removed and the regular primers are added for PCR amplification of the target gene.

21. In this example, 10 μL cDNA was used for the barcoding step and the DNA was eluted in the same volume of nuclease-free water after purification. Should the purified barcoded DNA be more diluted than the input cDNA, just use a greater volume for the following PCR step.
22. The sequence of the adapter primer corresponds to the adapter sequence of the adapter-barcode primer.
23. Typically, a molar ratio of 3:1 of insert to vector is recommended. If assuming a 200 bp PCR product and the use of 1 μL of pGEM®-T vector (50 ng), one would need ~10 ng for appropriate cloning conditions.
24. For an initial assessment of m5C rates, choose 10 clones for further analysis. If methylated residues are detected, more clones can be analyzed to increase statistical power. Lower counts of clones may complicate analysis and could lead to misinterpretation of data.
25. If you detect Cs with low methylation rates, it becomes more important to analyze higher numbers of clones to obtain greater statistical reliability of the data. Ideally, clones from different biological replicates should be analyzed.

Acknowledgment

The research was funded by the Austrian Science Fund (FWF): P27024-BBL.

References

1. Grosjean H (2009) Nucleic acids are not boring long polymers of only four types of nucleotides: a guided tour. In: Grosjean H (ed) DNA and RNA modification enzymes: structure, mechanism, function and evolution. LandesBioscience, Austin
2. Motorin Y, Lyko F, Helm M (2010) 5-methylcytosine in RNA: detection, enzymatic formation and biological functions. Nucleic Acids Res 38:1415–1430. doi:10.1093/nar/gkp1117

3. Carell T, Brandmayr C, Hienzsch A et al (2012) Structure and function of noncanonical nucleobases. Angew Chem Int Ed Engl 51:7110–7131. doi:10.1002/anie.201201193
4. Cowling VH (2010) Regulation of mRNA cap methylation. Biochem J 425:295–302. doi:10.1042/BJ20091352
5. Dominissini D, Moshitch-Moshkovitz S, Schwartz S et al (2012) Topology of the human and mouse m6A RNA methylomes revealed by m6A-seq. Nature 485:201–206. doi:10.1038/nature11112
6. Meyer KD, Saletore Y, Zumbo P et al (2012) Comprehensive analysis of mRNA methylation reveals enrichment in 3′ UTRs and near stop codons. Cell 149:1635–1646. doi:10.1016/j.cell.2012.05.003
7. Squires JE, Patel HR, Nousch M et al (2012) Widespread occurrence of 5-methylcytosine in human coding and non-coding RNA. Nucleic Acids Res 40:5023–5033. doi:10.1093/nar/gks144
8. Khoddami V, Cairns BR (2013) Identification of direct targets and modified bases of RNA cytosine methyltransferases. Nat Biotechnol 31:458–464. doi:10.1038/nbt.2566
9. Hussain S, Sajini AA, Blanco S et al (2013) NSun2-mediated cytosine-5 methylation of vault noncoding RNA determines its processing into regulatory small RNAs. Cell Rep 4:255–261. doi:10.1016/j.celrep.2013.06.029
10. Carlile TM, Rojas-Duran MF, Zinshteyn B et al (2014) Pseudouridine profiling reveals regulated mRNA pseudouridylation in yeast and human cells. Nature. doi:10.1038/nature13802
11. Frommer M, McDonald LE, Millar DS (1992) A genomic sequencing protocol that yields a positive display of 5-methylcytosine residues in individual DNA strands. Curr Issue 89(5): 1827–1831
12. Schaefer M, Pollex T, Hanna K, Lyko F (2009) RNA cytosine methylation analysis by bisulfite sequencing. Nucleic Acids Res 37:e12. doi:10.1093/nar/gkn954
13. Shapiro R, Servis RE, Welcher M (1970) Reactions of uracil and cytosine derivatives with sodium bisulfite. J Am Chem Soc 92(2): 422–424
14. Hayatsu H, Wataya Y, Kai K, Iida S (1970) Reaction of sodium bisulfite with uracil, cytosine, and their derivatives. Biochemistry 9: 2858–2865
15. Amort T, Soulière MF, Wille A et al (2013) Long non-coding RNAs as targets for cytosine methylation. RNA Biol 10:1003–1008. doi:10.4161/rna.24454
16. Hayatsu H, Shiragami M (1979) Reaction of bisulfite with the 5-hydroxymethyl group in pyrimidines and in phage DNAs. Biochemistry 18:632–637
17. Schaefer M, Hagemann S, Hanna K, Lyko F (2009) Azacytidine inhibits RNA methylation at DNMT2 target sites in human cancer cell lines. Cancer Res 69:8127–8132. doi:10.1158/0008-5472.CAN-09-0458
18. Blanco S, Dietmann S, Flores JV et al (2014) Aberrant methylation of tRNAs links cellular stress to neuro-developmental disorders. EMBO J 33:2020–2039. doi:10.15252/embj.201489282

Chapter 9

Transcriptome-Wide Detection of 5-Methylcytosine by Bisulfite Sequencing

Thomas Amort, Xueguang Sun, Daria Khokhlova-Cubberley, and Alexandra Lusser

Abstract

While low-throughput RNA bisulfite sequencing is the method of choice to assess the methylation status of specific cytosines in candidate RNAs, the combination of bisulfite treatment of RNA with today's high-throughput sequencing techniques opens the door to methylation studies at nucleotide resolution on a transcriptome-wide scale. Below we describe a protocol for the transcriptome-wide analysis of total or fractionated poly(A)RNA in cells and tissues. Although the nature of the bisulfite sequencing protocol makes it comparably easy to translate from a low to a high-throughput approach, several critical points require attention before starting such a project. We describe a step-by-step protocol for planning and performing the experiment and analyzing the data.

Key words Bisulfite sequencing, RNA methylation, 5-Methylcytosine, High throughput, Transcriptome, Next-generation sequencing

1 Introduction

Despite the long-standing knowledge about different modified bases in high abundant noncoding ribosomal and transfer RNA [1], the presence of 5-methylcytosine (m5C) in coding RNA has long remained elusive. However, already in the 1970s, it has been reported that m5C constitutes a small fraction in the transcripts of hamster BKH-21 and HeLa cells [2, 3], although the techniques used in these studies did not allow for the detection of the precise locations of the modified C. The recent adaptation of bisulfite sequencing (BS-seq) for RNA has now enabled the search for methylated C on a transcriptome-wide level [4]. In acidic pH, bisulfite reacts with unmethylated C resulting in deamination to uracilsulfonate, which can be further converted to uracil in basic pH [5, 6]. Since this reaction takes place on a much slower scale with methylated or hydroxymethylated C [7], subsequent sequencing of the reaction products leads to the selective identification of

Alexandra Lusser (ed.), *RNA Methylation: Methods and Protocols*, Methods in Molecular Biology, vol. 1562,
DOI 10.1007/978-1-4939-6807-7_9,

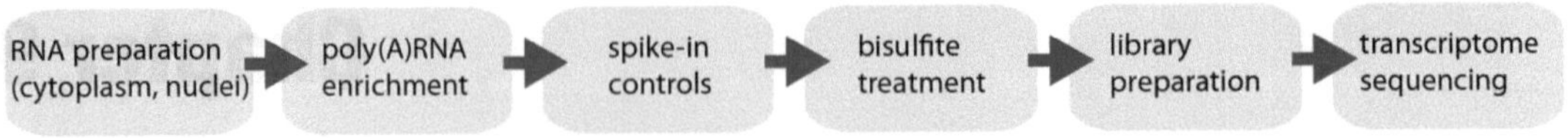

Fig. 1 Schematic overview of the experimental setup for transcriptome-wide RNA-BS-Seq

m5C by its appearance as C, while all unmethylated Cs are read as thymine (T). BS treatment of RNA combined with high-throughput transcriptome sequencing, therefore, has the potential to circumvent the drawbacks of chromatography- and mass spectrometry-based methods, in that it does not require the purification of relatively large amounts of specific RNAs, and it allows for the identification of m5C at nucleotide resolution. However, analyses based on BS treatment need to be carefully controlled, since the efficiency of the reaction is strongly affected by secondary structure, the presence of RNA-protein complexes, as well as excessive RNA degradation (*see* also Chapter 8 in this volume). In addition, reaction conditions need to be established that will lead to efficient conversion of all unmethylated Cs but at the same time will not deaminate methylated Cs.

Here, we describe a protocol for the transcriptome-wide analysis of methylated C in poly(A)RNA (BS-RNA-Seq; Fig. 1). To this end, the input RNA requires depletion of the abundant rRNA and tRNA species, as their presence severely affects the detection limit for the much rarer mRNAs and lncRNAs. Enrichment of poly(A) RNA can be achieved by the use of oligo(dT) beads or by the removal of ribosomal RNA using kits like RiboZero/RiboMinus (although in the latter case, tRNAs will still be present). Depending on the research question, further enrichment can be achieved by purifying the RNA from specific cellular fractions, such as cytoplasm or nuclei [8]. The purified poly(A)RNA is then subjected to BS treatment, purified again and ligated to adapters for subsequent sequencing by a deep sequencing method of choice. In the current protocol, we describe an approach using the Illumina HiSeq™ platform. To control for the efficiency of BS treatment, it is advisable to add to the sample small amounts of non-methylated control RNA, which can be generated by in vitro transcription. The obtained sequences are then mapped to an in silico C → T converted reference genome or transcriptome using, for example, an analysis pipeline as described in [9] and in Chapters 10 and 11 of this volume.

2 Materials

Take care while handling RNA. Degradation will result in a significant loss of input. Keep samples on ice unless otherwise stated. Use RNase-free plastics, filter tips, nuclease-free water, and wear gloves all the time.

2.1 Controls for BS-RNA-Seq

1. Nuclease-free water: Add 1 mL DEPC per 1 L of water and stir for 1 h at room temperature to completely dissolve DEPC. Autoclave twice at 121 °C for 20 min to remove DEPC completely.
2. Forward primer with add-on T7 promoter sequence: 10 μM in nuclease-free water.
3. Reverse primer: 10 μM in nuclease-free water.
4. dNTP solution: 5 mM dATP, 5 mM dTTP, 5 mM dCTP, 5 mM dGTP in nuclease-free water.
5. *Taq* Polymerase: 5 U/μL.
6. *Taq* Polymerase Buffer (10×).
7. MEGAscript® T7 Transcription Kit (Ambion) containing Enzyme Mix, 10× Reaction Mix, 75 mM ATP, 75 mM CTP, 75 mM GTP, 75 mM UTP, TURBO DNase (2 U/μL), 7.5 M Lithium Chloride Precipitation Solution and nuclease-free water.
8. Isopropanol.
9. 70 % Ethanol.

2.2 Agarose Gel Electrophoresis

1. Agarose gel: 1–1.5 % agarose in 1× TBE buffer with 0.5 μg/mL ethidium bromide.
2. 10× TBE buffer: Dissolve 108 g Tris, 55 g boric acid, and 9.3 g Na_2EDTA (pH 8) in 1 L distilled water. Stir on a magnetic stirrer until completely dissolved and store at room temperature. Use 0.5× TBE buffer supplemented with 0.5 μg/mL ethidium bromide for gel electrophoresis.
3. Ethidium bromide stock solution: 10 mg/mL in nuclease-free water. Mix well and store at 4 °C in the dark.

2.3 Isolation of Nuclei

1. 1 M Tris–HCl, pH 8: In a 1 L glass beaker, add 121.1 g Tris to 800 mL water, dissolve completely, and chill to 4 °C. Adjust pH at 4 °C with HCl and fill up to 1 L with water.
2. Nuclear extraction buffer (NE buffer): 0.32 M sucrose, 5 mM $CaCl_2$, 3 mM magnesium acetate, 0.1 mM EDTA, 10 mM Tris–HCl, pH 8, 0.1 % Triton X-100 (w/v). Prepare 10 mL of the buffer as indicated (enough for five isolations) and store at 4 °C. Before use, add 0.1 mM PMSF and 1× proteinase inhibitors.
3. Sucrose cushion buffer: 1.8 M sucrose, 3 mM magnesium acetate, 10 mM Tris–HCl, pH 8. Store at 4 °C.

2.4 RNA Isolation and Enrichment

1. TRIzol®.
2. Chloroform.
3. 75 % ethanol.

4. Isopropanol.
5. DNaseI: 2 U/μL.
6. DNaseI Reaction Buffer (10×).
7. Dynabeads® mRNA purification kit (Ambion) containing Dynabeads® oligo(dT)$_{25}$ (magnetic), Binding Buffer, Washing Buffer, and Elution Buffer.

2.5 Bisulfite Treatment

1. EZ RNA methylation kit (Zymo Research) containing RNA Conversion Reagent, RNA Binding Buffer, RNA Wash Buffer, RNA Desulphonation Buffer, DNase/RNase-Free Water, and Zymo-Spin IC Columns (*see* **Note 1**).
2. 100 % ethanol.

2.6 High-Throughput Sequencing Library Preparation

1. ScriptSeq V2 RNA-Seq Library Preparation Kit from Illumina (SSV21124) containing cDNA Primer, Fragmentation Solution, cDNA Synthesis PreMix, 100 mM DTT, StarScript AMV Reverse Transcriptase, Finishing Solution, Terminal Tagging PreMix, DNA Polymerase, Forward PCR Primer.
2. DNA Clean & Concentrator™-5 from Zymo Research containing DNA Binding Buffer, DNA Wash Buffer, DNA Elution Buffer, and Spin Columns.
3. One *Taq* 2× Master Mix with Standard Buffer (New England Biolabs).

2.7 Library Purification

1. Agencourt AMPure XP magnetic beads (Beckman Coulter).
2. 80 % ethanol.

2.8 Library Quantification

1. Kapa Library Quantification Kit—Complete Kit (Universal; Kapa Biosystems) containing KAPA SYBR® FAST qPCR Master Mix 2×, Primer Premix 10×, Library Quantification DNA Standard 1 Illumina, Library Quantification DNA Standard 2 Illumina, Library Quantification DNA Standard 3 Illumina, Library Quantification DNA Standard 4 Illumina, Library Quantification DNA Standard 5 Illumina, Library Quantification DNA Standard 6 Illumina.
2. If the kit is used for the first time, add the 10× Primer Premix (1 mL) to the bottle of 2× KAPA SYBR® FAST qPCR Master Mix (5 mL). Mix thoroughly using a vortex mixer. Record the date on which the primers were added to the qPCR Master Mix. KAPA SYBR® FAST qPCR Master Mix with primers is stable through 30 freeze-thaw cycles, and should be stored protected from light at −20 °C when not in use.
3. 1 M Tris–HCl, pH 8.
4. DNA Dilution Buffer: 10 mM Tris–HCl, pH 8.0 (25 °C), 0.05 % Tween® 20 (w/v). Store at room temperature.

2.9 Instrumentation

1. Thermal cycler.
2. UV/Vis spectrophotometer.
3. Magnetic stand for 1.5 mL reaction tubes.
4. Vortex mixer.
5. Microcentrifuge.
6. Agarose gel electrophoresis chamber: clean thoroughly to avoid contamination with other nucleic acids.
7. Shaking incubator.
8. Heat block.
9. Fume hood.
10. Ultracentrifuge with swing-out rotor.
11. Ultracentrifuge centrifugation tubes 13 × 51 mm (5 mL).
12. CryoPREP™ extraction system (Covaris).
13. TissueTUBE™ TT1XT (Covaris).
14. Real-time PCR system.
15. 2200 TapeStation (Agilent).
16. HiSeq 1500 (Illumina).

3 Methods

All pipetting steps can be carried out at room temperature unless otherwise stated. Be aware that for statistical analysis three replicates are required.

3.1 Prepare Controls for RNA-Seq

3.1.1 Template Preparation for In Vitro Transcription of Spike-In Control RNA

1. For the spike-in control, design PCR primers to amplify a fragment of *E. coli* 16S rRNA. Include a T7 promoter overhang at the 5′-end of the forward primer (T7 promoter sequence 5′-GAAATTAATACGACTCACTATAGGG-3′) (*see* **Notes 2** and **3**).
2. Set up the PCR reaction by mixing 2.5 μL *Taq* Polymerase Buffer, 1 μL dNTPs, 1 μL forward primer with T7 promoter, 1 μL reverse primer, and 1 μL cDNA template in a 0.2 mL thin-walled PCR tube. Add nuclease-free water to a final volume of 24.5 μL and finally add 0.5 μL Taq polymerase.
3. Perform PCR amplification in a thermal cycler with the following program:
   (a) 95 °C for 30 s.
   (b) 95 °C for 30 s.
   (c) xx °C for 20 s (*see* **Note 4**).
   (d) 68 °C for 30 s.

(e) Repeat **steps (b)–(d)** five times.

(f) Repeat **steps (b)–(d)** for an additional 30 cycles using an annealing temperature (**step (c)**) corresponding to your primer (overhang included) (*see* **Note 5**).

(g) 68 °C for 10 min.

(h) Hold at 4 °C.

4. Load the entire PCR reaction mixed with loading dye along with a suitable DNA ladder on an agarose gel, run for 30 min at 100 V in 0.5× TBE buffer with ethidium bromide, and check the ethidium bromide-stained PCR product using a UV detection system.
5. Excise the PCR product using a clean blade wearing eye protection and purify the DNA using a commercial gel purification kit.
6. Determine concentration of the purified DNA by measuring absorbance at 260 nm in a UV/Vis-spectrophotometer (*see* **Note 6**).

3.1.2 In Vitro Transcription of Spike-In Control RNA

The assembly of an in vitro transcription reaction follows the manufacturer's instructions provided with the MegaScript T7 kit.

1. At room temperature combine 2 μL of each nucleoside (ATP, CTP, GTP, UTP) with 2 μL of 10× reaction buffer and up to 1 μg (*see* **Note 6**) of template including the T7 promoter generated by PCR in the previous steps. Add 2 μL of the enzyme mix provided in the kit and add nuclease-free water to a final volume of 20 μL.
2. Mix the reaction by pipetting and collect liquid by brief centrifugation in a microcentrifuge.
3. Incubate at 37 °C for 2 h to overnight (*see* **Note** 7).
4. After in vitro transcription, remove the DNA template by the addition of 1 μL TURBO DNase to the reaction.
5. Mix well by pipetting and incubate at 37 °C for 15 min.
6. For RNA purification, stop the reaction by the addition of 30 μL nuclease-free water and 30 μL LiCl solution.
7. Mix by slow vortexing or pipetting, collect the mixture by short centrifugation and chill at −20 °C for 1 h.
8. Centrifuge at 4 °C for 15 min at maximum speed in a microcentrifuge. You should now see a white RNA pellet. Remove the supernatant and avoid disturbing the pellet.
9. Wash the pellet once by the addition of 1 mL 70 % ethanol. Invert the tube several times and centrifuge as in **step 8**. Remove the supernatant.
10. Air dry the pellet in a heat block at 45 °C and dissolve in 50 μL nuclease-free water.

11. Check in vitro transcribed RNA on an appropriate agarose gel and assess RNA concentration and quality by measuring absorbance at 260 and 280 nm in a UV/Vis-spectrophotometer. Highly pure RNA should give an A_{260}/A_{280} ratio of ~2.

3.1.3 In Vitro RNA Folding

1. Dilute 1 μg of the in vitro transcribed RNA in 50 μL nuclease-free water.
2. Heat in a thermo block to 70 °C for 15 min.
3. Switch off the thermo block and leave the RNA in the switched off incubator until cooled to room temperature (*see* **Note 8**).

3.2 RNA Isolation and Poly(A) Enrichment

3.2.1 Isolation of Total RNA

Isolate the RNA with a procedure of your choice, e.g., TRIzol® reagent or RNA extraction kit following the manufacturer's suggestions (*see* **Note 9**). We describe below the purification of total RNA from mouse brain tissue using the TRIzol® method. Always work in a fume hood if using TRIzol® or chloroform.

1. Dissect a mouse brain, rinse in ice cold PBS, and flash freeze in liquid nitrogen.
2. Pulverize the tissue in liquid nitrogen using a mortar and pestle or use an instrument like the CryoPREP™ from Covaris.
3. In the latter case, place the frozen tissue into a tissue tube, put into the CryoPREP™, and pulverize two times with impact setting 6.
4. To 50–100 mg of pulverized frozen tissue add 1 mL TRIzol® and mix thoroughly by vortexing.
5. Incubate the sample for 5 min at room temperature, then add 0.2 mL chloroform per mL TRIzol® used, and shake vigorously on a vortex mixer for at least 15 s (*see* **Note 10**).
6. Incubate at room temperature for 15 min. Centrifuge at 12,000 × *g* for 15 min at 4 °C.
7. Transfer upper aqueous phase to a new 1.5 mL reaction tube and add 0.5 mL isopropanol per mL of TRIzol® to precipitate RNA.
8. Incubate for 10 min at room temperature.
9. Centrifuge at 12,000 × *g* for 15 min at 4 °C, remove supernatant, and wash the pellet with 1 mL 75 % ethanol.
10. Centrifuge at 12,000 × *g* for 10 min (*see* **Note 11**).
11. Remove supernatant, air-dry the RNA pellet for 5 min, and dissolve in 100 μL RNase-free water.
12. To remove residual DNA by DNaseI digestion (*see* **Note 12**), add 11 μL of DNaseI buffer and 2 μL of DNaseI to the RNA sample, mix carefully and incubate at 37 °C for 20 min (*see* **Note 13**).
13. Repurify the RNA using any commercial kit that uses spin column purification and elute with nuclease-free water.

14. Determine concentration and purity of the RNA sample by measuring absorbance at 260 and 280 nm in a UV/Vis-spectrophotometer. Highly pure RNA should give an A_{260}/A_{280} reading of ~2 (*see* **Note 14**).

3.2.2 Isolation of Nuclear RNA

1. To 300 mg frozen pulverized tissue (*see* Subheading 3.2.1, **step 2**), add 2 mL NE buffer and dounce on ice at least 25 times—for cells less douncing is recommended (10×) (*see* **Note 15**).
2. To a 13 × 51 mm (5 mL) ultracentrifuge tube, add 3 mL sucrose cushion buffer to generate a sucrose cushion.
3. Carefully pipette the homogenized sample on the top of the sucrose cushion (*see* **Note 16**).
4. In an ultracentrifuge, centrifuge at 240,000 × *g* for 2 h at 4 °C in a swing-out rotor.
5. Take away supernatant and discharge—be careful to not disturb the pellet.
6. Use the pelleted nuclei for RNA isolation as described in Subheading 3.2.1 starting with **step 4** (*see* **Note 17**).

3.2.3 Poly(A) RNA Enrichment

Enrich for poly(A)RNA using an appropriate technique like oligo(dT)$_{25}$ magnetic beads. In the following protocol the procedure for Dynabeads®—mediated enrichment is described.

1. Dilute 75–100 μg of purified RNA (either total RNA or nuclear RNA) in 100 μL DEPC water.
2. In a thermoblock, heat to 65 °C for 2 min to denature secondary structures.
3. Place immediately on ice.
4. Transfer 1 mg (200 μL) of well-resuspended Dynabeads® to a 1.5 mL reaction tube. Place the tube on a magnet and wait until all Dynabeads® have migrated to the wall of the tube.
5. To equilibrate the beads take off the supernatant by careful pipetting. Remove the tube from the magnet and add 100 μL Binding buffer directly to the beads. Put the tube back on the magnet and remove the supernatant as stated before. Again remove the tube from the magnet.
6. Add 100 μL Binding buffer to the Dynabeads®.
7. Add the RNA from **step 1** to the Dynabeads®/Binding Buffer suspension. Mix thoroughly by pipetting and rotate on a roller mixer for 10 min at room temperature.
8. Place the tube on a magnet and wait at least 30 s until the solution is clear. Remove the supernatant carefully with a pipette.
9. Remove the tube from the magnet and wash the beads twice with 200 μL Washing buffer. Remove the supernatant between the washing steps using the magnet.

10. To elute the bound poly(A)RNA, add 100 μL of DEPC water directly to the Dynabeads®. Heat to 70 °C for 2 min and place the tube immediately on the magnet.
11. Wait until the solution is clear, then transfer the eluted RNA to a new tube (*see* **Note 18**).
12. With the eluted RNA, repeat **steps 2–9** (*see* **Note 19**).
13. After the second round of enrichment elute RNA by directly pipetting 20 μL of DEPC water onto the Dynabeads® and mix by careful pipetting.
14. Heat to 70 °C for 2 min and place the tube immediately on the magnet.
15. Transfer the eluted RNA (volume 20 μL) to a new reaction tube.
16. Measure RNA concentration with a spectrophotometer. 1 μg of pure poly(A)RNA is sufficient for subsequent bisulfite treatment and library preparation (*see* **Notes 20** and **21**).

3.3 Bisulfite Treatment

1. Use 1 μg of the isolated poly(A)RNA and spike-in at a mass ratio of 1:20,000–1:40,000 the in vitro transcribed and folded control RNA from Subheading 3.1 in a final volume of 20 μL nuclease-free water (*see* **Note 22**).
2. Mix by pipetting with 130 μL bisulfite solution (conversion reagent) provided in the EZ RNA methylation kit in a 200 μL thin-walled PCR reaction tube. Briefly spin sample to ensure that there are no droplets at the side or the cap of the tube (critical step) (*see* **Note 23**).
3. Incubate the tube in a thermal cycler using the following program:
   (a) 70 °C for 5 min (denaturation).
   (b) 64 °C for 45 min (conversion).
   (c) *Optional*: Repeat **steps (a)** and **(b)** once or twice to ensure complete denaturation of the RNA and optimal efficiency of the bisulfite reaction (*see* **Note 24**).
4. Place a spin column provided in the kit into a collection tube and add 250 μL of RNA Binding Buffer to the column.
5. Load the deaminated RNA sample from **step 3** into the spin column containing the RNA Binding Buffer and mix by pipetting up and down (*see* **Note 25**).
6. Add 400 μL of 100 % ethanol to the sample/Binding Buffer mixture in the column. Close the cap and mix immediately by inverting the column several times.
7. Centrifuge at full speed (≥10,000 × *g*) for 30 s. Discard the flow-through.
8. Add 200 μL RNA Wash Buffer to the column and centrifuge at full speed for 30 s.

9. Add 200 μL of RNA Desulphonation Buffer to the column and let stand at room temperature (20–30 °C) for 30 min. After the incubation, centrifuge at full speed for 30 s. Discard the flow-through.
10. Add 400 μL RNA Wash Buffer to the column and centrifuge at full speed for 30 s. Repeat the wash step and discard the flow-through.
11. Centrifuge the column in the emptied collection tube at full speed for at least 2 min to remove residual liquid. Transfer the column carefully from the collection tube into an RNase-free tube provided in the kit.
12. Elute RNA by directly pipetting 10 μL of RNase-free water onto the column matrix. Let stand at room temperature for 1 min. Centrifuge at full speed for 1 min.
13. Determine the concentration of the RNA sample by measuring absorbance at 260 and 280 nm in a UV/Vis-spectrophotometer. Highly pure RNA should give an A_{260}/A_{280} reading of ~2 (*see* **Note 26**).

3.4 Library Preparation from Bisulfite-Converted Poly(A)RNA for Deep Sequencing

The described protocol follows the instructions for library preparation using ScriptSeq V2 RNA-Seq Library Preparation Kit with a few changes specified in the corresponding notes.

3.4.1 cDNA Synthesis

1. In a 0.2 mL thin-walled PCR tube combine 100 ng bisulfite-treated RNA (Subheading 3.3) and 2 μL cDNA primer and fill up with nuclease-free water to a total volume of 11 μL.
2. Mix by gently flicking the tube.
3. Incubate at 65 °C for 5 min in a thermal cycler with heated lid.
4. Place the tube on ice.
5. Prepare the cDNA Synthesis Master Mix (*see* **Note 27**). For a single reaction combine 1 μL Fragmentation Solution (*see* **Note 28**), 3 μL cDNA Synthesis Premix, 0.5 μL DTT, 0.5 μL StarScript AMV Reverse Transcriptase.
6. Gently but thoroughly mix the cDNA Synthesis Master Mix by pipetting ten times.
7. Add 5 μL of the cDNA Synthesis Master Mix to each RNA/cDNA-Primer reaction from **step 4** on ice and mix gently but thoroughly by pipetting.
8. Place into a thermal cycler and run the following program:
   (a) 25 °C for 5 min.
   (b) 42 °C for 20 min.
   (c) 37 °C Pause/Hold.

9. Remove reactions immediately from the thermal cycler when they are cooled to 37 °C, add 1.0 μL of Finishing Solution and mix gently but thoroughly by pipetting. Return reactions to the thermal cycler and run:
   (a) 37 °C for 10 min.
   (b) 95 °C for 3 min.
   (c) 25 °C Hold.
10. In the last step, allow the reactions to cool to 25 °C. While the samples are in the thermal cycler, start with preparing the reagents for the next step (Subheading 3.4.2, **steps 1** and **2**) to ensure a gapless transition.

3.4.2 Synthesize 3′-Tagged cDNA

1. The Terminal-Tagging PreMix is a viscous solution. Mix it thoroughly before use (*see* **Note 29**). On ice, prepare the Terminal Tagging Master Mix (*see* **Note 27**). For a single reaction combine 7.5 μL Terminal Tagging Premix and 0.5 μL DNA Polymerase.
2. Thoroughly mix by pipetting several times.
3. Remove the cDNA reactions from the thermal cycler (Subheading 3.4.1, **step 10**) and add 8.0 μL of the Terminal Tagging Master Mix (*see* **Note 30**).
4. Gently but thoroughly mix the reactions by pipetting.
5. In a thermal cycler incubate the reactions at 25 °C for 15 min followed by denaturation at 95 °C for 3 min and subsequent cooling to 4 °C.

3.4.3 Purification of cDNA

Use spin column-based techniques for purification of the cDNA like recommended in the kit. The protocol here is described for the DNA Clean and Concentrator-5 kit.

1. Add 350 μL of DNA binding buffer into the spin columns.
2. Add 25 μL of water to each reaction to bring the volume in each cDNA tube to 50 μL(*see* **Note 31**).
3. Transfer the cDNA dilution into the columns with the DNA binding buffer and mix very well by pipetting (at least six times).
4. Spin at 15,000 × *g* for 30 s, discard the flow-through.
5. Add 700 μL of DNA wash buffer to the columns.
6. Spin at 16,000 × *g* for 30 s, discard the flow-through.
7. Add 400 μL of DNA wash buffer to the columns.
8. Spin at 15,000 × *g* for 2 min.
9. Transfer the columns into a clean 1.5 mL reaction tube. Add 23 μL DNA elution buffer directly onto the spin column membrane and incubate at room temperature for 5 min.

10. Spin at 16,000 × *g* for 30 s.
11. Transfer eluted cDNA into 0.2 mL thin-walled PCR tubes for amplification of the library.

3.4.4 Library Amplification and Barcoding

1. To the purified 23 μL cDNA sample, add 1 μL of appropriate index primer (*see* **Note 32**).
2. Prepare a master mix for amplification of several samples (*see* **Note 27**). For a single reaction combine 25 μL OneTaq 2× Master Mix and 1 μL cDNA Forward PCR Primer.
3. Add 26 μL of the amplification master mix to each cDNA/index primer sample (24 μL) from **step 1** and mix by pipetting.
4. Place tubes in a thermal cycler and run the following program:
   (a) 94 °C for 30 s.
   (b) 94 °C for 30 s.
   (c) 55 °C for 30 s.
   (d) 68°C for 3 min.
   (e) Repeat **steps (b)–(d)** 14 times to a total of 15 cycles.
   (f) 68 °C for 5 min.
   (g) Hold at 4 °C (*see* **Note 33**).

3.4.5 Purification of the RNA-Seq Library

Start the procedure with a 1× AMPure XP purification scheme (1:1 ratio (v/v) of AMPure XP beads to cDNA library) since it will yield a ScriptSeq library with fragments of >200 nts.

1. Warm the AMPure XP beads to room temperature.
2. Transfer each amplified and barcoded library (50 μL) to a separate 1.5 mL tube.
3. Important! Vortex the AMPure XP beads until they are a homogeneous suspension.
4. Add 50 μL of the beads to each sample.
5. Mix thoroughly by gently pipetting the entire volume of each tube 15 times.
6. Incubate the samples at room temperature for 15 min. During the incubation period prepare 400 μL of freshly diluted 80 % ethanol at room temperature for each sample.
7. Place the samples in a magnetic stand (do not remove until **step 13**) at room temperature and wait for at least 5 min, until the liquid appears clear.
8. Remove and discard the supernatant from each tube using a pipette. Some liquid may remain in each tube (~5 μL) and this is fine as it is crucial to not disturb the beads.

9. With the sample remaining on the magnetic stand, add 200 μL of the 80 % ethanol from **step 6** to each tube without disturbing the beads.
10. Incubate the samples at room temperature for at least 30 s, then remove and discard all of the supernatant. Take care not to disturb the beads.
11. Repeat **steps 9** and **10** one more time for a total of two 80 % ethanol washes.
12. Allow the samples to air-dry with opened lid on their magnetic stands for 15 min at room temperature.
13. Add 20 μL of nuclease-free water to each tube and remove 1.5-mL tubes from their magnetic stand.
14. Thoroughly resuspend the beads by gently pipetting 15 times.
15. Incubate the samples at room temperature for 2 min.
16. Place the samples on the magnetic stand at room temperature for at least 5 min, until the liquid appears clear.
17. Transfer the clear supernatant, which contains the RNA-Seq library, from each tube to an appropriate collection tube.
18. Assess the quality of libraries on Agilent TapeStation using D1000 DNA ScreenTape. The size distribution of libraries can also be assessed using 2100 BioAnalyzer and a High Sensitivity DNA Chip (*see* **Note 34**).

3.5 Library Quantification

Quantify the generated libraries by qPCR with KAPA Library Quantification Kit for Illumina platforms. In the following a slightly modified protocol from the kit is provided. Ensure that all components of the KAPA Library Quantification Kit are completely thawed and thoroughly mixed (*see* **Note 35**).

1. Using DNA Dilution Buffer, prepare the following serial library dilutions in triplicates: 1:1000–1:2000–1:4000–1:8000.
2. Vortex and spin down each library dilution before proceeding with the next serial dilution (*see* **Note 36**).
3. Determine the total number of reactions that will be performed for the appropriate number of replicates of each of the following reactions:
   (a) Six DNA Standards (each standard in triplicate).
   (b) Each dilution of every library to be assayed (each dilution in triplicate).
4. Prepare the required volume of qPCR master mix using the following reaction setup: 16 μL of 2× KAPA SYBR® FAST qPCR Master Mix containing the 10× Primer Premix and 4 μL of PCR-grade water.
5. Mix and briefly centrifuge the reagent master mix.

6. Dispense the appropriate volume of the master mix into each PCR tube or well.
7. Dispense 4 μL of each DNA Standard into the appropriate tubes/wells.
8. Dispense 4 μL of each dilution of a library.
9. Cap tubes or seal the PCR plate, and transfer to the qPCR instrument.
10. Perform qPCR with the following cycling protocol:
    (a) 95 °C for 5 min.
    (b) 95 °C for 30 s.
    (c) 60 °C for 45 s.
    (d) Repeat **steps (b)** and **(c)** 35 times.
11. Annotate the DNA Standards as outlined below (*see* **Note 37**):
    (a) DNA Standard 1—20 pM.
    (b) DNA Standard 2—2 pM.
    (c) DNA Standard 3—0.2 pM.
    (d) DNA Standard 4—0.02 pM.
    (e) DNA Standard 5—0.002 pM.
    (f) DNA Standard 6—0.0002 pM.
12. Review amplification curves and the Ct scores for replicate data points (DNA Standards and libraries), and exclude obvious outliers. Replicate data points should differ by ≤0.2 cycles (*see* **Note 38**).
13. Exclude all library dilutions that fall outside the dynamic range of the assay, i.e., that return an average Ct score lower than that of Standard 1 or higher than that of Standard 6. If all the dilutions of a library fall outside the standard curve, re-quantify a more appropriate dilution of the library.
14. Generate a standard curve by using the instrument software.
15. Review the standard curve to ensure that the following criteria are met:
    (a) The average ΔCt value between DNA Standards is in the range of 3.1–3.6.
    (b) The calculated reaction efficiency is in the range of 90–110 % (*see* **Note 39**).
    (c) $R^2 \geq 0.99$.
16. If the standard curve does not meet these criteria, calculated library concentrations will not be reliable, and the assay must be repeated.
17. Use the standard curve to convert the average Ct score for each dilution of every library that was assayed to average concentration (in pM).

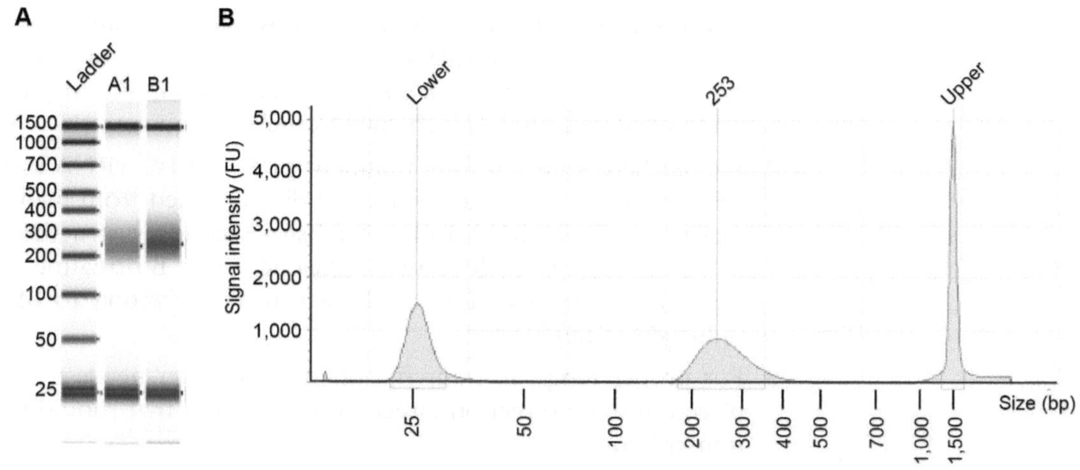

Fig. 2 (**a**) TapeStation D1000 gel electrophoresis image of two libraries (A1, B1) prepared from 100 ng of bisulfite-treated poly(A)RNA. (**b**) Histogram results of *panel A, lane A1*. "*Lower*" and "*Upper*" designate lower and upper markers, respectively

18. Multiply the average concentration calculated for each dilution of every library with the appropriate dilution factor to calculate the concentration of the undiluted library.

With the protocol described here a library with fragment size of ~180–400 bp as assessed by D1000 DNA SreenTape (Fig. 2) can be generated while the concentration of the libraries will range from 18 to 33 pM.

3.6 Sequencing Using Illumina HiSeq 1500 Platform

Multiplexed libraries were clustered at 11 pM for sequencing. Although sequencing parameters depend on the specific experimental question, we recommend the generation of 100 bp single end reads. This will be sufficient for reliable mapping to mammalian reference genomes or transcriptomes. As the extent of methylation of a particular C in a specific mRNA type can vary from molecule to molecule, it is important to obtain a high number of reads. We found that >70 million reads per replicate sample are required for a reasonable survey of the cytosine-methylated transcriptome in mouse.

3.7 Data Analysis

The processing of the obtained sequencing data is described in Chapter 10 of this volume.

4 Notes

1. We find that using commercial kits makes it easier to generate reproducible results—bisulfite treatment by adapting DNA bisulfite kits is possible [10] but at least in our hands prone to

artifacts. Although using kits, it is still possible to modify several steps to improve performance as stated in the protocol below. We recommend testing each kit for deamination efficiency by using positive controls, if changing crucial parameters.

2. We use here a 552 nt long fragment of *E. coli* 16S rRNA as a spike-in control as it can be readily distinguished from mammalian samples. rRNA forms extensive secondary structures and is therefore suitable to assess efficiency of the BS-treatment, since secondary structures are known to be refractory to BS-induced C deamination.
3. For proper template transcription the promoter needs to be placed in sense direction directly on the strand that should be transcribed.
4. Use an annealing temperature calculated for the template-specific sequence of the forward primer without the T7 promoter sequence for the initial five cycles as stated in the protocol; then raise the temperature to match the entire primer sequence to increase specificity of primer annealing.
5. Use a suitable tool for the calculation of primer melting temperatures, especially if primers differ that much in length. We found http://tmcalculator.neb.com to give very good results.
6. For in vitro transcription usually 500 ng–1 μg of template is needed. If one PCR reaction does not yield enough product, just perform several reactions and combine them. Gel purification is needed to remove enzymes, unincorporated nucleotides, and buffer components.
7. Overnight incubation of the reaction gives the maximum amount of RNA.
8. Slow cooling of the RNA solution will take about 3–4 h depending on the thermo block used.
9. When using Trizol, purified RNA should be additionally purified through spin columns, since RNA quality is critical for the outcome of the experiment. Alternatively, many commercially available kits for RNA isolation give excellent results.
10. Initial incubation for 5 min will ensure complete dissociation of RNA-protein complexes. This is important, since bisulfite treatment will not work, if RNA is protected by bound proteins.
11. If you wish to pause your procedure, this is an optimal point, since precipitated RNA in ethanol is protected from degradation and can be easily stored at 4 °C for several weeks.
12. It is important to remove all contaminating DNA, since it is typically not possible to distinguish at later steps, if identified potential m5C sites originate from DNA or RNA.

13. Take care with DNase I, since it is sensitive to physical denaturation. Do not vortex and just mix gently by inverting the tube or pipetting up and down.
14. Absence of DNA in the purified RNA sample can be assessed by PCR amplification of an endogenous transcript using RNA without reverse transcription as a template.
15. 300–500 mg mouse brain tissue is used in this procedure. The amount of tissue/cells needed has to be determined with respect to the RNA amount needed for further poly(A) enrichment and/or bisulfite sequencing.
16. To avoid a sharp border between the sample and the cushion, which could keep the nuclei from entering and passing the cushion, we recommend disturbing the border with a pipette tip by gently pricking several times the cushion solution.
17. By using 300–500 mg mouse brain tissue we could purify about 70 μg of nuclear RNA. This amount is enough to perform poly(A) selection and enrich sufficient polyadenylated RNA for subsequent bisulfite treatment and sequencing.
18. Proper preparation of input RNA is inevitable to achieve high-quality results. The presence of genomic DNA or high abundant noncoding RNA (rRNA and tRNA) will interfere with sequencing depth and lead to loss of coverage of low abundant and low expressed messenger RNAs. Therefore, the sample should be subjected to at least two rounds of poly(A)-enrichment.
19. The Dynabeads® can be reconstituted and reused to minimize the cost of poly(A)-selection. Please refer to the manual for additional information.
20. To assess the depletion of contaminating rRNA and tRNA in the purified sample, we recommend running a small aliquot on an Agilent BioAnalyzer RNA Chip. However, if this equipment is not available, a denaturing agarose gel electrophoresis following standard methodology with subsequent visual inspection of ethidium bromide-stained RNA can be performed. Expect to see a smear along the gel lane and no presence of distinguishable bands for 28S and 18S rRNA. Since analysis of successful poly(A)RNA enrichment by gel electrophoresis requires considerably more RNA than use of a BioAnalyzer, be prepared to increase the amount of input material for RNA isolation and scale up all steps accordingly.
21. Assuming 98 % of total RNA is going to be unwanted rRNA and tRNA and therefore will get lost during the enrichment procedure, by using 100 μg total RNA as starting material one should end with approximately 1–1.5 μg pure poly(A)RNA.
22. Do not exceed the suggested RNA amount; too much input RNA will not be efficiently deaminated and will lead to artifacts and difficult to interpret data.

23. Complete mixing of the RNA sample with the conversion solution is crucial for efficient deamination. Droplets can harbor RNA molecules that are protected from deamination and will lead to artifacts in the subsequent analysis.

24. Reaction and deamination of non-methylated Cs with the bisulfite reagent is highly dependent on the structure of the RNA. For the reaction to occur efficiently, the RNA needs to be in a single-stranded form. Hence, to ensure the absence of any local secondary structures, denaturation and conversion steps can be repeated. Thereby, the repeated denaturation step not only increases the chances for melting of secondary structures, the deamination achieved during the first and/or second cycle in addition progressively weakens secondary structures. Caution is necessary, however, not to exaggerate denaturation, since this will exacerbate the already significant degradation of RNA that occurs at high temperature. It should be noted that m5C or hm5C are not inert to reacting with bisulfite, and prolonged incubation with the reagent will ultimately lead to deamination of also the methylated Cs resulting in the loss of the signal upon sequencing. Should this occur, the bisulfite reagent can be diluted with nuclease-free water (e.g., to 80 or 50 % bisulfite solution) and/or the optional repetition of the reaction cycles can be omitted. In any case, it is necessary to test reactions with different conditions (e.g., different numbers of treatment cycles and/or different bisulfite concentrations) to achieve optimal results. To optimize bisulfite-treatment conditions we recommend performing a PCR-based analysis of the spike-in control RNA according to the protocol described in Chapter 8 of this volume.

25. Make sure that the RNA is completely mixed with the Binding Buffer to avoid loss during centrifugation.

26. For subsequent library preparation at least 100 ng of input RNA at a concentration of 12 ng/μL is needed. We suggest using 1 μg RNA as input for the bisulfite conversion reaction. Less input is possible as far as after desulphonation and purification enough RNA can be recovered. Consider that some RNA is lost during the procedure due to degradation, during binding to the columns as well as through insufficient elution from the column.

27. Prepare enough for all samples and take pipetting errors into account by preparing 10 % extra Master mix (multiply all volumes by the number of reactions and by 1.1).

28. Although BS-treated RNA is already fragmented to a size of ~200 nt, RNA Fragmentation Solution is added to supplement Mg^{2+} in the cDNA synthesis reaction.

29. Although the use of wide bore pipet tips when pipetting the Terminal Tagging PreMix and the Terminal Tagging Master

Mix is recommended in the kit protocol, we use regular pipet tips and mix 15–20 times by pipetting.

30. Do not use P10 pipet tips to add the Terminal Tagging Master Mix as the bores are too small. Use tips with larger bores.
31. The ratio of Binding Buffer to cDNA should be 7:1.
32. The index primer acts as a reverse primer. Index primers are designed as described in Appendix 3 of ScriptSeq v2 RNA-Seq Library Preparation protocol from Epicentre/Illumina.
33. The amplification can be performed overnight; amplified libraries can be stored at −20 °C and purified later.
34. Library preparation is considered successful if library fragment size ranges from ~170 to 400–500 bp, and the concentration of the library is >10 nM, as assessed on D1000 DNA Screen Tape from Agilent.
35. Libraries can be multiplexed either before quantification by qPCR with KAPA Library Quantification Kit for Illumina platforms from Kapa Biosystems or after quantification by qPCR. To multiplex libraries, the same amount of each library should be combined based on the concentrations of libraries. If multiplexing libraries prior to quantification by qPCR, it is recommended to measure the concentration of libraries with Qubit ds DNA HS assay kit (ThermoFisher Scientific), since Qubit assay is more precise than TapeStation.
36. If the concentration of a library (or multiplexed libraries) is >20 nM or >20 ng/μL, dilute the library (or multiplexed libraries) to ≈20 nM or ≈20 ng/μL with DNA elution buffer (10 mM Tris–HCl, pH 8.5, 0.1 mM EDTA) before preparing serial dilutions.
37. Note that the specified values correspond to the concentrations of the DNA Standards, and not the final DNA concentration in each reaction. It is not necessary to convert these to the actual concentrations in the reaction, as long as the same volume of template (DNA Standard, diluted library) is used in all reactions.
38. If the data set contains many outliers, results are unlikely to be reliable.
39. Given an optimal efficiency the PCR product has increased 1.8- to 2.2-fold per cycle, and the slope of the standard curve is between −3.1 and −3.6).

Acknowledgment

Funding in A.L's lab is provided by the Austrian Science Fund (FWF): P27024-BBL.

References

1. Grosjean H (2009) Nucleic acids are not boring long polymers of only four types of nucleotides: a guided tour. In: Grosjean H (ed) DNA and RNA modification enzymes: structure, mechanism, function and evolution. LandesBioscience, Austin
2. Dubin DT, Taylor RH (1975) The methylation state of poly A-containing messenger RNA from cultured hamster cells. Nucleic Acids Res 2:1653–1668
3. Salditt-Georgieff M, Jelinek W, Darnell JE, Furuichi Y, Morgan M, Shatkin A (1976) Methyl labeling of HeLa cell hnRNA: a comparison with mRNA. Cell 7: 227–237
4. Squires JE, Patel HR, Nousch M, Sibbritt T, Humphreys DT, Parker BJ, Suter CM, Preiss T (2012) Widespread occurrence of 5-methylcytosine in human coding and non-coding RNA. Nucleic Acids Res 40:5023–5033. doi:10.1093/nar/gks144
5. Hayatsu H, Wataya Y, Kai K, Iida S (1970) Reaction of sodium bisulfite with uracil, cytosine, and their derivatives. Biochemistry 9:2858–2865
6. Shapiro R, Servis RE, Welcher M (1970) Reactions of uracil and cytosine derivatives with sodium bisulfite. J Am Chem Soc 92(2): 422–424
7. Hayatsu H, Shiragami M (1979) Reaction of bisulfite with the 5-hydroxymethyl group in pyrimidines and in phage DNAs. Biochemistry 18:632–637
8. Amort T, Soulière MF, Wille A, Jia X-Y, Fiegl H, Wörle H, Micura R, Lusser A (2013) Long non-coding RNAs as targets for cytosine methylation. RNA Biol 10:1003–1008. doi:10.4161/rna.24454
9. Rieder D, Amort T, Kugler E, Lusser A, Trajanoski Z (2015) meRanTK: methylated RNA analysis ToolKit. Bioinformatics. doi:10.1093/bioinformatics/btv647
10. Schaefer M, Pollex T, Hanna K, Lyko F (2009) RNA cytosine methylation analysis by bisulfite sequencing. Nucleic Acids Res 37:e12. doi:10.1093/nar/gkn954

Chapter 10

Analysis of High-Throughput RNA Bisulfite Sequencing Data

Dietmar Rieder and Francesca Finotello

Abstract

Methylation of the 5-cytosine (m^5C) is a common but not well-understood RNA modification, which can be detected by sequencing of bisulfite-treated transcripts (RNA-BSseq). In this Chapter, we discuss computational RNA-BSseq data analysis methods for transcriptome-wide identification and quantification of m^5C.

Key words RNA-BSseq, RNA methylation, m^5C, Differential methylation, 5-methylcytosine, Transcription, RNA modification, Bisulfite conversion, High-throughput sequencing, Data analysis, RNA, Modification, (Cytosine-5) methylation, Sodium bisulfite

1 Introduction

In RNA more than 100 chemically distinct posttranscriptional modifications are known [1], roughly two-thirds of which are methylations [2]. In particular, methylation (m^5C) and hydroxymethylation (hm^5C) of the carbon 5 atom of cytosine are a interesting examples, because they occur in both classes of nucleic acids, DNA and RNA, suggesting a universal utilization of these modifications. In DNA, they are known to affect the activity state of genomic regions; in RNA, however, its role is less clear.

In the past, m^5C was mainly studied on ribosomal RNAs (rRNAs) and transfer RNAs (tRNAs) and was associated with structural and metabolic stabilization [3]. m^5C was also found in the long noncoding RNAs (lncRNAs) HOTAIR and XIST, where it has the ability to interfere with RNA-protein interaction [4]. Moreover, evidence from studies on mutations in RNA cytosine 5 methyltransferases (enzymes promoting methylation) suggests a role in RNA quality control, RNA stability, protein translation, as well as immunity and consequently also in disease etiology [5].

However, surprisingly little is known about the general abundance, significance, and function of m^5C in poly(A)RNAs. To

Alexandra Lusser (ed.), *RNA Methylation: Methods and Protocols*, Methods in Molecular Biology, vol. 1562,
DOI 10.1007/978-1-4939-6807-7_10,

address this question it is important to identify and compare m^{5}Cs on a transcriptome-wide scale preferably at single base pair resolution.

Recently, studies using high-throughput methods to globally evaluate RNA cytosine methylation showed that m^{5}C occurs throughout the human and mouse transcriptomes, in coding and noncoding RNAs [6–8].

RNA m^{5}C modifications can be detected by bisulfite treatment, which converts all unmethylated cytosines into uracils, leaving methylated cytosines unaffected. An alternative technique is the enrichment of RNAs covalently bound to RNA-methyltransferases (RMT) that were trapped on their substrate either by the presence of 5-azacytidine (Aza-IP) [7, 9] or by mutating the catalytic center of the respective RMT (miCLIP) [8]. When these methods are combined with high-throughput sequencing it is possible to obtain either global (RNA-BSseq) or enzyme-specific (Aza-IP or miCLIP) m^{5}C-modified transcriptomes.

Like all high-throughput sequencing approaches, these techniques generate millions to hundred millions of short sequencing reads, which need to be processed by specialized analysis pipelines and software tools to finally extract the relevant information. In this Chapter, we will discuss computational methods to identify candidate methylated cytosines at single base pair resolution from RNA-BSseq data.

2 RNA-BSseq

2.1 Experimental Design and Data Features

A reliable and well-established method to detect cytosine methylation in DNA is based on treatment of DNA with bisulfite, which leads to deamination of all un-methylated cytosines in DNA transforming them into uracils, while methylated cytosines tend to be preserved. In mammalian genomes, methylated cytosines represent only a minor fraction of all cytosines and occur only at specific positions. So, after bisulfite treatment, only the methylated cytosines (about ~1 %) are preserved. The bisulfite-based m^{5}C detection method was recently adopted for RNA [10] and first commercially available kits are now available. When bisulfite conversion is applied to total- or specific RNAs (e.g., tRNA, rRNA, ncRNA, and mRNA), massive parallel sequencing of the resulting deaminated RNAs can be used to obtain transcriptome-wide methylation states.

Similar to standard RNA sequencing, RNA-BSseq typically results in millions of short read sequences (50–100 bp), which are stored in FASTQ formatted files. Standard RNAseq data is generated by sequencing cDNA, obtained by reverse transcription of RNA, and sequencing reads therefore consist of A, C, G, and T. However, due to the deamination process the RNA-BSseq reads differ from standard high-throughput RNAseq reads in the sense that they have very infrequent occurrences of C (or G in the reverse reads generated through paired-end sequencing).

To ensure a good coverage of the methylated sites in the transcriptome, fairly deep sequencing is required. Sequencing depths of about ~100 million single-end reads are typical for RNA-BSseq experiments [6, 7]. Paired-end sequencing experiments have not been published so far, but in pilot experiments using 80–100 million 100 bp read pairs, we have obtained results comparable to those from single-end libraries.

The maximum coverage of a certain genomic position after deduplication (i.e., removal of identical reads, *see* below) is equal to the length of the read. Thus, sequencing to a depth exceeding the maximum coverage is not useful [11]. When paired-end sequencing is used, the maximum useful coverage is higher since both reads of a pair need to map to locations identical to those of another read pair to be considered a duplicate of it. A further advantage of using paired end sequencing is that it can increase the efficiency of mapping to low complexity regions where one end may serve as anchor for the other [12]. The mappability is generally reduced for bisulfite-converted sequences since the alphabet is shortened (mostly three instead of four letters). However, for reads longer than 35 bases, the difference in mappability between the bisulfite reads and normal reads was found to be quite low [13]. In addition, bisulfite-treated RNAs undergo random degradation [10] and long fragments (>200 bp) are not very likely to occur in high numbers so in most cases sequencing reads longer than 100 bp will not lead to better results than shorter reads.

The bisulfite-induced deamination process is not very efficient in regions with complex secondary structure, and therefore it is advisable to include un-methylated control sequences that help to estimate the achieved deamination rate. Control sequences are usually highly structured in vitro transcribed spike-in RNAs from other organisms and have low similarity with sequences from the studied species. In addition to this technical control, it is important to perform replicate experiments to account for biological variance [12] and to increase the significance of the conclusions drawn from the resulting data, especially when studying differential methylation.

2.2 Quality Control

Depending on the specific library preparation protocol used for generating the sequencing library, stretches of nonconverted unmethylated cytosines or other biases in the sequencing reads might be present. Therefore, it is important to know as much as possible about how the data was generated, i.e., library preparation method, primers, and adapters used. Moreover, knowing the properties and problems of the sequencing technology, like tendency to homopolymer errors that come from over- or under-calling homopolymeric regions, may be helpful in interpreting unexpected results.

To spot potential problems, basic quality controls prior to analyzing the sequencing data need to be performed. Detecting biases or problems in the data at early stages of the analysis saves precious time and correct problem handling leads to more reliable results.

FastQC [14] can be used to extract different quality metrics and properties from raw sequencing reads. These metrics give a first impression of whether the data suffers from problems such as low base calling quality, adapter contaminations, unexpected over-represented sequences, low conversion rate (e.g., C to T < 0.9), positional C-biases (i.e., read positions with high nonconversion rates), or high percentage of sequence duplications (i.e., read sequences that are exact copies of each other). When specific issues are highlighted by the FastQC report, ad hoc preprocessing steps can be performed before the main analysis to reduce the impact of bias on the interpretation of the results. In the following subheadings, we present the major quality metrics reported by FastQC and discuss the related data issues and possible solutions.

2.2.1 Sequence Quality

The "Per base sequence quality" plot shows the base-call quality score distribution of all reads in a FASTQ file as box-plots across the individual read positions. The quality score Q is computed as Phred score:

$$Q = -10\log_{10}P$$

where P is the base-calling error probability. Thus, base calls with scores higher than 30 can be considered high quality as the probability that the corresponding base call is incorrect is <0.001. As shown in Fig. 1a, base-call quality scores of typical RNA-BSseq data show a slight decrease toward the 3′ end of sequencing reads. It is crucial that the RNA-BSseq reads have correct and high quality base calls to ensure correct mapping and correct methylation calls. Therefore, it is advisable to use only good quality parts of the data (*see* Fig. 1 and Subheading 2.3).

2.2.2 Position-Based Nucleotide Composition

The "Per base sequence content" plot shows the base composition at each position along the reads as fractions of the four nucleotides (ACGT, *see* Fig. 1c, d). It provides very helpful information about the success and efficiency of the bisulfite-induced deamination process and can help in uncovering potential contamination.

Ideally, the bisulfite treatment converts all unmethylated cytosines into uracils and leaves only methylated cytosines unaffected. In RNA-BSseq data from mammalian poly-A RNA the conversion rate is typically >99 %, so the proportion of C's at any position along the sequenced reads should range around 1 % (called "non-conversion rate"). However, this estimate may vary for different species and for different classes of RNAs (i.e., tRNAs). Especially with longer reads, a steady increase of the occurrence of cytosines at positions toward the 3′ end of the reads may occur (Fig. 1c).

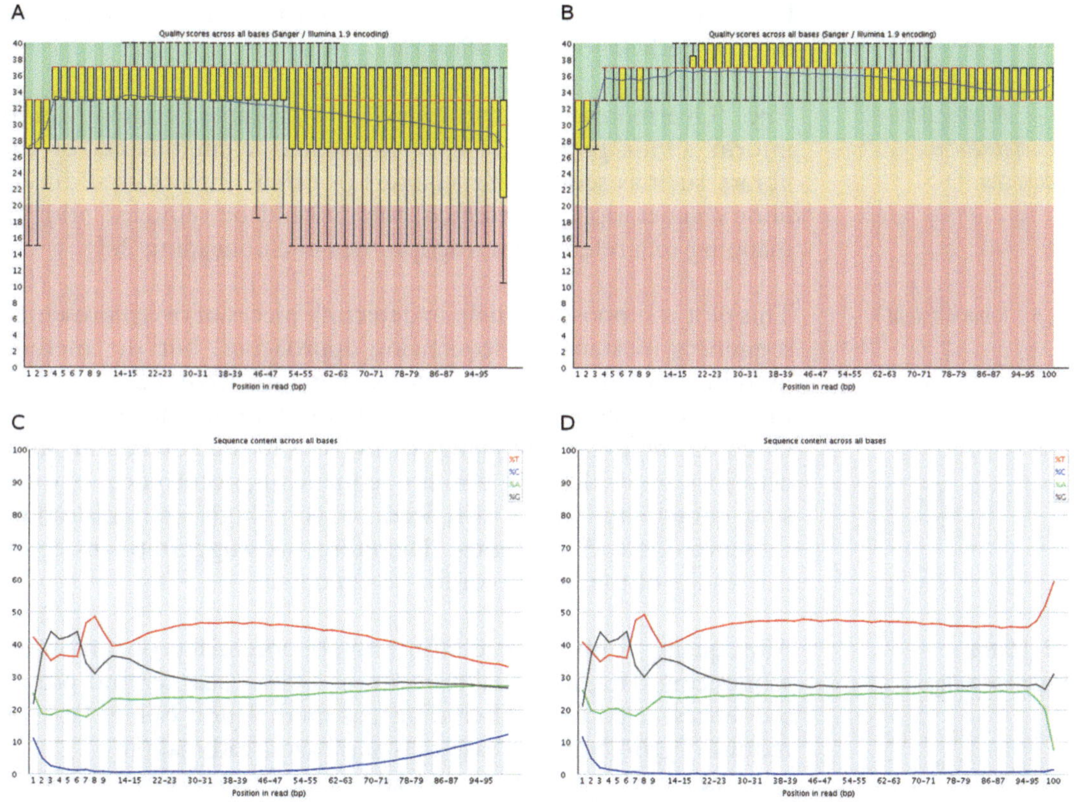

Fig. 1 (**a**, **b**) FastQC per base position quality scores from Illumina reads. Raw RNA-BSseq reads (**a**) typically show a slight decrease in quality toward the 3′ end. After stringent filtering and trimming the quality scores remain high throughout the entire read length. (**c**, **d**) FastQC per base sequence content plots from Illumina reads. Raw RNA-BSseq reads (**c**) typically show cytosine content around 1 % in efficiently deaminated RNA preparations. Adapter contamination can be spotted as increase in cytosine content toward the end of the reads (i.e., base positions > 50). After filtering and adapter removal the cytosine content remains stable at ~1 % throughout the entire read (**d**). In the shown example random oligomers were used for cDNA generation during library preparation; therefore, one can see strongly biased base contents in the first ten read positions

This is a strong indication of adapter presence since RNA methylation or unsuccessful deamination is not depending on positions within a sequencing read. In RNA-BSseq data from library preparation protocols using Illumina ScriptSeq, we noticed an increased cytosine content in the first ten positions of the reads, which most likely originates from the random oligomers used in the reverse transcription step during library preparation (Fig. 1c, d).

2.2.3 Sequence Duplication Levels

The sequence duplication levels plot gives information about the complexity of the sequencing library. A high duplication level (e.g., >50 %) indicates that the library has low diversity. Duplicate reads map to identical coordinates and arise from reverse transcription, PCR, or fragmentation biases during library preparation. They are problematic and lead to incorrect quantification of the methylation levels. Therefore, duplicate reads should be removed or ignored during methylation quantification.

2.2.4 Adapter Content/ Overrepresented Sequences

The "Overrepresented sequences" and "Adapter content" plots inform about potentially overrepresented contaminants of the library such as sequencing adapters or other primers used during library generation. If detected, they should be removed prior to mapping to a reference transcriptome (*see* Subheading 2.3).

2.2.5 Kmer Content

This plot is supposed to help in detecting pieces of overrepresented sequences in standard sequencing approaches, but the method used in generating the plot is not suitable for RNA-BSseq data. It calculates an observed over expected ratio for the individual Kmers based on the overall frequency of all bases in the library and, since cytosines are depleted in bisulfite sequencing, occurrences of cytosine containing Kmers will lead to very high and misleading Observerd/Expected ratios. Therefore, we suggest to disregard the results of the Kmer plot for BSseq data.

2.3 Data Preprocessing

After quality control and identification of potential problems in the sequencing data, low quality reads, as well as adapter and primer sequences, need to be filtered out, as methylation calling later on requires high-confident base calls and exact read mapping.

There are several tools available (e.g., trimmomatic, flexbar, fastx-toolkit, or cutadapt [15–18]) that can perform quality filtering (removing reads), quality trimming (removing read parts), and adapter removal. Typically, we first get rid of adapter contaminants at the 3′ end using stringent overlap settings (-O 1 in cutadapt) resulting in the removal of any base that overlaps with the adapter sequence. Then we use trimmomatic to trim the reads from 3′ to 5′ using a sliding window within which the average quality score of the sequenced base may not fall below the threshold of $Q > 22$, which was reported to be the ideal threshold maximizing the relative amount of "surviving" reads alignable on the reference for trimmomatic [19]. We further discard reads that have an average Phred score below 30 in more than 25 % of the bases. This filter removes reads that have potential low quality parts spread throughout the entire read extent. The remaining high confidence reads can then be aligned to a reference transcriptome (Fig. 1b, d).

For paired-end reads, the reads should remain properly paired after filtering, i.e., the order of the reads should be the same in the forward- and reverse read files. Tools like trimmomatic, flexbar, or cutadapt maintain read pairing when they are run in paired-end mode. If forward and reverse read files are not paired after filtering, correct pairing can be restored using the pairfq tool [20].

2.4 Read Alignment to Reference Genome or Transcriptome

In order to determine the methylation state of cytosines in RNA, quality filtered RNA-BSseq reads need to be aligned to a set of reference sequences, i.e., a reference transcriptome or genome.

If unmethylated in-vitro transcribed spike-in control sequences are included in the sequencing library, the reference sequences for these controls should also be added to the reference sequences.

Alignment of high-throughput sequencing reads is in general performed by short read mapping tools like Bowtie2 [21], BWA [22], and many others that are available to date [23]. For standard RNA sequencing, specialized splice-aware read mapping programs exist, which are capable of aligning reads to a genome even when they cross an exon-exon border, e.g., STAR [24], TopHat2 [25], or HiSat [26].

However, the special nature of the bisulfite-treated RNA reads demands for special aligning strategies. As discussed before, the bisulfite treatment converts all unmethlyated cytosines, which represent the great majority, into uracils whereas the small number of methylated cytosines remains unaffected. Thus, sequencing reads of RNA-BSseq libraries contain mostly A, G, and T (reverse reads of paired end sequencing: A, C, and T). Given the resulting differences with respect to the original sequence, an attempt of aligning such reads to a standard reference using a general short read aligner would fail or at best result in very poor alignment rates.

To achieve better alignment rates, it is best to first generate an in-silico C $\rightarrow$ T converted reference and then to align the bisulfite sequencing reads to it. However, there might be regions in transcripts where multiple methylated cytosines occur, resulting in poor or failed alignment to the fully converted reference because the number of unconverted cytosines in the reads might exceed the mismatch threshold of the aligning algorithm. To prevent this, also the reads themselves should be fully converted, so that every C will be a T. When the resulting reads are then aligned to the fully converted reference the alignment rate should increase. In order to retain the methylation information (nonconverted Cs) one needs to restore the original bisulfite-read sequence after alignment of the fully converted read. This requires custom programming since there is no tool available today that can perform the read restoration.

A number of DNA-BSseq alignment tools, which have to deal with reads comparable to those generated in RNA-BSseq, implement this or similar strategies [27–30]; however, they are not crafted for dealing with spliced reads arising from transcripts [31] and so may perform only suboptimal when applied to RNA-BSseq data.

An approach reported in one of the first transcriptome-wide cytosine methylation studies [7] is to solve the splicing problem by first generating all possible (known and hypothetic) transcripts based on a given genome and a corresponding splice-site annotation and then aligning the BSseq reads to this reference using a DNA-BSseq short read mapping program. The Useq package (http://useq.sourceforge.net/) may be used for generating such reference transcriptomes, which should additionally be combined with a separately generated exon-masked genome. Eventually, these combined sequences may be used for generating a bisulfite index for the DNA-BSseq short read aligner. After aligning the RNA-BSseq reads to this reference, the alignments in the resulting SAM file need to be transformed back to genomic coordinates using the SamTranscriptomeParser—another tool from the USeq package.

As one can imagine the entire strategy requires many manual steps and is time-consuming. Thus, we have developed a toolkit named meRanTK (http://icbi.at/software/meRanTK/) that is specifically tailored for analyzing RNA-BSseq data [32]. It includes meRanG, a tool that enables alignment of RNA-BSseq reads to a reference genome using splice aware short read mappers (STAR or HiSat2). As transcripts can originate from either strand of the genome, meRanG runs two parallel aligning processes to map reads to a C → T and to a G → A converted genome. meRanG determines the highest scoring unique alignments from both processes and—by taking into account that strand-specific library preparation protocols preserve the information about which strand an RNA molecule was originally transcribed from—only correctly oriented alignments are finally retained.

As mentioned above, bisulfite treatment of RNA causes strong fragmentation. Thus, when using paired-end sequencing, the two read pair mates—if originating from fragments shorter than twice the read length—can overlap. It is important to correctly deal with this situation when analyzing RNA-BSseq as it may lead to incorrect methylation rate estimations. meRanG can detect such overlaps and clips the 3′ ends of both read mates, each by half the extent of the overlap, so that bases in the genomic region covered by the read pair may be counted only once per aligned read pair in subsequent methylation calling steps. If tools incapable of resolving overlaps are used in RNA-BSseq analysis, an alternative approach is to merge/flatten the alignments after read mapping using tools like BamUtil (http://genome.sph.umich.edu/wiki/BamUtil).

RNA-BSseq can produce many reads that map equally well to multiple locations on the reference. Such multi-mapping reads are often not considered for further analysis. However, depending on the type of RNA (e.g., tRNAs) and the sequenced organism, they might still contain useful information. Using meRanG, we routinely store such multi-mapping reads to separate output files and analyze them independent of unambiguously mapped reads.

It is also useful to produce M-bias plots (e.g., implemented in meRanG) that may help detecting potential sequencing or library problems (*see* also Subheading 2.2). In such plots the relative fraction of Cs over total read bases is plotted at each mapped read position. In an unbiased dataset this plot should present a flat horizontal line at the height of the expected nonconversion rate, since cytosine methylation is expected to occur independently of the position in a read. Deviations from this expected line may indicate biased read positions to be excluded from methylation calling. However, such deviations might be expected for special RNA families like tRNAs, which have common methylated positions (e.g., C at position 38) [33], especially if the individual reads are covering the full length of the RNA molecule.

2.5 Methylation Calling

Once the bisulfite-reads are aligned to reference sequences, the methylation state can be extracted from pileups of the reads covering each cytosine position. Methylation calling usually takes into consideration criteria such as coverage and nonconversion rate. A minimum coverage of ten reads and a minimum nonconversion rate of 0.2 have been used in transcriptome wide m^5C studies [6, 7]. This means that a cytosine will be reported as methylated if it is covered by at least ten reads and at least 20 % of them have a nonconverted C at the given position. In addition to these minimal criteria, there are other parameters that may help to increase specificity. For example, it is important to take the individual base call qualities into account and only consider bases that have a high Phred score (e.g., > Q30).

When calling the methylation states, potential PCR duplicates should be filtered by defining a maximum allowed number of identical reads. Moreover, potential C-biased read positions as identified by C-bias plots or FastQC should be excluded from the analysis.

If spike-in control sequences are included in the data set, one can estimate the overall C → T conversion efficiency. The deamination rate of spike-in controls in successful experiments typically exceeds 0.99. Based on this rate, we can calculate a *p*-value of the methylation state (i.e., probability that the methylation call for the corresponding position is incorrect) [34] and the *p*-value of the methylation rate (i.e., probability that the reported methylation rate for the corresponding position is incorrect) [35]. If no conversion rate is available, a Fisher's exact test may be used for calculating the *p*-value for a given candidate methylated cytosine based on the baseline sequencing error similar to what is used in the variant caller VarScan 2 [36].

The here-described methylation calling method is implemented in meRanCall, the methylation caller of meRanTK. Besides reporting candidate methylated cytosines and their associated *p*-values, it also reports read counts, methylation rate, mutated positions, and additional information, such as sequence context around candidate methylated cytosines. Methylation data is stored in simple tab delimited text- and BED formatted files, which facilitates easy visualization using modern genome browsers like UCSC or IGV [37, 38].

2.6 Identification of Differentially Methylated Cytosines

In RNA methylation experiments, it is also interesting to compare methylation patterns in different tissues or physiological conditions.

Presence or absence of methylation at specific sites and fold-change of methylation rates between different conditions may be of interest. For statistical evaluation of methylation differences in single or multiple replicate experiments either a Fisher's exact test or a Cochran-Mantel-Haenszel test may be used. Differences in the library sizes should always be considered and corrected when methylation fold changes between different experimental conditions are calculated.

meRanTK includes meRanCompare a tool that facilitates the identification of differentially methylated cytosines in two experimental conditions. It can handle single or multiple replicates and reports statistical significance levels. meRanCompare takes into account different library sizes and estimates scaling factors, similar to the method described in DESeq2 [39]. It uses these scaling factors to scale the raw read counts at the candidate methylated cytosines to correct for the individual library sizes.

3 Conclusion

RNA-BSseq is a powerful high-throughput method for studying transcriptome-wide methylation states. Although there is still room for improvement, first studies have shown that it is able to identify known (i.e., on tRNAs [7]) as well as novel methylation sites and to provide new insight into general patterns of the m^5C RNA modification [6].

The extraction of meaningful and reproducible information from RNA-BSseq data requires high-quality sequencing reads of bisulfite-converted RNA fragments. For the analysis it is crucial that the deamination process completely converts unmethylated cytosines to uracils. Poor deamination leads to many false-positive methylation calls. Therefore, the conversion process should always be verified by including one or more control sequences. Experiments with deamination rates clearly below 0.99 should be excluded or repeated. Raw sequencing data should be carefully inspected to detect potential biases and contaminations. The alignment and methylation calling steps clearly benefit from stringently filtered, trimmed, and C-bias free reads.

Moreover, we suggest always carrying out RNA-BSseq experiments in three or more biological replicates as candidate sites that can be detected in at least two replicates give a higher confidence. We also suggest validating a number of potential methylation sites by independent methods such as meRIP [40], AzaIP [7], mCLIP [8], or mass spectrometry.

Until lately, only guidelines and strategies involving time-consuming and cumbersome steps that also demand manual programming were available; however, tools dedicated specifically for the analysis of RNA-BSseq data are now available [32]. Here, we suggested some practical strategies for experimental design of BSseq experiments, as well as data quality control, preprocessing, and analysis. These recent computational developments facilitate RNA-BSseq studies and may so help to shed more light on the nature and significance of m^{5}C RNA modifications.

References

1. Schaefer M (2015) Chapter fourteen – RNA 5-methylcytosine analysis by bisulfite sequencing. Methods Enzymol 560:297–329. doi: 10.1016/bs.mie.2015.03.007, ISSN 0076-6879, ISBN 9780128021927. http://www.sciencedirect.com/science/article/pii/S0076687915002335
2. Cantara WA, Crain PF, Rozenski J et al (2011) The RNA modification database, RNAMDB: 2011 update. Nucleic Acids Res 39:D195–D201. doi:10.1093/nar/gkq1028
3. Motorin Y, Lyko F, Helm M (2010) 5-methylcytosine in RNA: detection, enzymatic formation and biological functions. Nucleic Acids Res 38:1415–1430. doi:10.1093/nar/gkp1117
4. Amort T, Soulière MF, Wille A et al (2013) Long non-coding RNAs as targets for cytosine methylation. RNA Biol 10:1003–1008. doi:10.4161/rna.24454
5. Schaefer M (2015) Chapter fourteen – RNA 5-methylcytosine analysis by bisulfite sequencing. Methods Enzymol 560:297–329. doi: 10.1016/bs.mie.2015.03.007, ISSN 0076–6879, ISBN 9780128021927. http://www.sciencedirect.com/science/article/pii/S0076687915002335
6. Squires JE, Patel HR, Nousch M et al (2012) Widespread occurrence of 5-methylcytosine in human coding and non-coding RNA. Nucleic Acids Res 40:5023–5033. doi:10.1093/nar/gks144
7. Khoddami V, Cairns BR (2013) Identification of direct targets and modified bases of RNA cytosine methyltransferases. Nat Biotechnol 31:458–464. doi:10.1038/nbt.2566
8. Hussain S, Sajini AA, Blanco S et al (2013) NSun2-mediated cytosine-5 methylation of vault noncoding RNA determines its processing into regulatory small RNAs. Cell Rep 4:255–261. doi:10.1016/j.celrep.2013.06.029
9. Khoddami V, Cairns BR (2014) Transcriptome-wide target profiling of RNA cytosine methyltransferases using the mechanism-based enrichment procedure Aza-IP. Nat Protoc 9:337–361. doi:10.1038/nprot.2014.014
10. Schaefer M, Pollex T, Hanna K, Lyko F (2009) RNA cytosine methylation analysis by bisulfite sequencing. Nucleic Acids Res 37:e12. doi:10.1093/nar/gkn954
11. Lee J-H, Ang JK, Xiao X (2013) Analysis and design of RNA sequencing experiments for identifying RNA editing and other single-nucleotide variants. RNA 19:725–732. doi:10.1261/rna.037903.112
12. Sims D, Sudbery I, Ilott NE et al (2014) Sequencing depth and coverage: key considerations in genomic analyses. Nat Rev Genet 15:121–132. doi:10.1038/nrg3642
13. Storvall H, Ramsköld D, Sandberg R (2013) Efficient and comprehensive representation of uniqueness for next-generation sequencing by minimum unique length analyses. PLoS One 8:e53822. doi:10.1371/journal.pone.0053822
14. Andrews S (2010) FastQC: a quality control tool for high throughput sequence data. http://www.bioinformatics.babraham.ac.uk/projects/fastqc
15. Bolger AM, Lohse M, Usadel B (2014) Trimmomatic: a flexible trimmer for Illumina sequence data. Bioinformatics 30:2114–2120. doi:10.1093/bioinformatics/btu170
16. Dodt M, Roehr JT, Ahmed R, Dieterich C (2012) FLEXBAR-flexible barcode and adapter processing for next-generation sequencing platforms. Biology 1:895–905. doi:10.3390/biology1030895
17. Gordon A (2009) FASTX-Toolkit FASTQ/A short-reads pre-processing tools. http://hannonlab.cshl.edu/fastx_toolkit/index.html.
18. Martin M (2011) Cutadapt removes adapter sequences from high-throughput sequencing reads. EMBnet.journal 17:10–12. doi:10.14806/ej.17.1.200
19. Fabbro CD, Scalabrin S, Morgante M, Giorgi FM (2013) An extensive evaluation of read trimming effects on Illumina NGS data analysis. PLoS One 8:e85024. doi:10.1371/journal.pone.0085024
20. sestaton/Pairfq. In: GitHub. https://github.com/sestaton/Pairfq. Accessed 22 Feb 2016

21. Langmead B, Salzberg SL (2012) Fast gapped-read alignment with Bowtie 2. Nat Methods 9:357–359. doi:10.1038/nmeth.1923
22. Li H, Durbin R (2009) Fast and accurate short read alignment with Burrows-Wheeler transform. Bioinformatics 25:1754–1760. doi:10.1093/bioinformatics/btp324
23. Bao S, Jiang R, Kwan W et al (2011) Evaluation of next-generation sequencing software in mapping and assembly. J Hum Genet 56:406–414. doi:10.1038/jhg.2011.43
24. Dobin A, Davis CA, Schlesinger F et al (2013) STAR: ultrafast universal RNA-seq aligner. Bioinformatics 29:15–21. doi:10.1093/bioinformatics/bts635
25. Kim D, Pertea G, Trapnell C et al (2013) TopHat2: accurate alignment of transcriptomes in the presence of insertions, deletions and gene fusions. Genome Biol 14:R36. doi:10.1186/gb-2013-14-4-r36
26. Kim D, Langmead B, Salzberg SL (2015) HISAT: a fast spliced aligner with low memory requirements. Nat Methods 12:357–360. doi:10.1038/nmeth.3317
27. Kunde-Ramamoorthy G, Coarfa C, Laritsky E et al (2014) Comparison and quantitative verification of mapping algorithms for whole-genome bisulfite sequencing. Nucleic Acids Res 42:e43. doi:10.1093/nar/gkt1325
28. Krueger F, Andrews SR (2011) Bismark: a flexible aligner and methylation caller for Bisulfite-Seq applications. Bioinformatics 27:1571–1572. doi:10.1093/bioinformatics/btr167
29. Lim J-Q, Tennakoon C, Li G et al (2012) BatMeth: improved mapper for bisulfite sequencing reads on DNA methylation. Genome Biol 13:R82. doi:10.1186/gb-2012-13-10-r82
30. Xi Y, Li W (2009) BSMAP: whole genome bisulfite sequence MAPping program. BMC Bioinformatics 10:232. doi:10.1186/1471-2105-10-232
31. Finotello F, Di Camillo B (2015) Measuring differential gene expression with RNA-seq: challenges and strategies for data analysis. Brief Funct Genomics 14:130–142. doi:10.1093/bfgp/elu035
32. Rieder D, Amort T, Kugler E et al (2015) meRanTK: methylated RNA analysis ToolKit. Bioinformatics. doi:10.1093/bioinformatics/btv647
33. Tuorto F, Liebers R, Musch T et al (2012) RNA cytosine methylation by Dnmt2 and NSun2 promotes tRNA stability and protein synthesis. Nat Struct Mol Biol 19:900–905. doi:10.1038/nsmb.2357
34. Lister R, Pelizzola M, Dowen RH et al (2009) Human DNA methylomes at base resolution show widespread epigenomic differences. Nature 462:315–322. doi:10.1038/nature08514
35. Barturen G, Rueda A, Oliver JL, Hackenberg M (2013) MethylExtract: high-quality methylation maps and SNV calling from whole genome bisulfite sequencing data. F1000Res 2:217. doi:10.12688/f1000research.2-217.v2
36. Koboldt DC, Zhang Q, Larson DE et al (2012) VarScan 2: somatic mutation and copy number alteration discovery in cancer by exome sequencing. Genome Res 22:568–576. doi:10.1101/gr.129684.111
37. Karolchik D, Kuhn RM, Baertsch R et al (2008) The UCSC Genome Browser Database: 2008 update. Nucleic Acids Res 36:D773–D779. doi:10.1093/nar/gkm966
38. Thorvaldsdóttir H, Robinson JT, Mesirov JP (2013) Integrative genomics viewer (IGV): high-performance genomics data visualization and exploration. Brief Bioinform 14:178–192. doi:10.1093/bib/bbs017
39. Love MI, Huber W, Anders S (2014) Moderated estimation of fold change and dispersion for RNA-seq data with DESeq2. Genome Biol 15:550. doi:10.1186/s13059-014-0550-8
40. Meyer KD, Saletore Y, Zumbo P et al (2012) Comprehensive analysis of mRNA methylation reveals enrichment in 3′ UTRs and near stop codons. Cell 149:1635–1646. doi:10.1016/j.cell.2012.05.003

Chapter 11

Statistical Methods for Transcriptome-Wide Analysis of RNA Methylation by Bisulfite Sequencing

Brian J. Parker

Abstract

For the transcriptome-wide detection and quantification of the 5-methylcytosine (m^5C) methylation modification of RNA, one experimental approach is via bisulfite conversion. In this chapter we discuss statistical methods, and a corresponding computational pipeline, to perform transcriptome-wide differential m^5C methylation analysis between RNA samples, specialized for this assay.

Key words RNA methylation, Differential methylation, 5-methylcytosine, Epitranscriptomics, Bisulfite conversion, High-throughput sequencing

1 Introduction

For transcriptome-wide detection of m^5C methylation in RNA, a variety of protocols have been developed including immunoprecipitation-based methods (meRIP [1], aza-IP [2], miCLIP [3]), and approaches based on bisulfite treatment [4], as commonly used previously for DNA methylation and later modified for application to RNA. Bisulfite treatment of RNA leads to unmethylated cytosines being deaminated to uracil (represented as T in the sequenced libraries), while m^5C sites remain as cytosine (C in the sequenced libraries). So the level of m^5C methylation at a base position can be estimated by the proportion of C to T mismatches when mapping the bisulfite-converted reads back to the reference genome.

Bisulfite conversion experimental protocols for high-throughput sequencing (HTS) of RNA have been described in detail elsewhere [5–7], and is referred to as bsRNA-seq in this chapter. While this bisulfite conversion method is well-understood when applied to DNA, and standard bioinformatic tools are available for analysis in this case, here we describe a computational protocol and statistical models for detecting differential m^5C

Alexandra Lusser (ed.), *RNA Methylation: Methods and Protocols*, Methods in Molecular Biology, vol. 1562,
DOI 10.1007/978-1-4939-6807-7_11, © Springer Science+Business Media LLC 2017

methylation sites optimized for RNA, including splicing of mature mRNAs.

This bioinformatic protocol has been used in the analysis of human (HeLa and prostate) cell lines [8] and plant (*Arabidopsis*) [9].

We assume that the context for this analysis is a differential methylation study following knockdown of a methyltransferase (e.g., m^5C methyltransferases such as NSUN2 or TRDMT1) compared with wild type (WT) controls. However, the linear model underlying the method described is fully general and has been successfully applied to varying experimental designs including calling global m^5C sites in WT, and factorial designs using multiple methyltransferase knockdowns and simultaneous manipulations of environmental conditions.

2 Materials

The computational pipeline described here commences with the output of high-throughput sequencing of the bisulfite-converted RNA library, producing read sequences typically in FASTQ file format [10]. The pipeline described here has been successfully applied to both 150 bp paired end Illumina sequencing data and 50 bp single end SOLiD sequencing technologies, but can be generalized to any similar sequencing protocol. The following description focusses on Illumina sequencing.

In the following description we assume an experimental design with libraries from two or more biological conditions, for example methyltransferase knockdown versus wild type experiments, but arbitrarily complex experimental designs are supported. We assume that the experimental design incorporates sufficient biological replicates from each of the groups or conditions, with the number of replicates determining the overall power of the analysis. For example, three replicates in each condition would be typical; a more formal power analysis may be possible for some studies. In the following we refer to these individual experimental units/biological replicates as "RNA samples."

The following software packages are required to be installed:

- Trimmomatic adapter removal software [11] from http://www.usadellab.org/cms/index.php?page=trimmomatic.
- fastQC read quality software from http://www.bioinformatics.babraham.ac.uk/projects/fastqc/.
- Bowtie read mapping software [12, 13] from http://bowtie-bio.sourceforge.net/bowtie2/index.shtml (Bowtie2 required for Illumina, Bowtie for SOLiD sequencing).
- Tophat spliced read mapping software [14] from https://ccb.jhu.edu/software/tophat/index.shtml.

- Modified versions of Bismark [15] and/or Bsolana [16] DNA methylation calling programs, modified for RNA methylation, from https://github.com/parker-lab/rna_methylation/.
- R statistical package [17] from https://www.r-project.org/.
- Limma [18] differential expression analysis pipeline from https://bioconductor.org/packages/release/bioc/html/limma.html, and modified voom [19] function source code from https://github.com/parker-lab/rna_methylation/.
- IGV browser [20] from https://www.broadinstitute.org/igv/.
- Scripts implementing the entire pipeline described here, with example datasets, are available from https://github.com/parker-lab/rna_methylation/.

3 Methods

An overview of the computational pipeline is given in Fig. 1. We describe both the underlying statistical theory required to

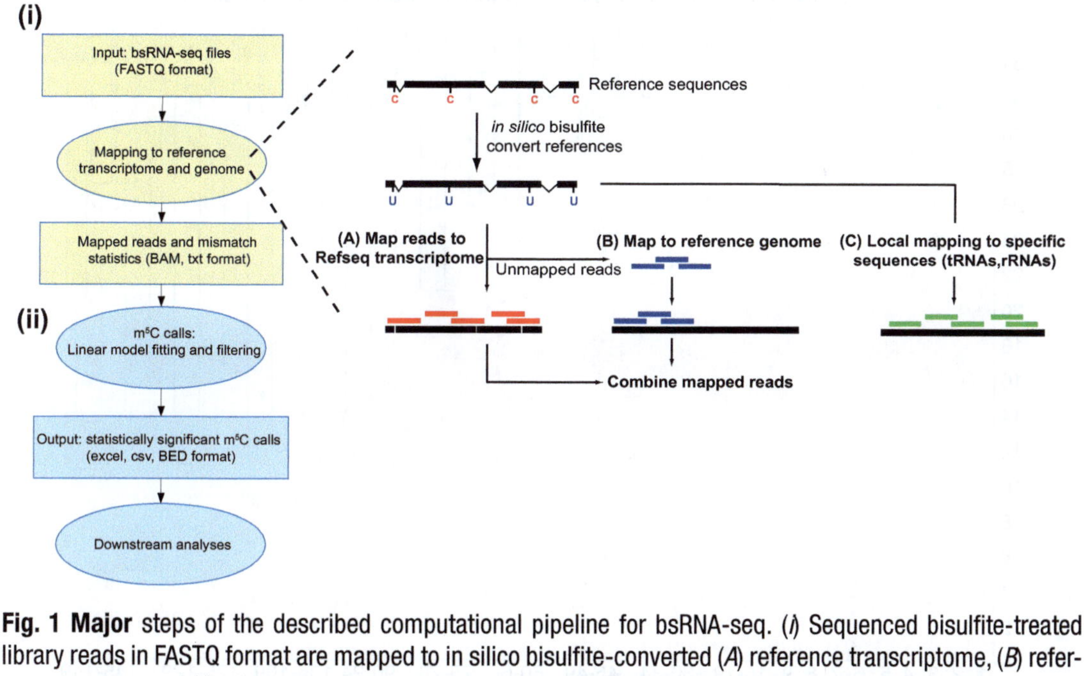

Fig. 1 Major steps of the described computational pipeline for bsRNA-seq. (*i*) Sequenced bisulfite-treated library reads in FASTQ format are mapped to in silico bisulfite-converted (*A*) reference transcriptome, (*B*) reference genome, and (*C*) other reference sequences of interest. Mismatch statistics relative to the references are then calculated. Output files at this stage are BAM files of mapped reads and text files summarizing the per-base mismatch statistics against these references. (*ii*) To detect statistically significantly differentially methylated sites, a linear model is fit at each candidate C base, based on the C to T mismatch proportion. Statistically significant called m^{5}C sites with a bounded false discovery rate are saved, with genomic location, as final output in Excel, CSV, and BED files

understand and adjust the protocol, and the applied tools needed for the analysis. The computational pipeline consists of the following steps.

3.1 Library Quality Assessment

The FASTQ files should be initially checked for basic quality metrics to ensure the library preparation produced RNA is of high quality and sufficient complexity. fastQC is a widely used program for this purpose and its output plots are described in detail at the url above. *See* Fig. 2 for an example.

3.2 Adapter Trimming

Adapter sequences ligated to the ends of RNA fragments to facilitate sequencing should be trimmed from the raw reads. Trimmomatic is a software suitable for Illumina sequencing. The adapter sequence used needs to be matched to the library protocol, and standard sequences are available from the Trimmomatic web site. Typical Trimmomatic parameters used are: ILLUMINACLIP:illuminaClipping.fa:4:30:10:1:true LEADING:3 TRAILING:20 SLIDINGWINDOW:4:20 MINLEN:36.

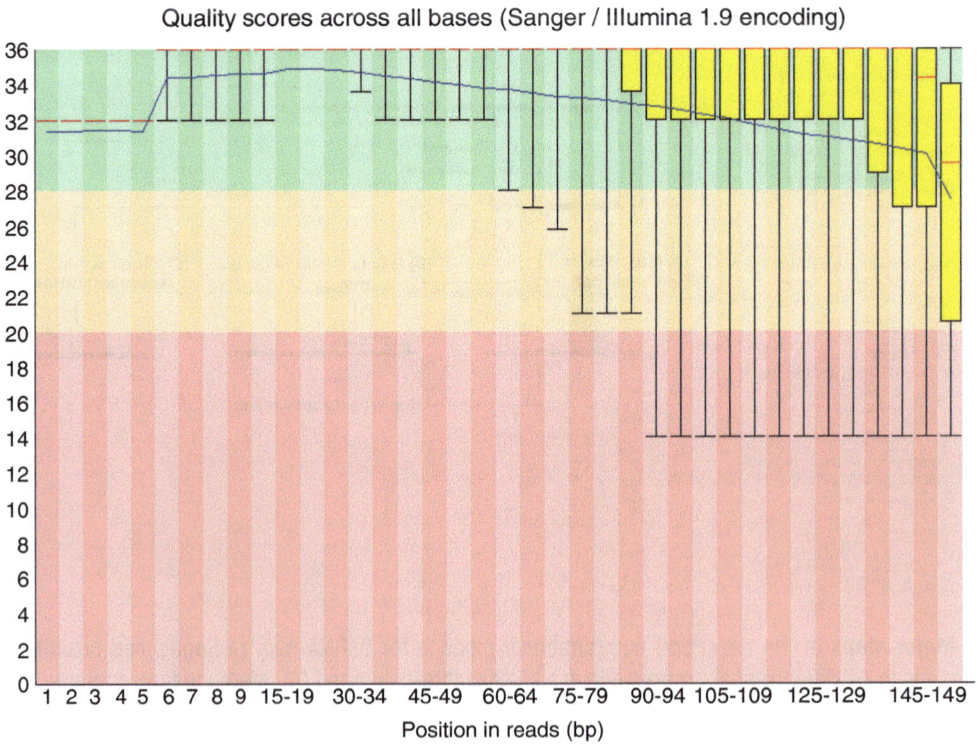

Fig. 2 FastQC quality checks on the sequenced library. FastQC implements multiple checks of RNA and library quality for Illumina sequencing. In this example plot, the quality scores for base calls across read length are plotted. In this high-quality library, there is only slight fall-off in base quality at the 3′ read ends before adaptor trimming

These parameters ensure stringent 3′ base quality trimming is performed, to miminize artifactual mismatches at the lower quality 3′ read end. Bases at the 5′ end that show a Phred quality score < 3 and bases at the 3′ end that show a Phred quality score < 15 are removed. Also, if any four base window within the read has a Phred quality score of < 15, then the rest of the read is trimmed. For paired end reads, the parameters enable palindromic mode which trims the adapter from 5′ and 3′ ends simultaneously, allowing single base precision in removing adapter sequences.

3.3 Global Mapping of Bisulfite-Converted Sequencing Data

Read mapping is performed with the Bismark or B-SOLANA bisulfite assay mapping programs, which were originally designed for DNA m^5C and which we have subsequently modified for application to RNA methylation. Bismark and B-SOLANA were designed for Illumina and SOLiD sequencing, respectively. The main modification made is to call the Tophat [14] spliced read mapper from within these programs to handle junction reads across splice sites in mature RNA sequences. Using Tophat, reads are first mapped to a Refseq reference transcriptome matching the species studied, in order to detect known transcript isoforms, followed by mapping to the appropriate species reference genome (*see* Fig. 1) to detect unannotated transcripts including mRNAs and ncRNAs. Both Bismark and B-SOLANA handle bisulfite-treated reads by mapping the data to in silico-converted reference transcriptomes and genomes, which have C's converted to T's to match the bisulfite-induced conversions of unmethylated bases; Bismark also in silico converts the read sequences. See the respective references for additional technical details.

To ensure maximal specificity of predicted sites, only uniquely mapped reads are retained, where "uniquely mapped" is defined by requiring a Bowtie2 mapping quality (MAPQ) score of ≥ 20 for Illumina sequencing. For SOLiD sequencing, Bowtie1 is required for which MAPQ scores are not available, in which case "uniquely mapped" is defined as reads with a single best mapping location.

Reference genome and transcript annotations for model organisms can be downloaded from the Illumina iGenomes datasets at https://support.illumina.com/sequencing/sequencing_software/igenome.html.

3.4 Local Mapping of Sequencing Data

Mapping of highly repetitive sequences of particular interest, such as tRNA and rRNA, can be incomplete in the previous steps due to the stringent multi-mapping control and as these sequences may not be fully represented in the reference genome. For such specific regions of interest, and also specific experimental control sequences (such as R-Luc, *see* ref. [6]), the sequencing data can additionally be mapped to specific complete reference sequences rather than to the reference genome. For example, in human studies to map mature tRNA sequences, sequencing reads were mapped to the full set of Rfam tRNA annotations in the hg19 human reference that matched a tRNA database (tRNAdb [21]), with the in silico addition of the non-templated tRNA extension CCA bases to the 3′ sequence ends. A single representative entry was selected for identical tRNA repeat sequences. Sequencing reads were also mapped to the rRNA tandem repeat sequences, including human ribosomal DNA complete repeating unit (18S rRNA, 5.8S rRNA, 28S rRNA) and human 5S DNA from Genbank.

3.5 Additional Checks of Mapped Data

The output files produced by the previous steps are: (1) BAM files (see https://www.broadinstitute.org/igv/BAM), storing the mapped read information, produced by Tophat, and (2) text files summarizing the number of reads overlapping each C base position in the reference genome and the counts of each form of conversion—C to T, C to G, C to A—in the reads overlapping that coordinate, estimated by comparing the read sequence against the reference genome (Figs. 3 and 4).

The BAM files can be directly viewed in a genome browser such as IGV to visually check sites of interest, including the sites called in later steps. Methylated loci can be distinguished by a substantial fraction of reads showing C to T conversions. *See* Fig. 5 for an example. The total number of C to T mismatches should dominate, *see* Fig. 3, which provides a quick check of the preceding steps.

3.6 Statistical Calling of Candidate m⁵C Sites

Next, a linear model is fit across the RNA samples, separately at each site, to detect statistically significant differential methylation between conditions. Such a linear model supports arbitrary experimental designs and comparisons of interest between biological conditions (see the limma user guide at https://www.bioconductor.

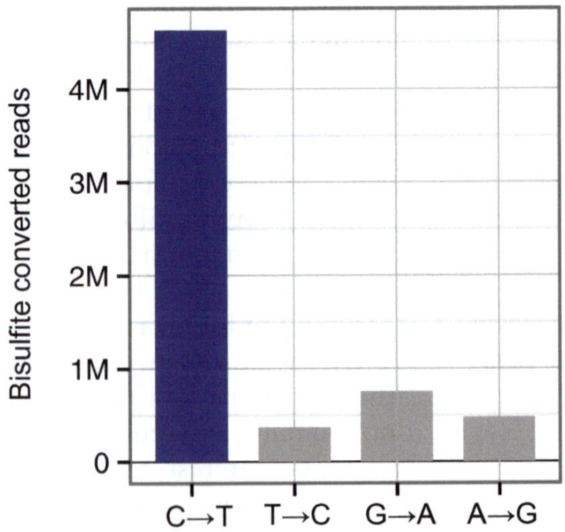

Fig. 3 Bar plot of distribution of all transitions of bases. As expected, C to T conversions dominate in this high-quality bisulfite-treated data

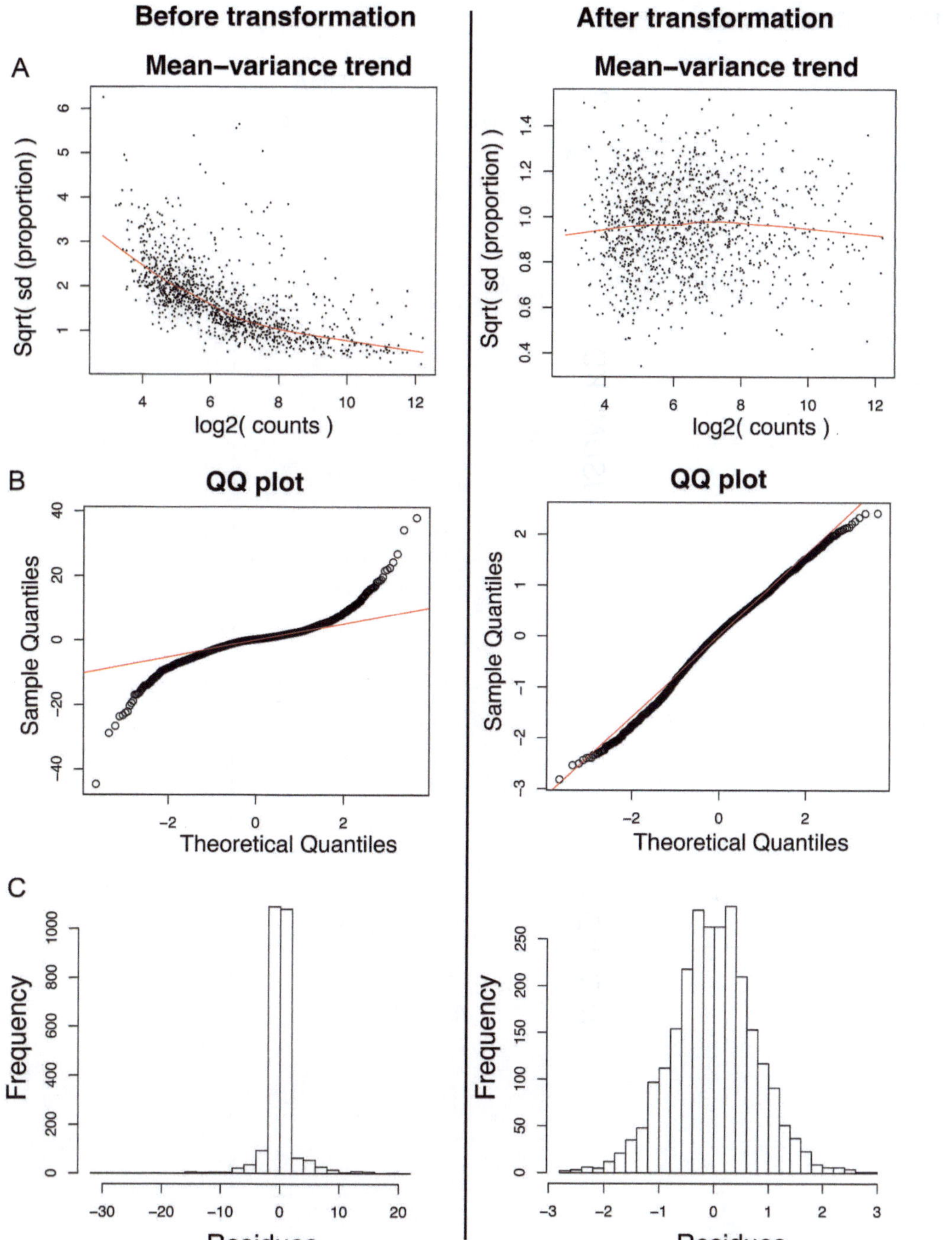

Fig. 4 Diagnostic plots for linear model assumptions. *Left subfigures* show plots of the raw data, before transformation or normalization, and *right subfigures* show corresponding plots after transforming and normalizing the data. Subfigures (**b**) and (**c**) show the distribution, across all candidate C bases, of the residuals between the data and the model fit. After data transformation the residuals are normally distributed: the quantile–quantile plot against a theoretical normal curve shows a good fit, and the histogram of the residuals shows a symmetric, approximately normal distribution. By contrast, the raw data show a skewed distribution with wide tails. Subfigure (**a**) for the raw data shows there is heterogeneity of variance with a substantial mean-variance trend, whereas the transformed data shows improved mean-variance trend. The *red line* shows a quadratic curve fit which is used to adjust weights in the model to further correct for any residual mean-variance trend, using the voom method

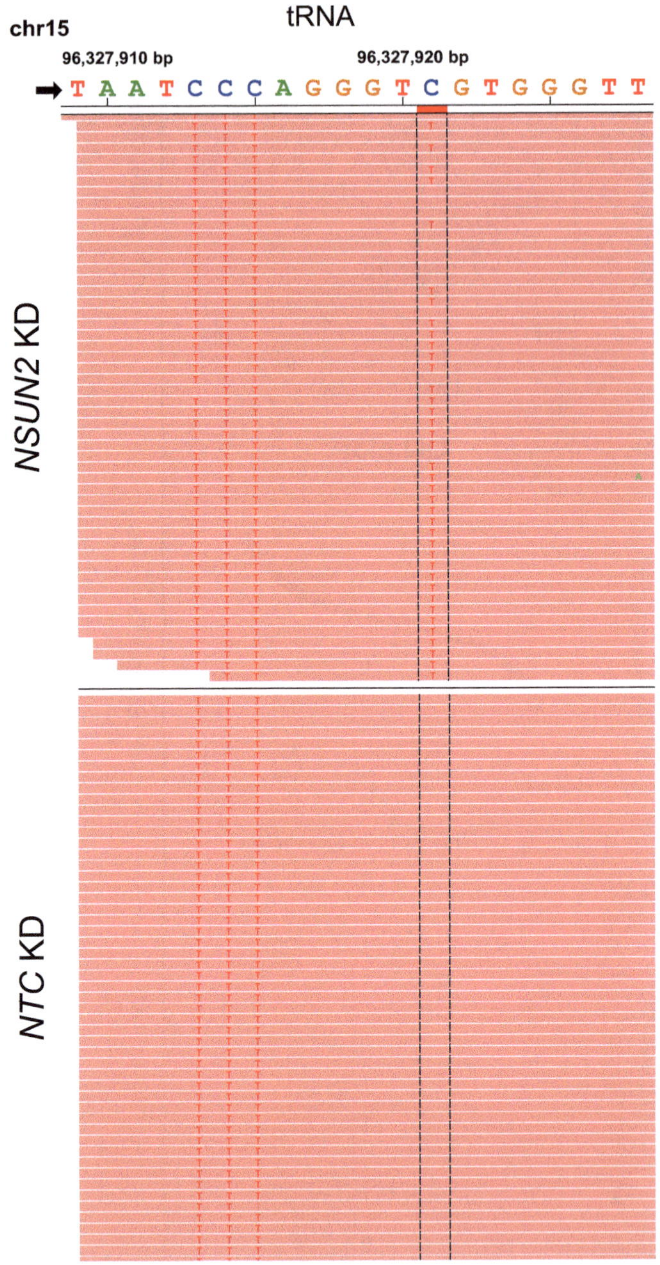

Fig. 5 **IGV** genome browser screenshot showing mapped reads in HeLa after bisulfite conversion at a tRNA locus on the + strand. The *bottom subfigure* shows mapped reads for a non-targetting control (NTC) RNA sample, compared with the *upper subfigure* which shows a methyltransferase (NSUN2) knockdown RNA sample. Note that the first three C positions show almost 100% C to T conversion in both the control and knockdown RNA samples, whereas the fourth C is not converted in the control RNA sample but a large fraction of reads are converted in the methyltransferase KD RNA sample. This is indicative of NSUN2-mediated m^5C methylation of the fourth C. Genomic coordinates are displayed at *top*

org/packages/3.3/bioc/vignettes/limma/inst/doc/usersguide.pdf for examples). As the distribution of the raw data from this assay does not directly match the assumptions needed to apply such a linear model, it is first transformed so as to meet these model assumptions, and per-site weights are computed to correct for increased variance due to low read counts, as follows.

Proportion statistic: The primary statistic used for detection of candidate m^5 C sites is a proportion statistic $p = \frac{C}{C+T}$, where C is the number of reads showing (unconverted) cytosines and T is the number of reads showing T mismatches, representing converted cytosines. The data is transformed into an approximate Gaussian distribution by a log arcsin transformation [22], $p' = \log_2 \frac{360}{2\pi} \cdot \arcsin\sqrt{p}$, which was determined empirically across multiple studies (*see* **Note 2**).

Non-specific filtering: Initial non-stringent filtering is performed across all RNA samples, both experimental and control, to exclude clearly non-candidate sites as follows. Lowly expressed candidate m^5C sites are excluded by requiring ≥ 5 reads across all RNA samples used in the model. For the detection of differential m^5C sites, initial filtering of low methylation sites across the RNA samples was performed by requiring significantly $\geq 1\%$ methylation (significance measured by binomial test, $p < 0.05$) in at least n RNA samples in the dataset, where n is the size of the control group of RNA samples. For overall calling of candidate m^5C sites (see below) all RNA samples require at least $> 0\%$ methylation.

Statistical model: A weighted linear model is fit to the transformed data p' for differential and overall methylation calls using limma. The weights correct for mean-variance trend due to the discrete counts of the bsRNA-seq assay. These weights are estimated by quadratic fit of mean variance of p' versus total read counts over each site using the voom method [19]. Moderation borrows information across sites to improve per-site variance estimates [23]. Optionally, the transformed data can be normalized before model fitting to remove technical differences between RNA samples such as varying bisulfite conversion efficiency. This is performed using a cyclic loess normalization option in the voom step above.

Differential m^5C site calling: In methyltransferase knockdown studies, methyltransferase-controlled m^5C sites are detected by defining an appropriate contrast matrix between the groups of interest, e.g. the methyltransferase knockdown RNA samples versus the control RNA samples (see the limma user guide, and example experimental designs provided at https://github.com/parker-lab/rna_methylation/).

Overall m^5C site calling: To call sites showing m^5C % above background levels, regardless of methyltransferase status, a model comparison is used to test whether the methylation % of the control group of RNA samples is significantly above background methylation %, at each base. In detail, the linear model design matrix is set for a one-sample moderated *t*-test against a fixed methylation level Δ, with Δ set to 0% methylation by default, but configurable to a small positive value to define a minimum effect-size above background for the test.

Multiple testing correction: The per-base *p*-values are adjusted to correct for the multiple hypothesis testing across all sites by the Benjamini and Hochberg [24] or Storey [25] methods. After this adjustment, a fixed threshold can be defined to limit the overall study false discovery rate (FDR) of the m^5C calls.

3.7 Post-filtering

Post-filtering is performed to exclude potentially artifactual candidate m^5C sites. Candidate m^5C sites are required to have a homogeneous C to T conversion with few C to A or C to G mismatches ($C + T$ is required to be $\geq 95\%$ of bases) to exclude calls at sites of high read error rate or unannotated SNPs. Also, calls overlapping known single nucleotide polymorphisms (SNPs), as annotated in dbSNP, and RNA editing sites, e.g. from the DARNED editing database [26], are excluded.

If simultaneous RNA-seq data is available for the identical cell line or tissue used for bsRNA-seq, it can be utilized to detect potential artifactual calls due, e.g., to differential somatic mutations, non-canonical editing sites, and unannotated SNPs: base positions where the genomic reference does not match the corresponding RNA-seq base call for $>90\%$ reads are excluded.

Once the final statistically significant set of differentially methylated sites is defined with a bounded false discovery rate, the candidate m^5C sites are output as a table in Excel and CSV formats, ordered by the fold change of methylation fraction, suitable for further processing (*see* **Note 1**). Also, a BED file (https://www.broadinstitute.org/igv/BED) with the called sites is produced, suitable for viewing in IGV. Optionally, for experimental validation, a more stringent set of candidate m^5C sites can be defined as sites that exhibit a substantial effect size, for example $\geq$ M% mean methylation across all control RNA samples for some threshold M.

3.8 Experimental Validation by Amplicon Sequencing

Candidates m^5C sites of interest can be validated by extremely deep MiSeq sequencing of an amplicon flanking the site, *see* ref. [5] for details of the experimental protocol. These validations can be computationally processed as described in Subheading 3.4 above, using the amplicon sequences as the mapping reference. Scripts and example data are provided at https://github.com/parker-lab/rna_methylation/.

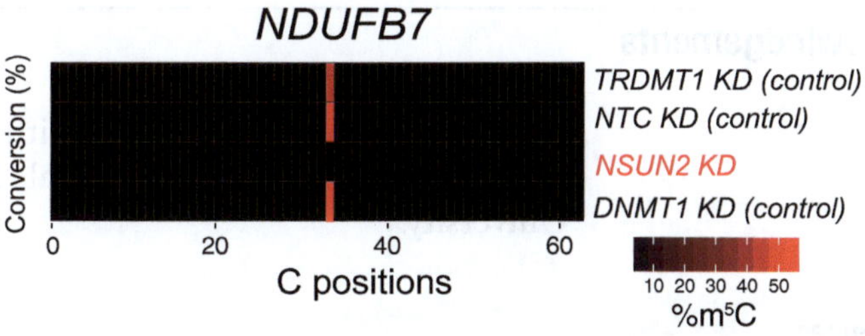

Fig. 6 Experimental validation by deep amplicon sequencing of site called from transcriptome-wide data. Columns of heatmaps show C bases in the amplicons flanking the called m^{5}C site; rows show different RNA samples representing methyltransferase knockdown and non-targetting control (NTC) RNA samples; color indicates estimated m^{5}C methylation %. In this result, an NSUN2-mediated loci in the mRNA of the gene *NDUFB7* is shown

4 Notes

1. *Downstream analyses.* The above computational pipeline will result in a final list of called sites, including the genomic coordinates of each methylation call, the methylation fold change, and the adjusted p-value, which can then be used to select candidates for experimental follow-up and downstream computational analyses. These downstream analyses include: overall genomic distribution, distribution plots around genomic landmarks, metagene plots, sequence and structural motif enrichment and logo generation, overlap with known ncRNAs, gene ontology and gene set enrichment analyses, etc. See the methods sections of the provided references for additional details of these example downstream analyses. Scripts for generating heatmaps over flanking regions of called m^{5}C sites (*see* Fig. 6) and other visualizations are available at https://github.com/parker-lab/rna_methylation/.
2. *Diagnostic checks of statistical model assumptions.* Running the above pipeline step from the provided script at https://github.com/parker-lab/rna_methylation/ will generate diagnostic plots used to check the model assumptions. In particular, that the residuals between the model fit and transformed data are approximately normally distributed, and that the variance is homogeneous or any remaining mean-variance trend is correctly modelled. *See* Fig. 4 for details. The transformation described above has been successfully applied to human and Arabidopsis bisulfite-converted RNA samples. As an alternative, a complementary log-log transformation was also found to perform well. If the diagnostic plots are not adequate for a particular experimental setup, a more general Box–Cox transformation, and other parameters, can be configured in the provided script, see the provided help files for details.

Acknowledgements

This work was in collaboration with the lab of Thomas Preiss at the John Curtin School of Medical Research, Australian National University.

References

1. Edelheit S, Schwartz S, Mumbach MR, Wurtzel O, Sorek R (2013) Transcriptome-wide mapping of 5-methylcytidine RNA modifications in bacteria, archaea, and yeast reveals m5C within archaeal mRNAs. PLoS Genet 9(6):e1003,602. doi:10.1371/journal.pgen.1003602
2. Khoddami V, Cairns BR (2013) Identification of direct targets and modified bases of RNA cytosine methyltransferases. Nat Biotechnol 31(5):458–464. doi:10.1038/nbt.2566
3. Hussain S, Sajini AA, Blanco S, Dietmann S, Lombard P, Sugimoto Y, Paramor M, Gleeson JG, Odom DT, Ule J, Frye M (2013) NSun2-mediated cytosine-5 methylation of vault noncoding RNA determines its processing into regulatory small RNAs. Cell Rep 4(2):255–261. doi:10.1016/j.celrep.2013.06.029
4. Schaefer M, Pollex T, Hanna K, Lyko F (2009) RNA cytosine methylation analysis by bisulfite sequencing. Nucleic Acids Res 37(2):e12. doi:10.1093/nar/gkn954
5. Sibbritt T, Shafik A, Clark SJ, Preiss T (2016) Nucleotide-level profiling of m^{5}C RNA methylation. Methods Mol Biol 1358:269–284. doi:10.1007/978-1-4939-3067-8_16
6. Squires JE, Patel HR, Nousch M, Sibbritt T, Humphreys DT, Parker BJ, Suter CM, Preiss T (2012) Widespread occurrence of 5-methylcytosine in human coding and non-coding RNA. Nucleic Acids Res 40(11): 5023–5033
7. Schaefer M (2015) RNA 5-methylcytosine analysis by bisulfite sequencing. Methods Enzymol 560:297–329. doi:10.1016/bs.mie.2015.03.007
8. Sibbritt T, Wen J, Squires JE, Shafik A, Beveridge NJ, Statham AL, Humphreys DT, Clark SJ, Parker BJ, Preiss T (2016) The RNA:m5C methyltransferase NSUN2 methylates a broad range of coding and non-coding RNAs (in preparation)
9. David R, Burgess A, Parker BJ, Li J, Pulsford K, Sibbritt T, Preiss T, Searle I (2016) Transcriptome-wide mapping of RNA 5-methylcytosine in Arabidopsis mRNAs and ncRNAs. The Plant Cell (in press).
10. Cock PJA, Fields CJ, Goto N, Heuer ML, Rice PM (2010) The Sanger FASTQ file format for sequences with quality scores, and the Solexa/Illumina FASTQ variants. Nucleic Acids Res 38(6):1767–1771. doi:10.1093/nar/gkp1137
11. Bolger AM, Lohse M, Usadel B (2014) Trimmomatic: a flexible trimmer for Illumina sequence data. Bioinformatics 30(15):2114–2120. doi:10.1093/bioinformatics/btu170
12. Langmead B, Trapnell C, Pop M, Salzberg SL (2009) Ultrafast and memory-efficient alignment of short DNA sequences to the human genome. Genome Biol 10(3):R25. doi:10.1186/gb-2009-10-3-r25
13. Langmead B, Salzberg SL (2012) Fast gapped-read alignment with bowtie 2. Nat Methods 9(4):357–359. doi:10.1038/nmeth.1923
14. Trapnell C, Pachter L, Salzberg SL (2009) Tophat: discovering splice junctions with rna-seq. Bioinformatics 25(9):1105–1111. doi:10.1093/bioinformatics/btp120
15. Krueger F, Andrews SR (2011) Bismark: a flexible aligner and methylation caller for bisulfite-seq applications. Bioinformatics 27(11):1571–1572. doi:10.1093/bioinformatics/btr167
16. Kreck B, Marnellos G, Richter J, Krueger F, Siebert R, Franke A (2012) B-SOLANA: an approach for the analysis of two-base encoding bisulfite sequencing data. Bioinformatics 28(3):428–429. doi:10.1093/bioinformatics/btr660
17. R Core Team (2015) R: A language and environment for statistical computing. R Foundation for Statistical Computing, Vienna. http://www.R-project.org/
18. Ritchie ME, Phipson B, Wu D, Hu Y, Law CW, Shi W, Smyth GK (2015) limma powers differential expression analyses for RNA-sequencing and microarray studies. Nucleic Acids Res 43(7):e47. doi:10.1093/nar/gkv007

19. Law CW, Chen Y, Shi W, Smyth GK (2014) voom: precision weights unlock linear model analysis tools for rna-seq read counts. Genome Biol 15(2):R29. doi:10.1186/gb-2014-15-2-r29
20. Thorvaldsdóttir H, Robinson JT, Mesirov JP (2013) Integrative Genomics Viewer (IGV): high-performance genomics data visualization and exploration. Brief Bioinform 14(2):178–192. doi:10.1093/bib/bbs017
21. Jühling F, Mörl M, Hartmann RK, Sprinzl M, Stadler PF, Pütz J (2009) tRNAdb 2009: compilation of tRNA sequences and tRNA genes. Nucleic Acids Res 37(Database issue):D159–162. doi:10.1093/nar/gkn772
22. Snedecor G, Cochran W (1980) Statistical methods, 7th edn. Iowa State University Press, Ames, IA
23. Smyth GK (2004) Linear models and empirical bayes methods for assessing differential expression in microarray experiments. Stat Appl Genet Mol Biol 3:Article3. doi:10.2202/1544-6115.1027
24. Benjamini Y, Hochberg Y (1995) Controlling the false discovery rate: a practical and powerful approach to multiple testing. J R Stat Soc Ser B (Methodol) 57(1):289–300
25. Storey JD (2002) A direct approach to false discovery rates. J R Stat Soc Ser B (Stat Methodol) 64(3):479–498
26. Kiran A, Baranov PV (2010) DARNED: a DAtabase of RNa EDiting in humans. Bioinformatics 26(14):1772–1776. doi:10.1093/bioinformatics/btq285

Part IV

Further RNA Modifications

Chapter 12

High-Throughput Mapping of 2′-O-Me Residues in RNA Using Next-Generation Sequencing (Illumina RiboMethSeq Protocol)

Virginie Marchand, Lilia Ayadi, Aseel El Hajj, Florence Blanloeil-Oillo, Mark Helm, and Yuri Motorin

Abstract

Detection of RNA modifications in native RNAs is a tedious and laborious task, since the global level of these residues is low and most of the suitable physico-chemical methods require purification of the RNA of interest almost to homogeneity. To overcome these limitations, methods based on RT-driven primer extension have been developed and successfully used, sometimes in combination with a specific chemical treatment. Nowadays, some of these approaches have been coupled to high-throughput sequencing technologies, allowing the access to transcriptome-wide data. RNA 2′-O-methylation is one of the ubiquitous nucleotide modifications found in many RNA types from bacteria, archaea, and eukarya. Here, we describe a reliable and optimized protocol based on alkaline fragmentation of total RNA coupled to a commonly used ligation approach followed by Illumina sequencing. We describe the methodology for detection and relative quantification of 2′-O-methylations with a high sensitivity and reproducibility even with a limited amount of starting material (1 ng of total RNA). Altogether this technique unlocks a technological barrier since it will be applicable for routine parallel treatment of biological and clinical samples to decipher the functions of 2′-O-methylations in pathologies.

Key words 2′-O-Methylation, High-throughput sequencing, RNA modification, Ribose methylation, Alkaline fragmentation

1 Introduction

One very common and widespread modification in RNA is a methylation of the ribose 2′-OH moiety, which was found in tRNA, rRNA, snRNA, mRNA as well as in piRNA and miRNA[1–5].

Three major approaches have been proposed for the detection of 2′-O-methylated residues in mature cellular RNAs. Some of them have been extensively used in the past (e.g., reverse transcription (RT) at low [dNTP] or RT on randomly fragmented RNAs) and allowed the establishment of 2′-O-methylation profiles for rRNAs from various species and also for some snRNAs [6, 7].

Alexandra Lusser (ed.), *RNA Methylation: Methods and Protocols*, Methods in Molecular Biology, vol. 1562,
DOI 10.1007/978-1-4939-6807-7_12,

Detection and even quantification of 2′-O-methylation in RNA was achieved using specific RNase H cleavage of 2′-O-methyl RNA-DNA hybrids and by dedicated ligation-based techniques [8, 9]. More recently, the combination of RT at low concentration of dNTPs ([dNTP]) with quantitative PCR was used for relative quantification of 2′-O-Me residues [10]. Similarly, a direct qRT-PCR detection method using a special mutant of thermostable DNA polymerase with high RT-activity was described recently [11]. However, all these methods are extremely laborious and time-consuming and only allow site-by-site detection and relative measurement of 2′-O-methylations. In some instances, bioinformatics approaches for 2′-O-Me mapping can also be employed, which resulted, for example, in the deduction of ribose methylation patterns of rRNAs in various archaea after careful identification of C/D-box sRNA guides [12].

A high-throughput approach for 2′-O-Me detection based on random RNA fragmentation under alkaline conditions and library preparation followed by next-generation sequencing (NGS) was proposed in the last year [13]. The phosphodiester bond at the 3′ of 2′-O-Me nucleotides is relatively resistant to nucleolytic cleavage, and thus the RNA fragments starting or ending at the nucleotide at the +1 position to the modified residue are excluded from a randomly fragmented library. However, the suggested approach was based on a rather nonconventional ligation procedure with homemade 3′-cyclophosphate modified adapters and particular RNA ligase mutant [13].

Here, we provide a detailed description and optimization of the RiboMethSeq protocol based on simple alkaline hydrolysis coupled to a standard library preparation protocol suitable for Illumina sequencing (Fig. 1). The RiboMethSeq protocol can be robustly applied to detection and even relative quantification of 2′-O-Me sites in highly abundant or enriched RNAs and has an acceptable performance with a very small amount of starting RNA material (~1 ng).

2 Materials

Prepare all solutions using RNase-free water. Wear gloves to prevent degradation of RNA samples by RNases.

2.1 RNA Quantification and Quality Assessment

2.1.1 RNA Quantification

1. UV-visible small volume spectrophotometer: any kind of UV visible spectrophotometer allowing measurements of 1 μL samples.
2. 1.5 mL RNase-free microcentrifuge tubes.
3. RNase-free water.

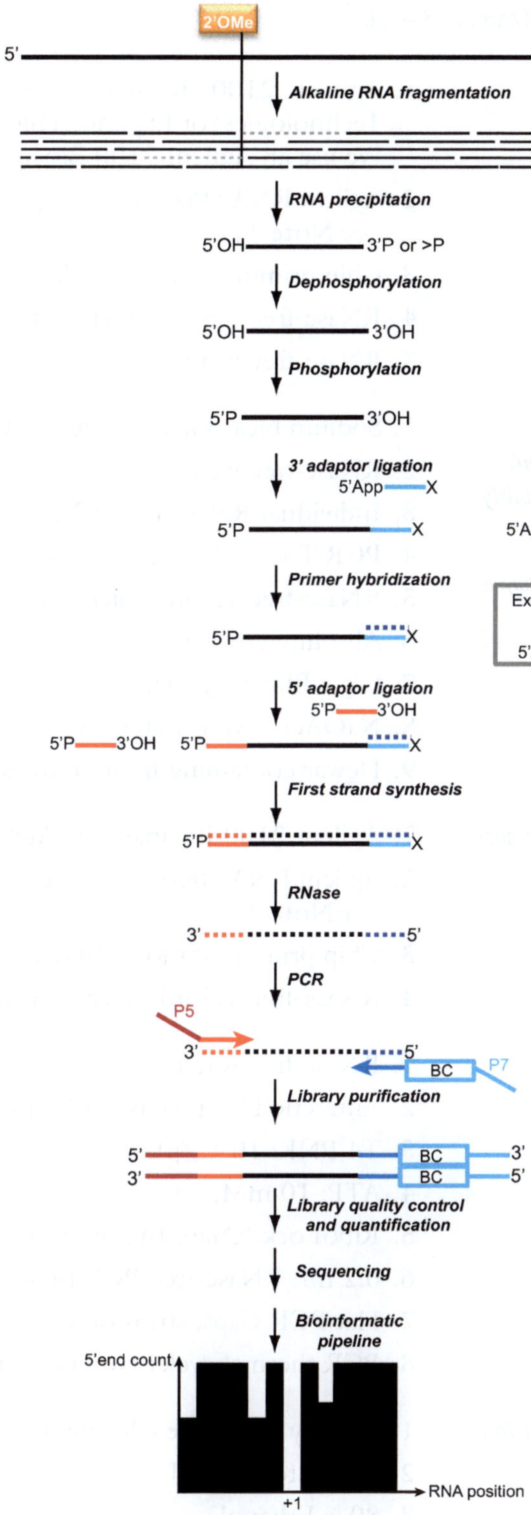

Fig. 1 General overview of the RiboMethSeq protocol. An RNA containing 2′-O-Me residues is randomly fragmented by alkaline hydrolysis, 5′- and 3′-ends are repaired and adapters are ligated to both extremities. After amplification and barcoding, amplicons are subjected to Illumina sequencing on HiSeq1000 in single-read mode. The 2′-O-Me residues protect the 3′-adjacent phosphodiester bond from cleavage, generating a typical gap in 5′-ends coverage profile

2.1.2 RNA Quality Assessment

1. Agilent 2100 Bioanalyzer or 2200 TapeStation (Agilent Technologies) or Experion (BioRad) or LabChip GX (Caliper). We use an Agilent 2100 Bioanalyzer.
2. Agilent RNA 6000 Pico kit (quantitative range 50–5000 pg/μL) (*see* **Note 1**).
3. Chip priming station (Agilent Technologies).
4. RNase-free 1.5 mL microcentrifuge tubes.
5. RNase-free water.

2.2 Alkaline Hydrolysis and RNA Fragmentation Quality Control

2.2.1 Alkaline Hydrolysis

1. Sodium bicarbonate buffer: 100 mM, pH 9.2.
2. RNase-free water.
3. Individual RNase-free 0.2 mL PCR tubes.
4. PCR Thermal cycler (we use Agilent SureCycler 8000).
5. RNase-free 1.5 mL microcentrifuge tubes.
6. Absolute ethanol.
7. GlycoBlue Coprecipitant: 15 mg/mL (Ambion).
8. NaOAc: 3 M in water, pH 5.2.
9. Dewar containing liquid nitrogen.

2.2.2 RNA Fragmentation Quality Control

1. Agilent 2100 Bioanalyzer (Agilent Technologies).
2. Agilent RNA 6000 Pico kit (quantitative range 50–5000 pg/μL) (*see* **Note 1**).
3. Chip priming station (Agilent Technologies).
4. RNase-free 1.5 mL microcentrifuge tubes.

2.3 End-Repair

1. RNase-free water.
2. Antarctic Phosphatase: 5 U/μL.
3. T4 PNK: 10 U/μL.
4. ATP: 10 mM.
5. RiboLock RNase Inhibitor: 40 U/μL.
6. 0.2 mL RNase-free PCR tubes, strips of 8.
7. Flat PCR Caps, strips of 8.
8. PCR thermal cycler (we use Agilent SureCycler 8000).

2.4 RNA Purification

1. RNeasy MinElute Cleanup kit.
2. Absolute Ethanol.
3. 80 % Ethanol.

2.5 Library Preparation Using NEBNext® Multiplex Small RNA Library Prep Set for Illumina®

1. NEBNext® Multiplex Small RNA Library Prep Set for Illumina® (set 1 or 2, New England Biolabs) (*see* **Note 2**).
2. 0.2 mL PCR tubes, strips of 8.
3. Flat PCR Caps, strips of 8.

2.6 Library Purification Using GeneJET PCR Purification Kit

1. GeneJET PCR Purification kit or equivalent.
2. 1.5 mL DNA low-binding tubes.

2.7 Library Quantification and Quality Assessment

2.7.1 Library Quantification

1. Any kind of fluorometer able to quantify DNA library with high sensitivity (e.g., Qubit® 2.0fluorometer).
2. Qubit® dsDNA HS Assay kit (0.2–100 ng).
3. Thin-walled polypropylene tubes of 500 μL compatible with the fluorometer (e.g., Qubit® Assay Tube or Axygen® PCR-05-C tubes).

2.7.2 Library Quality Assessment

1. Agilent 2100 Bioanalyzer (Agilent Technologies).
2. Agilent HS DNA kit (quantitative range 5–500 pg/μL).
3. Chip priming station (Agilent Technologies).
4. RNase-free 1.5 mL microcentrifuge tubes.

2.8 Library Sequencing

1. Any kind of Illumina sequencers (starting from Miseq to Hiseq).
2. Any appropriate sequencing kit for a single read length of 35–50 nt.

2.9 Bioinformatic Analysis

1. Unix (Linux) server (we are using Illumina Compute Dell server).
2. Adapter trimming software Trimmomatic (current version 0.36 http://www.usadellab.org/cms/?page=trimmomatic).
3. Alignment software Bowtie 2.0 (current version 2.2.9 http://bowtie-bio.sourceforge.net/bowtie2/index.shtml).
4. Microsoft Excel or Open(Libre)Office calc spreadsheet software.

3 Methods

3.1 RNA Quantification and Quality Assessment

Carry out all the procedures at room temperature.

3.1.1 RNA Quantification

1. On a Nanodrop 2000 start screen; select the "Nucleic Acid" application.
2. After the wavelength verification test, select the type of sample to measure, in this case "RNA."
3. Prepare the blank: the buffer/solution used for sample resuspension but without any trace of RNA (e.g., RNase-free water).

4. Load 1 μL of the blank solution to the bottom pedestal, lower the arm and click on the "Blank" button.
5. Wipe the upper and lower pedestal using a dry wipe and load 1 μL of one of your samples of interest to the bottom pedestal, lower the arm, and click "Measure."
6. Analyze the data obtained for your different RNA samples. For "pure" RNAs, the ratio A260/A280 should be 2, the ratio A260/A230 should be in the range of 1.8–2.2 (*see* **Notes 3** and **4**).

3.1.2 RNA Quality Assessment

1. Before starting the experiments, equilibrate all solutions of the kit at room temperature for at least 30 min in the dark. Vortex them and spin them down before use.
2. Transfer 550 μL of gel matrix (red cap vial) into a spin filter provided in the kit.
3. Centrifuge for 10 min at 1500 × *g* at room temperature.
4. Prepare 65 μL aliquots of the gel and store them at 4 °C for a maximum of 1 month.
5. Prepare the gel-dye mix by mixing 1 μL of RNA dye concentrate to a gel aliquot.
6. Centrifuge for 10 min at 13,000 × *g* at room temperature.
7. Dilute your RNA samples quantified on Nanodrop to 3–5 ng/μL with RNase-free water to be within the optimal range concentration of the assay.
8. Add 1 μL of your diluted RNA samples to 11 different 1.5 mL tubes already containing 5 μL of RNA marker (green cap vial) (*see* **Note 5**). Mix by pipetting up and down.
9. Mix 1 μL of the ladder (*see* **Note 6**) with 5 μL of RNA marker (green cap vial). Mix by pipetting up and down.
10. Prepare the chip priming station. Adjust the syringe clip to the highest top position.
11. Load 9 μL of the gel-dye mix in the well marked with a "G" surrounded by a black circle.
12. Close the chip priming station properly and press the plunger of the syringe until it is held by the clip.
13. Wait for 30 s and then release the clip.
14. Wait for 5 s until the plunger stops and pull it slowly back to the 1 mL position of the syringe.
15. Open the chip priming station and load 9 μL of the gel-dye mix in the 2 other wells marked "G."
16. Load 9 μL of the conditioning solution (white cap vial) in the well marked "CS."
17. Load 6 μL of the diluted ladder in the well marked with a ladder.

18. Load 6 μL of the diluted RNA samples in the wells marked 1–11.
19. Inspect the chip and make sure that no liquid spill is present on the edges of the wells.
20. Insert the chip in the Agilent 2100 Bioanalyzer and close the lid (*see* **Note 7**).
21. Select the following assay "Eukaryote Total RNA Pico series II" in the 2100 expert software screen.
22. Press "Start" to begin the chip to run (*see* **Note 8**).
23. After the run, immediately remove the chip and clean the electrodes with the electrode cleaner filled with 350 μL of RNase-free water.
24. Analyze the results of the chip (Fig. 2).

3.2 Alkaline Hydrolysis and RNA Fragmentation Quality Control

3.2.1 Alkaline Hydrolysis

1. Prepare one 1.5 mL tube per sample to be analyzed ("precipitation tube") containing 10 μL of NaOAc, 1 μL of glycoblue, and 1 mL of absolute ethanol for subsequent precipitation of the sample (store at −20 °C until further use).
2. Dilute your RNA samples to a concentration of 20 ng/μL with RNase-free water.
3. To individual PCR tubes, add 5 μL of each of your diluted RNA samples (*see* **Note 9**), keep on ice until further use.
4. Add 5 μL of bicarbonate buffer and mix by pipetting up and down.
5. Incubate in a thermal cycler preheated at 95 °C. Start a timer and incubate for 4–12 min (*see* **Note 10**).
6. Proceed with the next sample every 30 s.
7. Stop each reaction after the required time at 95 °C by spinning down the PCR microtube and add the whole sample into the corresponding 1.5 mL precipitation tube from **step 1**.
8. Mix by inverting the tube several times and throw it into liquid nitrogen.
9. Recover it from the liquid nitrogen and centrifuge your samples for 30 min at 4 °C at full speed in a microcentrifuge.
10. Take out the supernatant and make sure not to lose the pellet.
11. Wash with 600 μL of 80 % ethanol.
12. Centrifuge your samples for 10 min at 4 °C at full speed.
13. Take out the supernatant.
14. Centrifuge your samples for a short spin.
15. Remove any liquid left.
16. Incubate your samples with open lid for 2 min at 37 °C and 5 min at room temperature.
17. Resuspend the pellet with 20 μL of RNase-free water.

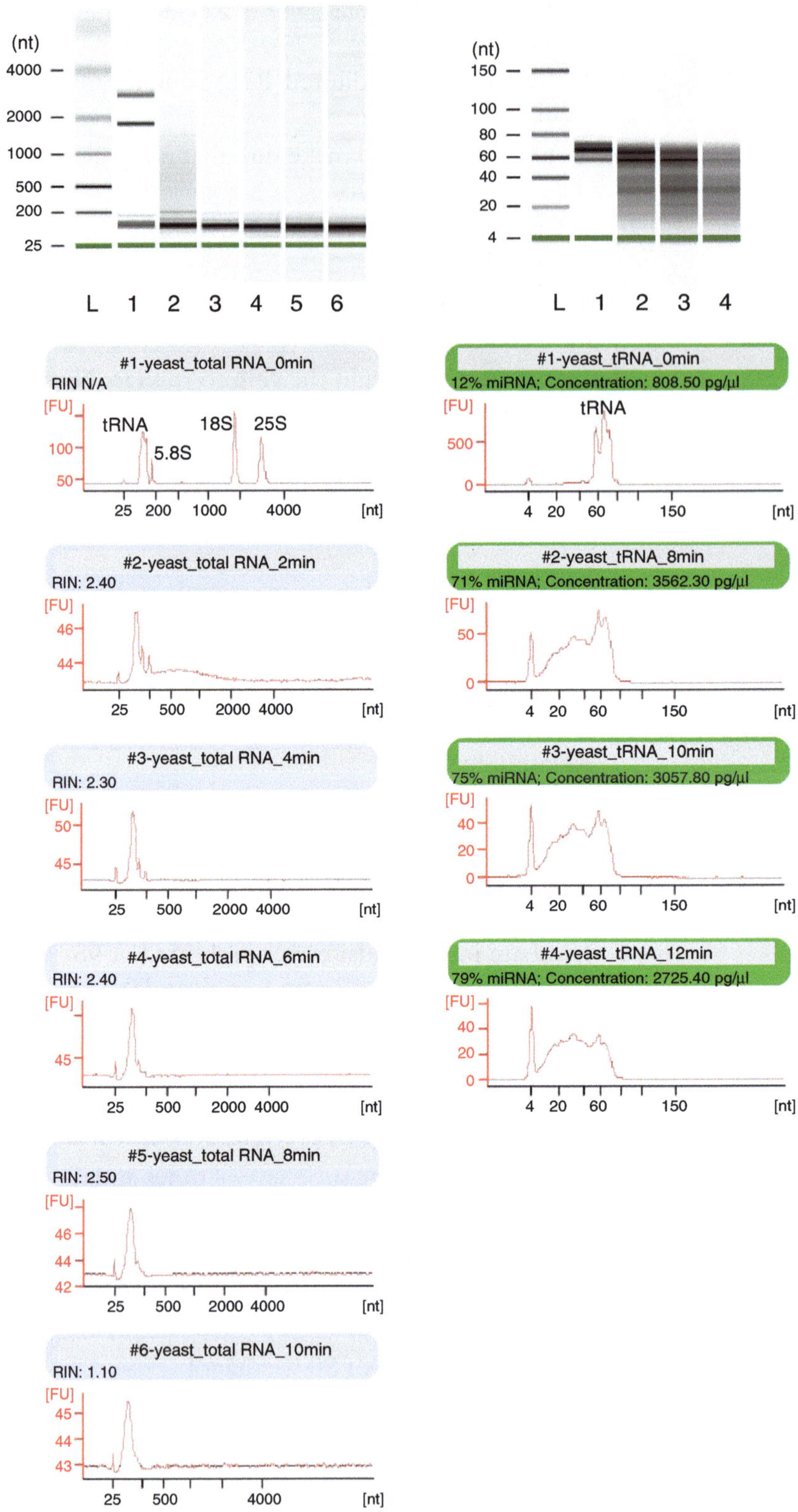

Fig. 2 Alkaline fragmentation and RNA fragmentation control. Representative capillary electrophoresis profiles of total yeast RNA (*left*, RNA 6000 Pico kit) and total tRNA fraction (*right*, Small RNA kit) fragmentation in alkaline conditions. *Top*—Digital gel-like representation of the Bioanalyzer 2100 profiles with appropriate size ladder. *Bottom*—Individual profiles for each sample, fragmentation times from 0 to 10 min for yeast total RNA (*left*) and 0 to 12 min for total tRNA fraction (*right*)

3.2.2 RNA Fragmentation Quality Control

1. Prepare your samples by mixing 5 μL of RNA marker (green cap vial) with 1 μL of your fragmented RNA samples.
2. With the rest of gel-dye mix, load 9 μL in the well marked "G" surrounded with a black circle and proceed as described in section "RNA quality control" from **steps 12** to **18** from Subheading 3.1.2.
3. Analyze the results obtained (Fig. 2).

3.3 End-Repair

1. Combine 16 μL of your treated RNA samples in a PCR tube with 2 μL of phosphatase buffer, 1 μL of RiboLock RNase Inhibitor, and 1 μL of Antarctic Phosphatase.
2. Mix by pipetting up and down.
3. Incubate the PCR tubes for 30 min at 37 °C and then for 5 min at 70 °C (to inactivate the phosphatase) and store for indefinite hold at 4 °C in a thermal cycler.
4. Add the following components to the previous mix: 17 μL of RNase-free water, 5 μL of PNK buffer, 5 μL of ATP (*see* **Note 11**), 1 μL of RiboLock RNase Inhibitor, 2 μL of PNK enzyme.
5. Incubate in a thermal cycler for 1 h at 37 °C and immediately proceed to the next step.

3.4 RNA Purification

All the reagents except ethanol used for RNA purification are part of RNeasy MinElute Cleanup kit.

1. Transfer the sample to a new 1.5 mL tube, and to adjust the final volume to 100 μL, add 50 μL of RNase-free water.
2. Add 350 μL of RLT buffer, mix by vortexing.
3. Add 675 μL of absolute ethanol and mix by inverting the tube up and down (*see* **Note 12**).
4. Transfer 700 μL of the sample to an RNeasy MinElute spin column (stored at 4 °C until use). Centrifuge for 30 s at 8000 × *g*.
5. Repeat the **step 4** with the rest of the sample. Then, add 500 μL of RPE buffer to the column. Centrifuge for 30 sec at 8000 × *g*.
6. Discard the flowthrough. Add 750 μL of 80 % Ethanol. Centrifuge for 2 min at 8000 × *g*.
7. Transfer the column to a new collection tube and centrifuge at full speed for 5 min with the lid open.
8. Transfer the column to a new 1.5 mL tube (provided with the kit). Add 10 μL of RNase-free water in the center of the column filter. Wait for 1 min.
9. Centrifuge at full speed for 1 min to elute. The recovered volume is about 9 μL.

3.5 Library Preparation Using NEBNext® Multiplex Small RNA Library Prep Set for Illumina®

1. Mix 6 μL of RNA sample with 1 μL of 3'SR adaptor (green cap vial) in a PCR tube.
2. Incubate for 2 min at 70 °C in a preheated thermal cycler. Transfer immediately to ice.
3. Add 10 μL of 3′ Ligation Buffer (green cap vial) and 3 μL of 3′ Ligation Enzyme (green cap vial).
4. Incubate for 1 h at 25 °C in a thermal cycler.
5. Add 4.5 μL of RNase-free water and 1 μL of SR RT primer (pink cap vial).
6. Incubate for 5 min at 75 °C, 15 min at 37 °C and 15 min at 25 °C.
7. Within the last 15 min of incubation, add 1.1 × n (n=number of samples) μL of the 5′ SR adapter (yellow cap vial) to an individual PCR tube (previously resuspended in 120 μL of RNase-free water and stored at −80 °C).
8. Denature the 5′SR adapter in a thermal cycler for 2 min at 70 °C and immediately place the tube on ice (*see* **Note 13**).
9. Add 1 μL of 5′SR adapter (previously denatured), 1 μL of 5′Ligation Reaction Buffer (yellow cap vial), and 2.5 μL of Ligase Enzyme Mix (yellow cap vial).
10. Incubate for 1 h at 25 °C in a thermal cycler.
11. Add the following components to the adapter ligated RNA mix from the previous step: 8 μL of First strand synthesis reaction buffer (red cap vial), 1 μL of Murine RNase inhibitor (red cap vial), 1 μL of ProtoScript II reverse transcriptase (red cap vial) and mix well by pipetting up and down.
12. Incubate for 1 h at 50 °C.
13. Immediately proceed to PCR amplification (*see* **Note 14**). Add the following components to the RT reaction mix from the previous step: 50 μL of LongAmp Taq Master Mix (blue cap vial), 2.5 μL of SR primer (blue cap vial), 2.5 μL of index primer (*see* **Note 15**), and 5 μL of RNase-free water. Mix well.
14. Perform the following PCR cycling conditions: 1 cycle of initial denaturation for 30 s at 94 °C, 12–15 cycles of denaturation 15 s at 94 °C, annealing 30 s at 62 °C, extension 15 s at 70 °C, 1 cycle of final extension for 5 min at 70 °C and store at 4 °C for indefinite hold.

3.6 Purification of the Library Using GeneJET PCR Purification Kit

1. Transfer the PCR mix to a 1.5 mL tube, and add 100 μL of binding buffer. Mix thoroughly.
2. Transfer the solution to the purification column. Centrifuge at full speed for 30 s. Discard the flowthrough.
3. Add 700 μL of wash buffer to the column and centrifuge at full speed for 30 s. Discard the flowthrough.

4. Centrifuge the empty column for 1 additional min.
5. Transfer the column to a clean 1.5 mL DNA low binding tube. Add 30 μL of Elution buffer to the center of the column membrane and centrifuge at full speed for 1 min.
6. Store the purified library at −20 °C until further use.

3.7 Library Quantification and Quality Assessment

3.7.1 Library Quantification

1. Before starting the experiments, incubate all solutions of the Qubit dsDNA HS assay kit at room temperature for at least 30 min. The kit provides the concentrated assay reagent, dilution buffer and prediluted standards.
2. Prepare the dye working solution by diluting the concentrated assay reagent 1:200 in dilution buffer. Prepare 200 μL of working solution for each sample and two additional standards.
3. Prepare the two standards annotated "C" and "D" by mixing 10 μL of standard with 190 μL of working solution.
4. Add working solution to 1 μL of RNA sample to obtain 200 μL in total.
5. Vortex the tubes for 2 s and incubate them for 2 min at room temperature.
6. Insert the tubes into the Qubit® 2.0 Fluorometer and proceed with measurements: on the home screen of the Qubit® 2.0 Fluorometer, choose the type of assay (e.g., "HS DNA") for which you want to perform a new calibration.
7. Press "Yes" to read new standards.
8. When indicated, insert the standard tube and press "Read." Standard #1 and #2 correspond to standards "C" and "D," respectively.
9. Once the calibration is done, insert each sample and press "Read" to make the measurements. Check that the value of your samples is within the assay's range, and press "Calculate Stock Conc" (*see* **Note 16**).

3.7.2 Quality Assessment

1. Before starting the experiments, incubate all solutions of the Agilent High Sensitivity DNA kit at room temperature for at least 30 min in the dark. Vortex them and spin them down before use.
2. Add 15 μL of High sensitivity DNA dye concentrate (blue cap vial) into a High Sensitivity DNA gel matrix vial (red cap vial).
3. Vortex for 10 s and transfer the gel-dye mix to the center of the spin filter.
4. Centrifuge for 10 min at 2240 × *g* (*see* **Note 17**).
5. Add 1 μL of each of your library to 11 different tubes of 1.5 mL already containing 5 μL of RNA marker (green cap vial). Mix by pipetting up and down.

6. Mix 1 μL of the ladder (yellow cap vial) with 5 μL of High sensitivity DNA marker (green cap vial). Mix by pipetting up and down.
7. Prepare the chip priming station. Adjust the syringe clip to the lowest top position.
8. Load 9 μL of the gel-dye mix in the well marked with a "G" surrounded by a black circle.
9. Close the chip priming station properly and press the plunger of the syringe until it is held by the clip.
10. Wait for 1 min and then release the clip.
11. Wait for 5 s. Until the plunger stops and pull it slowly back to the 1 mL position of the syringe.
12. Open the chip priming station and load 9 μL of the gel-dye mix in the 3 other wells marked "G".
13. Load 6 μL of the diluted ladder in the well marked with a ladder.
14. Load 6 μL of the diluted library samples in the wells labeled 1–11.
15. Insert the chip in the Agilent 2100 Bioanalyzer, close the lid, and select the following assay "High Sensitivity DNA" in the 2100 expert software screen.
16. Press "Start" to begin the chip to run.
17. After the run, immediately remove the chip and clean the electrodes with the electrode cleaner filled with 350 μL of RNase-free water.
18. Analyze the results of the chip (Fig. 3).

3.8 Library Sequencing

1. For sequencing, libraries are multiplexed and diluted to 8–12 pM final concentration.
2. Sequencing depth or coverage depends on the length of target RNA and can roughly be estimated at 1000 reads per RNA nt.
3. Sequencing length may vary from 35 to 50 nt in a single read mode.

3.9 Bioinformatic Analysis

1. Trim adapter sequences of raw reads (FastQ files) using Trimmomatic with the following parameters: java -jar trimmomatic-0.35.jar SE -phred33 input.fq.gz output.fq.gz ILLUMINACLIP:TruSeq3-SE:2:30:10 LEADING:30 TRAILING:30 SLIDINGWINDOW:4:15 AVGQUAL:30 MINLEN:17 (*see* **Note 18**) (Fig. 4).
2. Align the trimmed reads to the appropriate reference sequence (yeast or human rRNA for example) using bowtie2 with the following parameters: bowtie2 --sensitive -L 17 -x <bt2-idx>-U <r>-S.

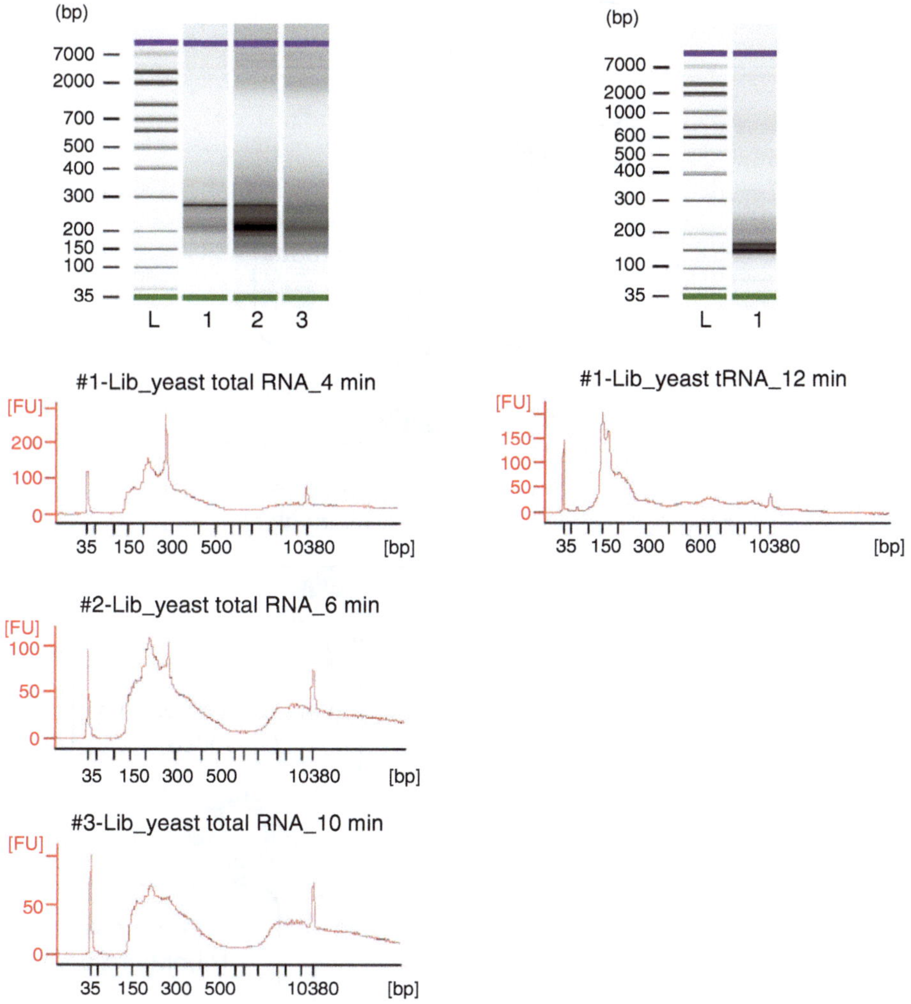

Fig. 3 Typical profiles obtained for sequencing libraries. Capillary electrophoresis analysis of sequencing libraries prepared from fragmented yeast total RNA (*left*) and total tRNA fraction (*right*). Separation is done using HS DNA kit, ladder shown at the *left*. Libraries were prepared from total RNA fragmented for 4 min, 6 min, and 10 min (*left*) and for total tRNA fraction fragmented for 12 min (*right*). Digital gel-like representations (*top*) and individual profiles for the corresponding libraries (*bottom*) are shown

3. Count the 5′-ends in the produced *.sam file using awk command: `awk'{print $4}' *.sam | sort | uniq -c | awk'{print $3,$2,$1,$4}' | sort -n`.
4. Import the file containing 5′-end counts into Excel or Calc.
5. Calculate ScoreMax for each position, make a ratio of number of 5′-reads ends between preceding and following position and calculate ScoreMAX as a ratio of a drop for a given position compared to the average for 12 neighboring positions (−6/+6).
6. Calculate the RiboMethScore, use the following formula: RiboMethScore = 1 − n_i/(0.5*(SUM(n_j*W_j)/SUM(W_j)+

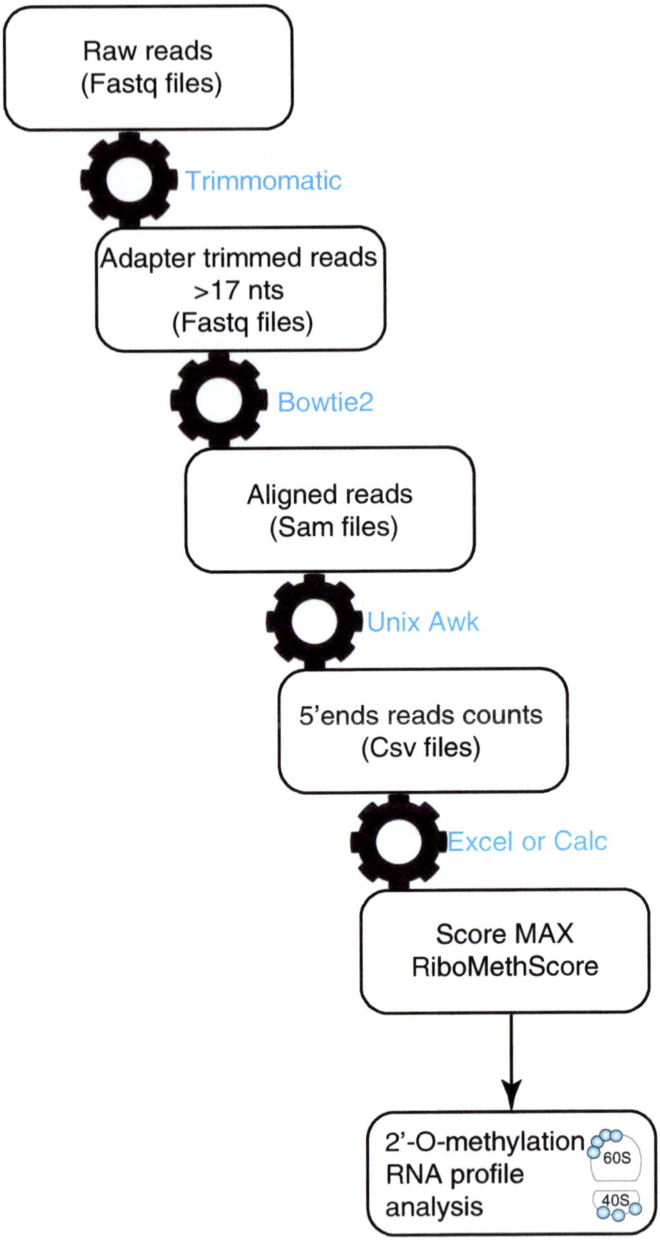

Fig. 4 Bioinformatics pipeline for the RiboMethSeq analysis. Raw reads after sequencing are trimmed using Trimmomatic to remove adapter sequences, aligned to the reference sequence by Bowtie2 and 5′-extremities of reads are counted. Calculations of ScoreMAX and RiboMethScore are performed using the spreadsheet software

SUM(n_k*W_k)/SUM(W_k)), where n_i – 5′-end count for a given position, j—varies from i−6 to i−1, k varies from i+1 to i+6, weight parameters are defined as following W_j = 1−0.1*(j−1) idem for W_k. Thus, W_j or W_k parameter varies from 0.5 for positions −6/+6 to 1 for positions −1/+1.

4 Notes

1. If the size of your RNA to be fragmented is below 200 nts, it is appropriate to use Agilent Small RNA kit (quantitative range 50–2000 pg/μL) following manufacturer's recommendations.
2. The kit NEBNext® Multiplex Small RNA Library Prep Set for Illumina® (set 1) (NEB, E7300S) includes a set of 12 barcoding primers (numbered 1–12) that will be used for multiplexing reactions during PCR amplification. There is also a version set 2 with primers (numbers 13–24). If you do not need these barcoding primers, you may order a similar kit without the primers and use any other source of barcoding primers (Illumina, Epicentre, NEB).
3. If your RNA sample is diluted with RNase-free water instead of 10 mM Tris-EDTA (TE) pH 8.0, the ratio A260/A280 may be below 2.0 due to the lower pH of water [14]. A ratio A260/A280 of 1.8 for samples diluted in RNase-free water is considered "pure" for RNA.
4. If your RNA sample is contaminated by phenol or chaotropic salts (e.g., guanidinium thiocyanate used in Trizol® extraction or other protocols), this will result in a ratio A260/A230 below 1.8. Another round of Phenol-Choroform-Isoamyl Alcohol (PCA) extraction and two successive steps of chloroform extraction followed by ethanol precipitation are in this case recommended before alkaline hydrolysis.
5. In case you are working with less than 11 samples, in the empty wells replace RNA with 1 μL of RNase-free water.
6. The ladder loaded in the Pico RNA chip is provided in a separate package and may be prepared before the experiment: spin down the tube and transfer 10 μL to a RNase-free tube. Heat for 2 min at 70 °C. Cool down on ice and add 90 μL of RNase-free water. Prepare 5 μL aliquots using the Safe-Lock PCR tubes provided in the kit and store them at −70 °C. Before use, thaw one tube and keep it on ice. The ladder is quite stable at −70 °C and may be used at least 4 months.
7. RNase contamination problems of the Bioanalyzer electrodes are very frequent and will affect the RNA integrity number of your samples. Therefore, if the Agilent 2100 Bioanalyzer is also frequently used to run DNA chips, it is strongly recommended to use a dedicated electrode cartridge only for RNA assays. In addition, we recommend for each chip to load an internal RNA control (total RNA preparation with a known RIN > 9). If you encounter contamination problems, soak the electrode cartridge into an RNaseZap® decontamination solution (Ambion) for at least 10 min, then rinse the electrodes with RNase-free water and let them dry out for at least one night.

8. The Agilent 2100 Bioanalyzer is very sensitive to vibrations and this may affect your results. Therefore, make sure that no vibrations will occur during the run.
9. The RNA quantity may be decreased to a minimal starting amount of 5–10 ng without considerably affecting coverage and calculation of the RiboMethScore. However, if you are aiming to discover new 2′-O-Me sites in an unknown RNA, we recommend using 100–250 ng as starting material.
10. If total RNA samples (from yeast, human, bacteria) are of good quality (RIN > 8), fragmentation time is around 8–10 min. However, this time may be decreased with RNA of poor quality (4–6 min) or may be increased with RNA species of higher stability (i.e. tRNAs) (12–14 min) (Fig. 2). We recommend testing three to four different times of fragmentation to define the appropriate conditions for hydrolysis. In general, for long RNAs, the optimal size distribution is around 50–100 nt, while for short RNAs (>200 nts), 20–50 nt is the appropriate size distribution.
11. We tried different ATP concentrations (1×10^{-3} M, 1×10^{-5} M, and 1×10^{-7} M) and lowering the [ATP] may affect the discovery of new methylation sites, due to incomplete phosphorylation of some 5′ends fragments. However, the RiboMethScores for all known rRNA positions were not affected.
12. Ethanol quantity is increased compared to the manufacturer's recommendations in order not to lose the small RNA fragments during the RNA binding to the silica membrane.
13. Do not leave the heated adapter on ice for more than 5–10 min before proceeding to the next step; this may impact your library preparation.
14. We recommend proceeding immediately with PCR amplification. However, if it is not possible, inactivate the RT by heating for 15 min at 70 °C and cool down the reaction at 4 °C for 1–3 h or safely store the reactions at −20 °C overnight.
15. Make sure to use only combinations of compatible primers for barcoding. Most Illumina sequencers use a green laser (or LED) to read G and T nucleotides and a red laser (or LED) to read A and C nucleotides. Within each sequencing cycle, at least one nucleotide for each color channel must be read in the index to ensure proper reading of the barcode sequence. Use as a reference the following guide (ScriptSeq™ Index PCR primers, Illumina) for the verification of barcode compatibility or check compatibility with Illumina Experimental Manager software.
16. This quantification step is crucial. Make sure to quantify all your libraries properly since an under- or overestimated quantification will interfere with subsequent sequencing reads proportion and quality.

17. The High Sensitivity DNA gel-dye mix is stable for 1 month at 4 °C protected from light.
18. MINLEN parameter can vary depending on the length of the analyzed RNA, 17 nt is suitable for rRNA, but it may be shorter for tRNAs, for example.

Acknowledgment

This work was supported by joint ANR-DFG grant HTRNAMod (ANR-13-ISV8-0001/HE 3397/8-1) to MH and YM, and AO Lorraine University-Lorraine Region "Aberrant RNA methylation in cancer" funding to YM.

References

1. Machnicka MA, Milanowska K, Osman Oglou O, Purta E, Kurkowska M, Olchowik A, Januszewski W, Kalinowski S, Dunin-Horkawicz S, Rother KM, Helm M, Bujnicki JM, Grosjean H (2013) MODOMICS: a database of RNA modification pathways--2013 update. Nucleic Acids Res 41(Database issue):D262–D267
2. Motorin Y, Helm M (2011) RNA nucleotide methylation. Wiley Interdiscip Rev RNA 2:611–631
3. Ji L, Chen X (2012) Regulation of small RNA stability: methylation and beyond. Cell Res 22:624–636
4. Ohara T, Sakaguchi Y, Suzuki T, Ueda H, Miyauchi K, Suzuki T (2007) The 3′ termini of mouse Piwi-interacting RNAs are 2′-O-methylated. Nat Struct Mol Biol 14:349–350
5. Tycowski KT, You ZH, Graham PJ, Steitz JA (1998) Modification of U6 spliceosomal RNA is guided by other small RNAs. Mol Cell 2:629–638
6. Maden BE (2001) Mapping 2′-O-methyl groups in ribosomal RNA. Methods 25: 374–382
7. Maden BE, Corbett ME, Heeney PA, Pugh K, Ajuh PM (1995) Classical and novel approaches to the detection and localization of the numerous modified nucleotides in eukaryotic ribosomal RNA. Biochimie 77:22–29
8. Yu YT, Shu MD, Steitz JA (1997) A new method for detecting sites of 2′-O-methylation in RNA molecules. RNA 3:324–331
9. Huang C, Karijolich J, Yu Y-T (2016) Detection and quantification of RNA 2′-O-methylation and pseudouridylation. Methods 103:68–76. doi:10.1016/j.ymeth.2016.02.003
10. Dong Z-W, Shao P, Diao L-T, Zhou H, Yu C-H, Qu L-H (2012) RTL-P: a sensitive approach for detecting sites of 2′-O-methylation in RNA molecules. Nucleic Acids Res 40:e157
11. Aschenbrenner J, Marx A (2016) Direct and site-specific quantification of RNA 2′-O-methylation by PCR with an engineered DNA polymerase. Nucleic Acids Res 44:3495–3502
12. Dennis PP, Tripp V, Lui L, Lowe T, Randau L (2015) C/D box sRNA-guided 2′-O-methylation patterns of archaeal rRNA molecules. BMC Genomics 16:632
13. Birkedal U, Christensen-Dalsgaard M, Krogh N, Sabarinathan R, Gorodkin J, Nielsen H (2015) Profiling of ribose methylations in RNA by high-throughput sequencing. Angew Chem Int Ed Engl 54:451–455
14. Wilfinger WW, Mackey K, Chomczynski P (1997) Effect of pH and ionic strength on the spectrophotometric assessment of nucleic acid purity. Biotechniques 22(474–6):478–481

Chapter 13

RiboMeth-seq: Profiling of 2′-*O*-Me in RNA

Nicolai Krogh, Ulf Birkedal, and Henrik Nielsen

Abstract

RiboMeth-seq is a sequencing-based method for mapping and quantitation of one of the most abundant RNA modifications, ribose methylation. It is based on a simple chemical principle, namely the several orders of magnitude difference in nucleophilicity of a 2′-OH and a 2′-*O*-Me. Thus, the method combines alkaline fragmentation and a specialized library construction protocol based on 5′-OH and 2′,3′ cyclic phosphate ends to prepare RNA for sequencing. The read-ends of library fragments are used for mapping with nucleotide resolution and calculation of the fraction of molecules methylated at the 2′-*O*-Me sites.

Key words 2′-*O*-Me, Ribose methylation, Box C/D snoRNA, Ribosome, Ribosomal RNA, RNA library, RNA-seq

1 Introduction

Ribose methylation (2′-*O*-Me) is one of the most abundant RNA modifications. It is found in all domains of life and in all major classes of RNA [1]. The modification is introduced by stand-alone enzymes (e.g., in *E. coli* rRNA) or by an RNA-guided mechanism. Here, box C/D sRNAs in Archaea or box C/D snoRNAs in Eukarya guide a generic methyl transferase (fibrillarin) to the target through base pairing and modify the nucleotide base paired to the fifth nucleotide upstream of the D (or D′) box [2, 3]. The ribose methylation confers chemical stability to the target RNA because the 2′-OH can no longer be activated for nucleophilic attack on the neighboring phosphodiester bond. It also has several structural implications; it promotes 3′-*endo* conformation of the sugar, blocks sugar edge interactions, and changes the hydration sphere around the 2′-*O* [4].

Ribose methylations are interesting to study both in terms of the modification itself and in terms of the guide RNAs. The role of the 2′-*O*-Me has been unclear in many cases because loss of individual modifications generally has no apparent phenotype when tested under standard growth conditions [5]. However, with

Alexandra Lusser (ed.), *RNA Methylation: Methods and Protocols*, Methods in Molecular Biology, vol. 1562,
DOI 10.1007/978-1-4939-6807-7_13,

recent high-resolution X-ray crystallography that reveals the structural context of 2′-*O*-Me sites in the ribosome [6, 7] and more efficient methods for monitoring 2′-*O*-Me such as RiboMeth-seq, this is likely to change. Ribose methylations in rRNA are frequently said to play a role in "ribosome biogenesis and translational fidelity" [2, 3]. In ribosome biogenesis, it may in fact be the guide RNA that plays the critical part by segregating the target sequence during folding of the rRNA at specific time-windows during folding thus acting as a chaperone. In this respect, it is interesting to note that the number of RNA-guided ribose methylations has increased tremendously during eukaryotic evolution and that the number of box C/D guide RNAs varies at least from 7 to 128 among species of Archaea [8]. Another interesting aspect of the biology of the guiding system is the many "orphan" guide RNAs with no apparent target RNA.

In order to facilitate the study of guide RNAs and 2′-*O*-Me, we developed a sequencing-based profiling method, RiboMeth-seq [9]. The method relies on the principle that 2′-*O*-methylated nucleotides are several orders of magnitude less sensitive to alkaline degradation than unmethylated nucleotides. Kiss et al. [10] already used this principle for studying individual sites in a primer extension format. We combined alkaline degradation with a cloning protocol for 2′,3′ cyclic phosphate ends developed by Schutz et al. [11]. The main steps of the protocol are described in Fig. 1. First, the RNA is partially degraded by incubation at alkaline pH at high temperature to obtain fragmentation that is unbiased with respect to RNA folding. Library fragments of 20–40 nucleotides are then gel-purified. Next, RNA oligonucleotide adapters are ligated to the ends of the library fragments using a tRNA ligase from *A. thaliana* that has an inactivating mutation in its kinase domain. The 3′ adapter has a 5′ phosphate and carries in its sequence a barcode and the sequence used as a primer site for sequencing. The 5′ adapter is made by in vitro transcription coupled to ribozyme cleavage (by a *cis*-encoded HDV ribozyme) and is thus synthesized to have a 2′,3′ cyclic phosphate end [12]. After enzymatic removal of the 2′ phosphates generated in the ligation step, the library fragments are subjected to standard sequencing on the Ion Proton platform. The advantages of this platform are that it is a bench top sequencer and that it can be used to sequence small amounts of library without an amplification step. However, the protocol can be adapted to other platforms as well. The sequencing reads are used for mapping onto a reference sequence and the first and last nucleotides of the library fragment sequence are recorded as 3′ and 5′ read-ends, respectively. The two sets of read-ends are merged by shifting the 5′ read-ends one nucleotide upstream such that both datasets refer to the same phosphodiester bond. A diagram showing read-end numbers along the sequence reveals the 2′-*O*-Me sites as positions with low number of read-ends on a background of unmethylated positions with many read-ends. This "negative

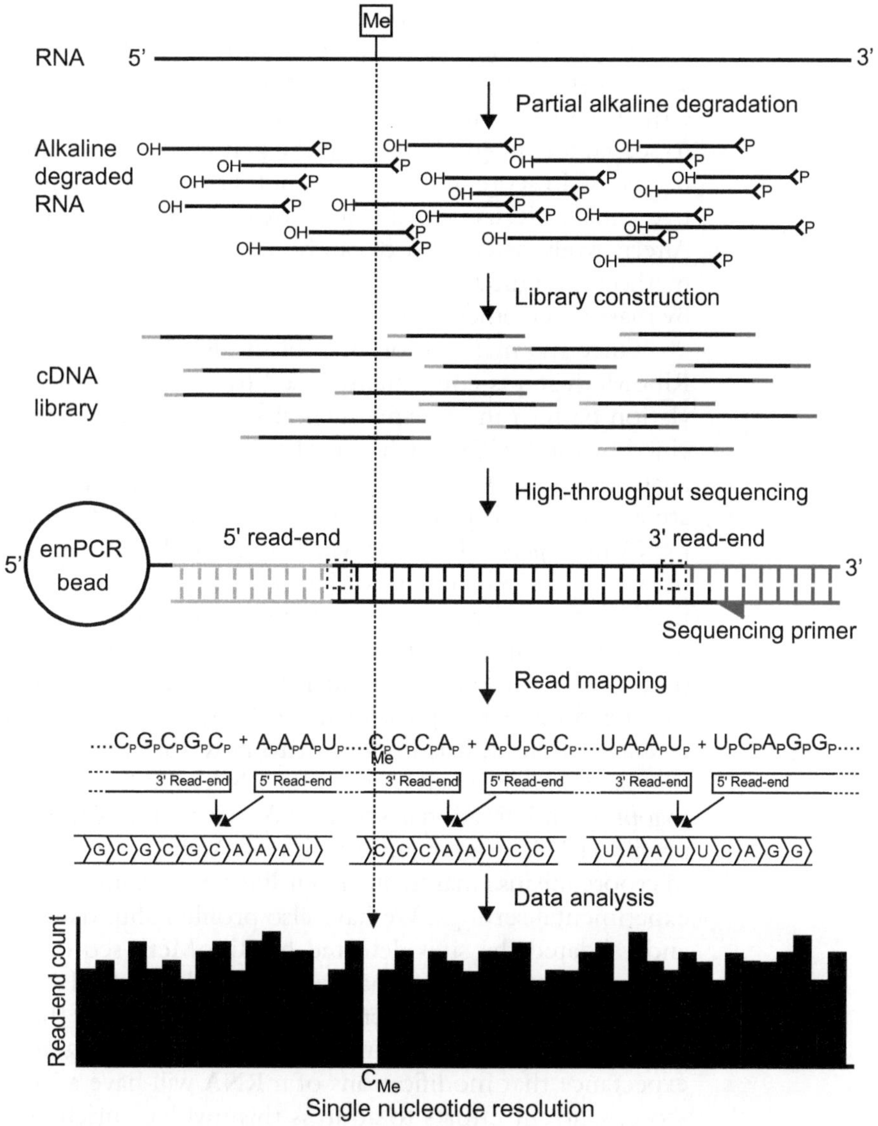

Fig. 1 Outline of RiboMeth-seq. See the text for explanation

image" of the 2′-*O*-Me profile of the RNA can then be converted into a peak diagram and a methylation score (the RMS score) expressing the fraction of molecules methylated at the queried position can be calculated. We usually perform RiboMeth-seq in technical triplicates for mapping and in biological triplicates for experimental studies.

The usefulness of RiboMeth-seq for de novo mapping of 2′-*O*-Me sites depends on the extent of modification. Sites that are fully or close to fully methylated will be revealed, but sites with a lower RMS score are hard to distinguish from background. We usually apply a threshold score of 0.75 (corresponding to 75 % of the molecules being methylated at the queried position). At this score, most detected sites will be true positives and the few false-positives are frequently due to base modifications or pseudouridylations. The most straightforward validation of the sites is to use the sequence of the putative site and the neighboring D box sequence to search for the guide RNA in the genome. The combination of a robust RMS score in triplicate samples and identification of a putative guide RNA constitutes strong evidence for 2′-*O*-sites. Alternatively, selected sites can be validated by any of a number of methods designed for 2′-*O*-Me detection at individual sites [13] or by mass spectrometry [14].

Once the methylation map of the RNA has been established, RiboMeth-seq is conveniently used to compare changes in methylation pattern in an experimental setting. The quantitation provided by the RMS score has not been systematically bench-marked against other methods for the simple reason that no other studies are available, except for a mass spectrometry study of 2′-*O*-Me in rRNA in *S. pombe* [15] for which RiboMeth-seq has not yet been performed. However, RiboMeth-seq is generally consistent with the literature with respect to 2′-*O*-Me mapping as well as in the few cases where quantitative information has been reported (e.g.,[16]). Although the quantitation aspect of RiboMeth-seq may need further improvement both experimentally and in terms of data handling, it is important to note that this is currently the only sequencing-based method for RNA modifications that deliver mapping and fractional scores. We have used RiboMeth-seq to profile rRNA from many species of Archaea, several eukaryotic microorganisms, mammalian cell lines and tissues, and in several experimental settings. We have also profiled abundant small RNAs and validated the sites detected by RiboMeth-seq in these RNAs by mass spectrometry. RiboMeth-seq analysis of mRNA remains a challenge due to the higher sequence complexity of this RNA class and the consequent demand for deeper sequencing as well as the expectance that modifications of mRNA will have a low fractional score. Current efforts to address this involve enrichment steps of methylated RNA prior to application of RiboMeth-seq.

2 Materials

All reagents and H_2O used must be nuclease free.

2.1 5′ Adapter Construction

1. Phusion HF buffer: 5× (New England Biolabs).
2. dNTP solution: 10 mM of each nucleotide.
3. PCR oligo1: 25 μM (*see* Table 1).
4. PCR oligo2: 25 μM (*see* Table 1).
5. Template DNA for in vitro transcription of the 5′ adapter: 20 ng/μL in H_2O, PCR oligo 1, and 2 (*see* Table 1).

Table 1
DNA and RNA oligos

Name	Sequence (5′-3′)
Template DNA	TAATACGACTCACTATAGGCCACTACGCCTCCGCTTTCCTCTCTATGGGC AGTCGGTGATGCCGGCCATGGTCCCAGCCTCCTCGCTGGCGGCCGGT GGGCAACATTCCGAGGGGACCGTCCCCTCGGTAATGGCGAATGG GACA
PCR oligo 1	TAATACGACTCACTATAGGCCACTACGCCTCCGC
PCR oligo 2	TGTCCCATTCGCCATTACCGAGGGGACGGTC
5′ adapter	pppGGCCACUACGCCUCCGCUUUCCUCUCUAUGGGCAGUCGGU GAU-(2′,3′)-p
3′ adapter 1	CAUUGUCUGAGUCGGAGACACGC
3′ adapter 2	UCUAGUCUGAGUCGGAGACACGC
3′ adapter 3	CAAGCUCUGAGUCGGAGACACGC
3′ adapter 4	CUCUUGCUGAGUCGGAGACACGC
3′ adapter 5	AUUGCGCUGAGUCGGAGACACGC
3′ adapter 6	AGUUAGCUGAGUCGGAGACACGC
3′ adapter 7	AAGUCCCUGAGUCGGAGACACGC
3′ adapter 8	CACUACCUGAGUCGGAGACACGC
3′ adapter 9	ACAGUACUGAGUCGGAGACACGC
Ion Proton oligo A	CCATCTCATCCCTGCGTGTCTCCGACTCAG
qPCR oligo 1	CCATCTCATCCCTGCGTG
qPCR oligo 2	CCACTACGCCTCCGC
Taqman probe	CCTCTCTATGGGCAGTCGGTGAT (5′-6-FAM, 3′-BHQ-1))

6. Phusion® High-Fidelity DNA polymerase: 2 U/μL (New England Biolabs) or similar high-fidelity, thermo-stable DNA polymerase.
7. Agarose gel: 2 %.
8. Phenol:chloroform:isoamyl alcohol 25:24:1 (PCI), pH 8.0: saturated with 10 mM Tris–HCl, 1 mM EDTA, pH 8.0 (Sigma Aldrich P2069, *see* **Note 1**).
9. KAc: 3 M, adjusted to pH 8.7 with NaOH.
10. 96 % and 70 % ethanol.
11. MEGAshortscript™ T7 Kit (Thermo Scientific) or similar.
12. Urea polyacrylamide gel (UPAG): 10 %, dimensions; L200 × W170 × D1 (in mm).

2.2 Phosphorylation of 3′ Adapter

1. PNK buffer A: 10× (e.g., Thermo Scientific).
2. ATP: 100 mM.
3. 3′ adapter: 100 μM in H_2O (*see* Table 1).
4. T4 polynucleotide kinase (PNK): 10 U/μL (e.g., Thermo Scientific).
5. PCI, pH 6.7: saturated with 10 mM Tris–HCl, 1 mM EDTA (Sigma Aldrich P3803, *see* **Note 1**).
6. Glycogen: 5 μg/μL.
7. NaAc: 3 M, adjusted to pH 6.0 with HCl.
8. 96 % and 70 % ethanol.

2.3 Alkaline Degradation

1. RNA: 100–200 ng/μL (>10 μg of RNA is recommended as input for triplicate libraries, *see* **Note 2**).
2. $NaHCO_3/Na_2CO_3$: 0.1 M, pH 9.9, prepared by titrating 10 mL of 0.1 M of $NaHCO_3$, pH 11.13 with 0.1 M of Na_2CO_3, pH 8.36 to pH 9.9.
3. Glycogen: 5 μg/μL.
4. KAc: 3 M, adjusted to pH 8.7 with NaOH.
5. 96 % ethanol.
6. Loading buffer (1×): 25 % urea, 5 mM EDTA, 0.5 mg/mL bromophenol blue, 0.5 mg/mL xylene cyanol.

2.4 Gel Purification of Alkaline Degraded RNA

1. Urea polyacrylamide gel (UPAG): 10 %, dimensions; 115 × 100 × 1 (L × W × B in mm).
2. Sybr®Gold diluted 10,000-fold in 1× TBE: 89 mM Tris base, 89 mM boric acid, 2 mM EDTA.
3. Plastic wrap.
4. PCI, pH 8.0: saturated with 10 mM Tris–HCl, 1 mM EDTA, pH 8.0 (Sigma Aldrich P2069, *see* **Note 1**).
5. Chloroform.
6. Glycogen: 5 μg/μL.
7. KAc: 3 M, adjusted to pH 8.7 with NaOH.
8. 96 % and 70 % ethanol.

2.5 3′ Adapter Ligation

1. ATP: 100 mM.
2. Phosphorylated 3’ adapter: 100 mM in H_2O.
3. RiboLock: 40 U/μL (Thermo Scientific), or similar RNase inhibitor.
4. tRNA ligation buffer (10×): 100 mM Tris–HCl, pH 7.5, 1 M KAc, 3 mM spermine, 60 mM $MgAc_2$, 5 mM DTT, 5 % Triton X-100. Prepare the buffer by addition of spermine and DTT from 0.1 M frozen stocks immediately before use.

5. Ligase (Ath-tRNLΔPNK): 4 μM (*see* **Note 3**).
6. PEG8000: 40 % in H_2O, store at −20 °C and thaw only once.
7. PCI, pH 8.0: saturated with 10 mM Tris–HCl, 1 mM EDTA, pH 8.0 (Sigma Aldrich P2069, *see* **Note 1**).
8. Chloroform.
9. Glycogen: 5 μg/μL.
10. NaAc: 3 M, adjusted to pH 6.0 with HCl.
11. 96 % and 70 % ethanol.

2.6 5′ Phosphorylation and 2′,3′-Cyclic Phosphate Removal

1. PNK buffer A: 10× (e.g., Thermo Scientific).
2. ATP: 100 mM.
3. PNK: 10 U/μL (e.g., Thermo Scientific).
4. Loading buffer (2×): 50 % urea, 10 mM EDTA, 1 mg/mL bromophenol blue, 1 mg/mL xylene cyanol.

2.7 Gel Purification of 3′ Ligation Products

1. Urea polyacrylamide gel (UPAG): 10 %, dimensions; L115 x W100 × B1.5 (in mm).
2. Sybr®Gold diluted 10,000-fold in 1× TBE: 89 mM Tris base, 89 mM boric acid, 2 mM EDTA.
3. Plastic wrap.
4. PCI, pH 6.7: saturated with 10 mM Tris–HCl, 1 mM EDTA (Sigma Aldrich P3803, *see* **Note 1**).
5. Chloroform.
6. Glycogen: 5 μg/μL.
7. NaAc: 3 M, adjusted to pH 6.0 with HCl.
8. 96 % and 70 % ethanol.

2.8 5′ Adapter Ligation

1. ATP: 100 mM.
2. 5′ adapter: 150 ng/μL in H_2O (*see* Table 1).
3. RiboLock: 40 U/μL (e.g., Thermo Scientific) or similar RNase inhibitor.
4. tRNA ligation buffer (10×): 100 mM Tris–HCl, pH 7.5, 1 M KAc, 3 mM spermine, 60 mM $MgAc_2$, 5 mM DTT, 5 % Triton X-100. Prepare the buffer by addition of spermine and DTT from 0.1 M frozen stocks immediately before use.
5. Ligase (Ath-tRNLΔPNK): 4 μM (*see* **Note 3**).
6. PEG8000: 40 %, store at −20 °C and thaw only once.
7. PCI, pH 6.7: saturated with 10 mM Tris–HCl, 1 mM EDTA (Sigma Aldrich P3803, *see* **Note 1**).
8. Glycogen: 5 μg/μL.
9. NaAc: 3 M, adjusted to pH 6.0 with HCl.

10. 96 % ethanol.
11. Loading buffer (1×): 25 % urea, 5 mM EDTA, 0.5 mg/mL bromophenol blue, 0.5 mg/mL xylene cyanol.

2.9 Gel Purification of 5′ Ligation Products

1. Urea polyacrylamide gel (UPAG): 10 %, dimensions; L115 × W100 × D1.5 (in mm).
2. Sybr®Gold diluted a 10,000-fold in 1× TBE: 89 mM Tris base, 89 mM boric acid, 2 mM EDTA.
3. Plastic wrap.
4. PCI, pH 6.7: saturated with 10 mM Tris–HCl, 1 mM EDTA (Sigma Aldrich P3803, *see* **Note 1**).
5. Chloroform.
6. Glycogen: 5 μg/μL.
7. NaAc: 3 M, adjusted to pH 6.0 with HCl.
8. 96 % and 70 % ethanol.

2.10 Removal of 2′ Phosphates

1. 10× FastAP buffer (Thermo Scientific).
2. Fast AP: 1 U/μL (Thermo Scientific) or similar alkaline phosphatase.
3. PCI, pH 6.7: saturated with 10 mM Tris–HCl, 1 mM EDTA (Sigma Aldrich P3803, *see* **Note 1**).
4. Chloroform.
5. Glycogen: 5 μg/μL.
6. NaAc: 3 M, adjusted to pH 6.0 with HCl.
7. 96 % and 70 % ethanol.

2.11 cDNA Synthesis

1. dNTP solution: 10 mM of each nucleotide.
2. Ion Proton oligo A (*see* Table 1).
3. First strand buffer: 5× (Invitrogen).
4. DTT: 0.1 M.
5. RiboLock: 40 U/μL (Thermo Scientific) or similar RNase inhibitor.
6. Superscript III: 200 U/μL (Invitrogen) or similar reverse transcriptase.

2.12 qPCR Quantitation of Libraries

1. Primer/probe master mix (10×): 5 μM qPCR oligo 1, 5 μM qPCR oligo 2, 1 μM Taqman probe (see **Table 1**).
2. FastStart Essential DNA Probes Master: 2× (Roche) or similar qPCR master mix.
3. Agarose gel: 3 %.

3 Methods

The protocol starts with preparation of the adapters (*see* Subheadings 3.1 and 3.2) and alkaline degradation of the sample RNA (*see* Subheadings 3.3 and 3.4). At this stage, the sample is split into three aliquots if the experiment involves technical triplicates. Then follows the library construction steps (*see* Subheadings 3.5–3.11) followed by assessment of the library by qRT-PCR (*see* Subheading 3.12). Finally, the library is sequenced (*see* Subheading 3.13) and the data analyzed (*see* Subheading 3.14).

3.1 5′ Adapter Construction by In Vitro Transcription (See Fig. 2)

1. Use PCR to amplify the template DNA which is used for in vitro transcription to construct the 5′ adapter by mixing the following in a 0.2 mL tube:

Component	Volume (μL)
H_2O	145
Phusion HF buffer	40
dNTP (dGTP, dATP,dTTP, dCTP)	4
PCR oligo 1	4
PCR oligo 2	4
Template DNA	1
Phusion® high-fidelity DNA polymerase	2

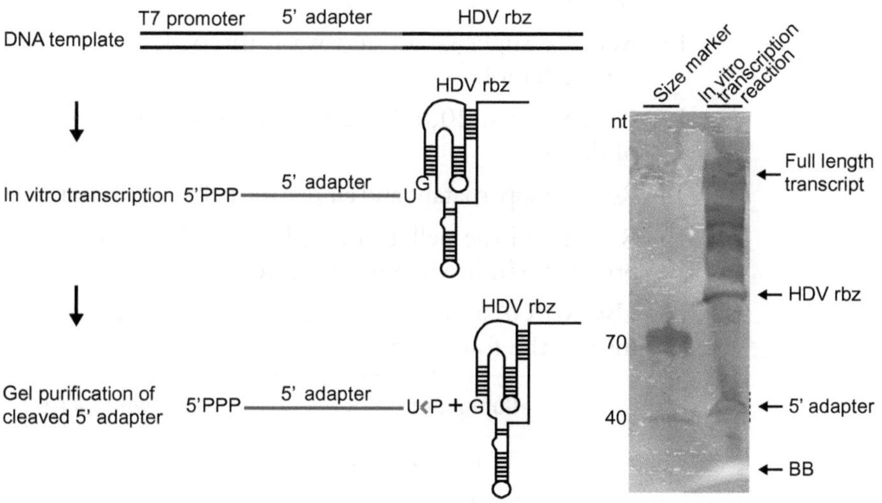

Fig. 2 Strategy for making 5′ adapters with cyclic phosphate ends. The schematic to the left shows the design of the template made by PCR, its transcription into an RNA that encodes the Hepatitis Delta Virus ribozyme (HDV rbz) in cis, and its self-cleavage to produce the 5′ adapter and the released ribozyme. The gel picture to the right shows an example of an experiment aimed at gel purification of the 5′ adapter. The yield of product can be increased by further incubation of the primary transcript. The position of the bromphenol blue (BB) from the loading buffer is marked on the gel

2. Mix by pipetting up and down and place the tube in a thermocycler programmed with the following settings:

Cycle step	Temperature (°C)	Time	Cycles
Initial denaturation	98	30 s	1
Denaturation Annealing and extension	98 72	7 s 15 s	30
Final extension	72	7 min	1
Store	12	Hold	1

3. Assess the quality of the PCR product by standard gel electrophoresis on a 2 % agarose gel. The correct size of the product is 145 bp.
4. To a new 1.5 mL tube add 200 μL of PCI, pH 8.0 and transfer the remainder of the PCR reaction to this tube.
5. Mix by vortexing for 5 s and separate the phases by centrifugation at 10,000 × *g* for 5 min at 4 °C.
6. Transfer the aqueous (top) phase to a new tube.
7. Add 200 μL of chloroform, vortex for 5 s.
8. Centrifuge at 10,000 × *g* for 5 min at 4 °C.
9. Transfer the aqueous (top) phase to a new tube.
10. Add 22 μL of KAc and 560 μL of ice-cold 96 % ethanol, vortex and precipitate the DNA by incubation on dry-ice for 15 min.
11. Recover the DNA by centrifugation at 20,000 × *g* for 30 min at 4 °C.
12. Remove supernatant and wash the pellet gently with 500 μL 70 % ethanol.
13. Centrifuge at 20,000 × *g* for 5 min at 4 °C to ensure recovery of the pellet.
14. Remove supernatant and air dry the pellet for 15 min with open lid.
15. Resuspend the pellet in 20 μL of H_2O and store at −20 °C or proceed within vitro transcription.
16. Use MEGAshortscript™ T7 Kit for in vitro transcription by mixing the following:

Component	Volume (μL)
T7 reaction buffer	5
T7 ATP	5
T7 CTP	5
T7 GTP	5
T7 UTP	5
Template DNA	20
T7 Enzyme Mix	5

17. Incubate for 2 h at 37 °C.
18. Add 50 μL loading buffer.
19. Denature the RNA by incubation for 1 min at 80 °C.
20. Separate the denatured RNA on a 10 % UPAG running at 300 V by loading 25 μL to each of four wells (*see* Fig. 2a).
21. Stain the gel by spreading a few mL of a 10,000-fold diluted Sybr®Gold in TBE onto the gel.
22. Incubate for 10 min.
23. Remove excess of Sybr®Gold by tipping the glass plate and cover the gel with plastic wrap.
24. Visualize the RNA by scanning the gel on a Typhoon scanner using the fluorescence mode and a laser in the range of (480–540 nm) (*see* **Note 4**).
25. Using ImageJ, print a 1:1 image of the gel and place it under the glass plate and align it with the gel.
26. Excise a gel slice containing the 5′ adapter (*see* Fig. 2a) and transfer the gel slice to a 1.5 mL tube.
27. Add 400 μL of H_2O and 400 μL of PCI, pH 8.0.
28. Elute the RNA by placing the tube at 4 °C in a rotating wheel at 25 rpm overnight.
29. Transfer the liquid to a new 1.5 mL tube.
30. Centrifuge at 10,000 × *g* for 5 min at 4 °C.
31. Transfer the aqueous (top) phase to a new tube.
32. Add 400 μL of chloroform, vortex for 5 s.
33. Centrifuge at 10,000 × *g* for 5 min at 4 °C.
34. Transfer the aqueous (top) phase to a new tube.
35. Add 3 μL of glycogen and 40 μL of KAc, vortex the tube for 5 s.
36. Add 1000 μL of ice-cold 96 % ethanol, vortex and precipitate the RNA by incubation on dry-ice for 15 min.
37. Centrifuge at 20,000 × *g* for 30 min at 4 °C.
38. Remove supernatant and wash the pellet gently with 500 μL of 70 % ethanol.
39. Centrifuge at 20,000 × *g* for 5 min at 4 °C.
40. Remove supernatant and air dry the pellet for 15 min with open lid.
41. Resuspend the 5′ adapter in 25 μL of H_2O by pipetting up and down.
42. Measure the concentration and adjust to 150 ng/μL with H_2O, store at −80 °C.

3.2 Phosphorylation of 3′ Adapter (See Note 5)

1. In a 1.5 mL tube combine:

Component	Volume (μL)
H_2O	62
PNK buffer A	10
ATP	6
3'adapter (RNA oligo)	20
PNK	2

2. Incubate for 30 min at 37 °C.
3. Place the tube on ice.
4. Extract the 3′ adapter by adding 100 μL of H_2O and 200 μL of PCI, pH 6.7, vortex for 5 s.
5. Centrifuge at 10,000 × *g* for 5 min at 4 °C.
6. Transfer the aqueous (top) phase to a new tube.
7. Add 200 μL of chloroform, vortex for 5 s.
8. Centrifuge at 10,000 × *g* for 5 min at 4 °C.
9. Transfer the aqueous (top) phase to a new tube.
10. Add 3 μL of glycogen and 22 μL of NaAc, vortex the tube for 5 s.
11. 560 μL of ice-cold 96 % ethanol, vortex and precipitate the RNA by incubation on dry-ice for 15 min.
12. Centrifuge at 20,000 × *g* for 30 min at 4 °C.
13. Remove supernatant and wash the pellet with 500 μL of 70 % ethanol.
14. Centrifuge at 20,000 × *g* for 5 min at 4 °C.
15. Remove supernatant and air dry the pellet for 15 min with open lid.
16. Resuspend the pellet in 40 μL of H_2O by pipetting up and down and store at −80 °C.

3.3 Alkaline Degradation of Sample RNA

1. To a 0.2 mL tube, add 100 μL of RNA (*see* **Note 2**) and denature at 90 °C for 2 min in a thermocycler.
2. Then add 100 μL of $NaHCO_3/Na_2CO_3$, pH 9.9 preheated to 90 °C to the RNA and incubate at 90 °C for 5 min to degrade the RNA into smaller library fragments (*see* **Note 6**).
3. To a new 1.5 mL tube add 3 μL of glycogen and 22 μL of KAc.
4. Add the degraded RNA and vortex the tube for 5 s.
5. Add 560 μL of ice-cold 96 % ethanol, vortex and precipitate the RNA by incubation on dry-ice for 15 min.
6. Recover the RNA by centrifugation at 20,000 × *g* for 10 min at 4 °C.

7. Remove supernatant and air dry the pellet for 15 min with open lid.
8. Resuspend in 10 µL loading buffer (*see* **Note 7**).

3.4 Gel Purification of Alkaline Degraded RNA

1. Denature the alkaline degraded RNA by incubation for 1 min at 80 °C and then place the tube on ice.
2. Fractionate the alkaline degraded RNA by gel electrophoresis on a 10 % UPAG running at 15 mA (*see* Fig. 3a).
3. Repeat **steps 21–24** in Subheading 3.1.
4. Excise a gel slice containing RNA in the size range of 20–40 nt (*see* Fig. 3a and **Note 8**) and transfer the gel slice to a 1.5 mL tube.
5. Add 400 µL of H_2O and 400 µL of PCI, pH 8.0 (*see* **Note 9**).
6. Repeat **steps 28–40** in Subheading 3.1.

3.5 3′ Adapter Ligation

1. To the dried pellet add the following on ice:

Component	Volume (µL)
H_2O	1.5
ATP	0.5
Phosphorylated 3′ adapter	2
RiboLock	0.5

2. Resuspend the pellet by pipetting up and down.
3. Add 1 µL of tRNA ligation buffer.
4. 1 µL of ligase (Ath-tRNLΔPNK).
5. Mix by pipetting up and down and avoid making many bubbles (*see* **Note 10**).
6. Add 3.5 µL of PEG8000.
7. Mix by pipetting up and down.
8. Incubate for 1 h at 30 °C.
9. Add 190 µL of H_2O and 200 µL of PCI, pH 6.7, vortex for 5 sec.
10. Repeat **steps 5–15** in Subheading 3.2.

3.6 5′ Phosphorylation and Removal of 2′,3′ Cyclic Phosphate

1. To the dried pellet add the following on ice:

Component	Volume (µL)
ATP	1
PNK buffer A	1
PNK	1

2. Resuspend the pellet by pipetting up and down.

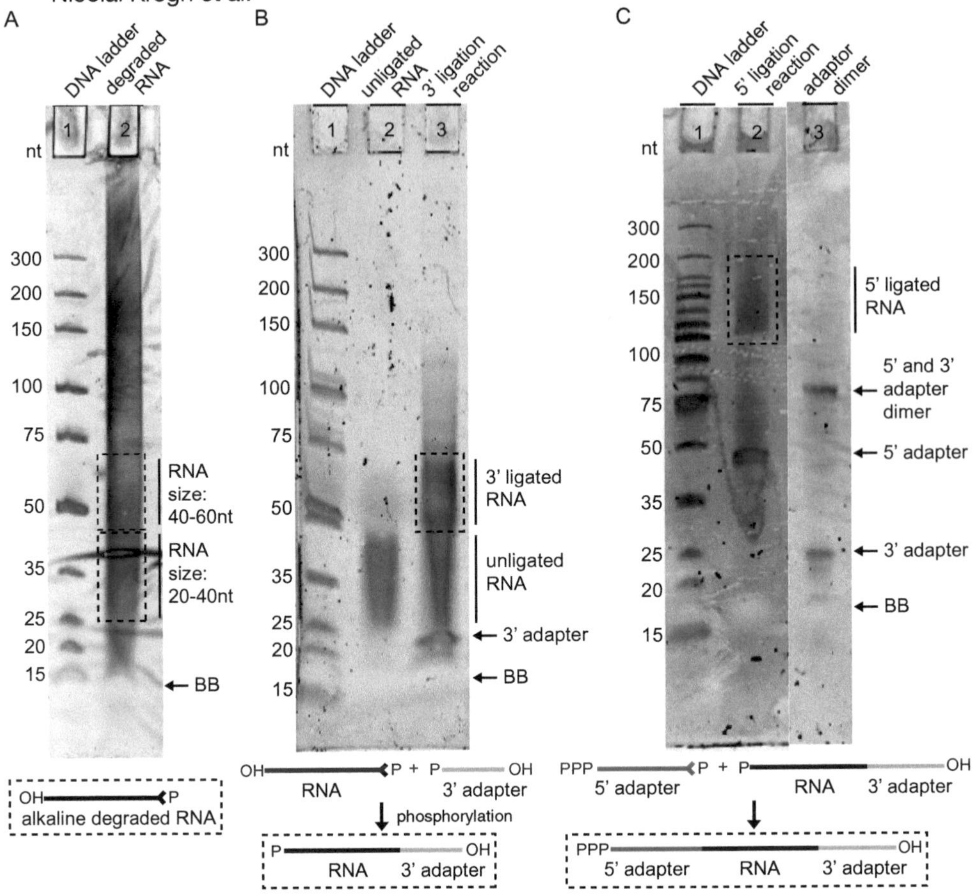

Fig. 3 Gel pictures showing the critical gel purification steps of library fragments (**a**), and first (**b**) and second (**c**) ligation steps. The panel below the gel pictures depicts the structure of RNA species that are being isolated. Note that RNA migrates approximately 10 % slower compared to DNA on a 10 % urea polyacrylamide gel. The migration lengths of the samples are illustrated by showing the position of the bromphenol blue (BB) from the loading buffer

3. Incubate for 30 min at 37 °C.
4. Add 10 μL of loading buffer.

3.7 Gel Purification of 3′ Ligation Products

1. Denature the 3′ ligated RNA by incubation for 1 min at 80 °C and then place the tube on ice.
2. Isolate the 3′ ligated RNA by gel electrophoresis on a 10 % UPAG running at 15 mA (*see* Fig. 3b).
3. Repeat **steps 21–24** in Subheading 3.1.
4. Excise a gel slice containing the 3′ ligated RNA (*see* Fig. 3b) and transfer the gel slice to a 1.5 mL tube.
5. Add 400 μL of H_2O and 400 μL of PCI, pH 6.7.
6. Elute the RNA by placing the tube at 4 °C in a rotating wheel at 25 rpm overnight.
7. Transfer the liquid to a new 1.5 mL tube.
8. Centrifuge at 10,000 × *g* for 5 min at 4 °C.

9. Transfer the aqueous (top) phase to a new tube.
10. Add 400 μL of chloroform, vortex for 5 s.
11. Centrifuge at 10,000 × *g* for 5 min at 4 °C.
12. Transfer the aqueous (top) phase to a new tube.
13. Add 3 μL of glycogen and 40 μL of NaAc, vortex the tube for 5 s.
14. Add 1000 μL of ice-cold 96 % ethanol, vortex and precipitate the RNA by incubation on dry-ice for 15 min.
15. Centrifuge at 20,000 × *g* for 30 min at 4 °C.
16. Remove supernatant and wash the pellet gently with 500 μL of 70 % ethanol.
17. Centrifuge at 20,000 × *g* for 5 min at 4 °C.
18. Remove supernatant and air dry the pellet for 15 min with open lid.

3.8 5′ Adapter Ligation

1. To the dried pellet add the following on ice (*see* **Note 11**):

Component	Volume (μL)
H_2O	0.5
ATP	0.5
5′ adapter	3
RiboLock	0.5

2. Repeat **steps 2–9** in Subheading 3.5.
3. Centrifuge at 10,000 × *g* for 5 min at 4 °C.
4. Transfer the aqueous (top) phase to a new tube.
5. Add 3 μL of 5 μg/μL glycogen and 22 μL of NaAc, vortex the tube for 5 s.
6. Add 560 μL of ice-cold 96 % ethanol, vortex and precipitate the RNA by incubation on dry-ice for 15 min.
7. Centrifuge at 20,000 × *g* for 30 min at 4 °C.
8. Remove supernatant and air dry the pellet for 15 min with open lid.
9. Resuspend in 10 μL 1×loading buffer.

3.9 Gel Purification of 5′ Ligation Products

1. Denature the 5′ ligated RNA by incubation for 1 min at 80 °C.
2. Place the tube on ice.
3. Separate the 5′ ligated RNA by gel electrophoresis on a 10 % UPAG running at 15 mA (*see* Fig. 3c).
4. Repeat **steps 21–25** in Subheading 3.1.

5. Excise a gel slice containing the 5′ ligated RNA (*see* Fig. 3c and **Note 12**) and transfer the gel slice to a 1.5 mL tube.
6. Repeat **steps 5–18** in Subheading 3.7.

3.10 Removal of 2′ Phosphates

1. To the dried pellet add the following on ice:

Component	Volume (μL)
H_2O	8
FastAP buffer	1
FastAP	1

2. Resuspend the pellet by pipetting up and down and incubate for 30 min at 37 °C.
3. Add 190 μL H_2O and 200 μL of PCI, pH 6.7 and vortex for 5 sec.
4. Repeat **steps 5–16** in Subheading 3.2.

3.11 cDNA Synthesis

1. To the dried pellet add the following on ice:

Component	Volume (μL)
H_2O	12
dNTP (dGTP, dATP,dTTP, dCTP)	1
Ion Proton oligo A	1

2. Resuspend the pellet by pipetting up and down.
3. Incubate for 1 min at 90 °C followed by 5 min at room temperature and then place on ice for 2 min.
4. Add the following to the tube:

Component	Volume (μL)
First strand buffer	4
DTT	1
RiboLock	0.5
Superscript III	0.5

5. Mix by pipetting up and down and incubate for 1 h at 50 °C.
6. Incubate for 15 min at 70 °C to inactive the Superscript III reverse transcriptase.
7. Make a 200 μL, 100-fold dilution of the library.
8. From the 100-fold dilution make three 200 μL, additional 10-fold dilutions to be used for qPCR quantitation.
9. Store the library at −20 °C or proceed to quantitation.

3.12 qPCR Quantitation of Libraries (See Note 13)

1. Add 2 μL of a 1000-fold diluted library to a 0.2 mL qPCR tube (*see* **Note 14**).

2. Along with the newly constructed library, analyze 2 μL of a 10-fold dilution series of a standard library with known concentration. This is used for generating a standard curve for quantitation of the library (*see* **Note 15**).
3. To each tube add the following prepared as a master mix:

Component	Volume (μL)
H_2O	2
Primer/probe master mix	1
FastStart Essential DNA Probes Master	5

4. Mix by pipetting up and down or flicking the tube.
5. Collect the contents at the bottom of the tube by a brief centrifugation.
6. Analyze the samples on a real-time PCR cycler the following program is for the LightCycler Nano (Roche) (*see* **Note 16**):

Cycle step	Temperature	Time	Cycles
Initial denaturation	95 °C	10 s	1
Denaturing Annealing and extension	95 °C 60 °C	10 s 30 s	45

3.13 Sequencing

The adapters listed in this protocol (*see* Table 1) are designed for the Ion Torrent/Ion Proton sequencing platform but can be redesigned to fit other platforms. If libraries are sequenced on an in-house Ion Proton sequencer, it is recommend to use the latest versions of the kits and the amount of library stated by the sequencing protocol.

As the very ends of the reads are used for the RMS analysis, it is important that the reads are not trimmed in any way prior to analysis. The current Torrent server v5.0.4 automatically removes the 3′ adapter sequence from the reads. Thus, it can no longer be determined whether a given library fragment was read through and the read-end therefore carried information or if the sequencing was terminated prematurely and the read-end therefore should be disregarded in the analysis. To circumvent this issue, the following steps should be taken when planning the sequencing run on the Torrent server:

In the final plan panel, under analysis parameters, select custom and in the fields Pre-BaseCaller for calibration and BaseCaller type in the option: --trim-adapter-cutoff=0.

3.14 Data Analysis

The data analysis basically consists of read trimming, read mapping, and counting of read-ends terminating at each position in the reference sequence. For trimming we find that a conservative approach demanding perfectly sequenced read-ends, barcode, and

3′ adapter is preferable to trimming based on quality scores, and thus we omit any quality trimming. The adapter trimming tool Cutadapt [17] may be used for both barcode splitting and 3′ adapter removal (*see* **Note 17**). As mentioned above, we demand high stringency on barcode and 3′ adapter, and only regard reads where all six barcode nucleotides and the first nine 3′ adapter nucleotides are read correctly.

For mapping, it is preferable to use a relevant transcriptome in fasta format as reference sequence. Alternatively, the genomic sequence may be used but in that case we find it most straightforward to include ribosomal and other highly abundant RNAs for which RMS analysis should be performed as independent fasta entries. The RMS analysis can then conveniently be confined to these sequences and the genomic mapping serves to reduce off-target mappings derived from related sequences. Also note that for a number of database genomes, e.g., the human hg19 the major ribosomal RNAs are omitted from the sequence. For the short, generally non-spliced reads generated by RiboMeth-seq, we find the read alignment tool Bowtie2 [18] with standard settings to be ideal for mapping (*see* **Note 17**).

For counting read-ends and calculating the RMS score it is necessary to use custom scripts to go through the sequence alignment map files. In our approach, we require the three distal nucleotides of each read-end to be perfectly aligned to the reference sequence when counting read-ends. The read count can be visualized as a simple column chart as depicted in the lower panel of Fig. 1. The read-end counts of each position should preferably be viewed in the context of the local surrounding read-end-coverage. One way to do this is to calculate the RMS score that considers the depression in read counts compared to a weighted average of the surrounding 12 positions [9]:

$$S_i = \max\begin{cases} 1 - \dfrac{n_i}{\frac{1}{2}\left(\dfrac{\sum_{j=i-\delta}^{i-1}\omega_j n_j}{\sum_{j=i-\delta}^{i-1}\omega_j} + \dfrac{\sum_{j=i+\delta}^{i-\delta}\omega_j n_j}{\sum_{j=i-1}^{i-\delta}\omega_j}\right)} \\ 0 \end{cases}$$

Where n_i = read count at position i and $\delta = 6$ (flanking window length). The weight ω varies with respect to distance:

j	−6	−5	−4	−3	−2	−1	0	1	2	3	4	5	6
ω	0.5	0.6	0.7	0.8	0.9	1	0	1	0.9	0.8	0.7	0.6	0.5

Base modifications on the Watson-Crick face of an RNA nucleotide may cause reverse transcriptase to misincorporate during cDNA synthesis. This property can be exploited to identify some base modifications by analyzing the reads using an approach similar to identifying single nucleotide polymorphisms (SNP) and investigate the misincorporation pattern [19]. The misincorporation may be subtle (for some modifications below 2 % distributed on the three alternative nucleotides, but still above the sequencing error background) and the threshold should therefore be lower than what is typically used for SNP detection. Numerous approaches exist for SNP detection and can be found elsewhere.

4 Notes

1. It is critical to use the listed phenol:chloroform:isoamyl alcohol to preserve the 2′,3′ cyclic phosphate on RNAs.
2. The RNA can be of different origin, e.g., whole cell, gel-purified rRNA, or nuclear RNA. In our experience, slight degradation of the RNA is not a problem, probably because degradation by, e.g., RNases and heavy metal ions leaves 2′,3′ cyclic phosphate and 5′ OH ends as with alkaline degradation in the first step of the protocol.
3. The enzyme is presently not commercially available but can be requested from the authors.
4. The gel can be scanned on any gel visualization equipment that allows printing of a 1:1 gel picture. Alternatively, the gel slices can be cut directly from the gel using a blue light source for visualization.
5. The 3′ adapter contains the barcode that should be unique for each library that is sequenced together on the same sequencing chip. The phosphorylation reaction presented here yields 3′ adapters sufficient for up to 20 library constructions. The reaction can be scaled or alternatively, the 3′ adapter can be purchased as a phosphorylated RNA oligo.
6. The incubation time presented here is optimized for RNA in the size range of ribosomal RNA. For RNA samples enriched for smaller RNAs the incubation time should be shorter.
7. Load appropriate size markers together with the RNA. We recommend proceeding to the gel purification of alkaline degraded RNA immediately to minimize opening of 2′,3′-cyclic phosphate on the RNA.
8. In cases where the RNA is very precious, we recommend to excise a gel slice in the size range of 40–60 nt and store it at −80 °C in a mixture of 400 μL of H_2O and 400 μL of PCI, pH 8.0. The RNA can be eluted from the gel slice and used for RiboMeth-seq analysis if the first experiment fails by starting at **step 6** in Subheading 3.4. The longer fragments may result in

slightly more library fragments being discarded in the data analysis and a slightly more pronounced problem with only read-ends from one side at the ends of the analyzed molecules.

9. It is very important to use PCI saturated with buffer at pH, 8.0 to prevent opening of the 2′,3′-cyclic phosphates.
10. For efficient ligation it is very important to have the pellet thoroughly resuspended and the RNA, adapter, and ligase mixed well before adding the 40 % PEG 8000.
11. As a control make a ligation reaction substituting H_2O with 0.5 μL of 50 μM phosphorylated 3′ adapter to generate adapter dimers that can be used as a marker on the following gel purification (*see* Fig. 3c).
12. It is very important to avoid cutting out any of the adapter dimers as this will result in fewer usable reads after the sequencing.
13. Alternatively, the libraries can be quantified using the Ion Library TaqMan™ Quantitation Kit (Thermo Scientific). The procedure listed here takes advantage of using the common qPCR reagents already used in many laboratories and is much cheaper compared to the alternative.
14. It is recommended to make the qPCR quantitation of three 1000-fold dilutions of the library to get a more accurate quantitation, hence a more even distribution of reads between libraries sequenced on the same chip.
15. The standard library can be generated by PCR amplification of a previous constructed library using qPCR oligo 1 and 2 (*see* Table 1). It can be quantitated using a Nanodrop or a Qubit.
16. After the qPCR analysis the quality of the library can be analyzed on a 3 % agarose gel. PCR products generated from the cDNA library will appear as a 100-150 nt smear compared to a DNA ladder, whereas PCR products generated from adapter dimers appear as a distinct band at approximately 75 nt. The visualization of the DNA in the agarose gel tends to overestimate the amount of adapter dimers as this is a band of one length compared to the smeary appearance of the library fragments.
17. The download files and installation manuals for Cutadapt and Bowtie2 can be found here: https://cutadapt.readthedocs.org/en/stable/

    http://bowtie-bio.sourceforge.net/bowtie2/index.shtml.

Acknowledgments

The work was supported by the Danish Research Council for Independent Research and the Danish Cancer Research Foundation.

References

1. Motorin Y, Helm M (2011) RNA nucleotide methylation. Wiley interdisciplinary reviews. RNA 2:611–631
2. Decatur WA, Fournier MJ (2002) rRNA modifications and ribosome function. Trends in biochemical sciences 27:344–351
3. Watkins NJ, Bohnsack MT (2012) The box C/D and H/ACA snoRNPs: key players in the modification, processing and the dynamic folding of ribosomal RNA. Wiley interdisciplinary reviews. RNA 3:397–414
4. Helm M (2006) Post-transcriptional nucleotide modification and alternative folding of RNA. Nucleic acids research 34:721–733
5. Decatur WA, Liang XH, Piekna-Przybylska D, Fournier MJ (2007) Identifying effects of snoRNA-guided modifications on the synthesis and function of the yeast ribosome. Methods in enzymology 425:283–316
6. Noeske J, Wasserman MR, Terry DS, Altman RB, Blanchard SC, Cate JH (2015) High-resolution structure of the *Escherichia coli* ribosome. Nature structural & molecular biology 22:336–341
7. Polikanov YS, Melnikov SV, Soll D, Steitz TA (2015) Structural insights into the role of rRNA modifications in protein synthesis and ribosome assembly. Nature structural & molecular biology 22:342–344
8. Dennis PP, Tripp V, Lui L, Lowe T, Randau L (2015) C/D box sRNA-guided 2'-O-methylation patterns of archaeal rRNA molecules. BMC genomics 16:632
9. Birkedal U, Christensen-Dalsgaard M, Krogh N, Sabarinathan R, Gorodkin J, Nielsen H (2015) Profiling of ribose methylations in RNA by high-throughput sequencing. Angewandte Chemie 54:451–455
10. Kiss-Laszlo Z, Henry Y, Bachellerie JP, Caizergues-Ferrer M, Kiss T (1996) Site-specific ribose methylation of preribosomal RNA: a novel function for small nucleolar RNAs. Cell 85:1077–1088
11. Schutz K, Hesselberth JR, Fields S (2010) Capture and sequence analysis of RNAs with terminal 2',3'-cyclic phosphates. RNA 16:621–631
12. Meyer M, Masquida B (2014) cis-Acting 5' hammerhead ribozyme optimization for in vitro transcription of highly structured RNAs. Methods Mol Biol 1086:21–40
13. Maden BE (2001) Mapping 2'-O-methyl groups in ribosomal RNA. Methods 25:374–382
14. Douthwaite S, Kirpekar F (2007) Identifying modifications in RNA by MALDI mass spectrometry. Methods in enzymology 425:3–20
15. Taoka M, Nobe Y, Hori M, Takeuchi A, Masaki S, Yamauchi Y, Nakayama H, Takahashi N, Isobe T (2015) A mass spectrometry-based method for comprehensive quantitative determination of post-transcriptional RNA modifications: the complete chemical structure of Schizosaccharomyces pombe ribosomal RNAs. Nucleic acids research 43:e115
16. Buchhaupt M, Sharma S, Kellner S, Oswald S, Paetzold M, Peifer C, Watzinger P, Schrader J, Helm M, Entian KD (2014) Partial methylation at Am100 in 18S rRNA of baker's yeast reveals ribosome heterogeneity on the level of eukaryotic rRNA modification. PloS one 9:e89640
17. Martin M (2011) Cutadapt removes adapter sequences from high-throughput sequencing reads. EMBnet.J 17(n. 1):10–12
18. Langmead B, Salzberg S (2012) Fast gapped-read alignment with Bowtie 2. Nat Methods 9:357–359
19. Ryvkin P, Leung YY, Silverman IM, Childress M, Valladares O, Dragomir I, Gregory BD, Wang LS (2013) HAMR: high-throughput annotation of modified ribonucleotides. RNA 19:1684–1692

Chapter 14

In Silico Identification of RNA Modifications from High-Throughput Sequencing Data Using HAMR

Pavel P. Kuksa, Yuk Yee Leung, Lee E. Vandivier, Zachary Anderson, Brian D. Gregory, and Li-San Wang

Abstract

RNA molecules are often altered post-transcriptionally by the covalent modification of their nucleotides. These modifications are known to modulate the structure, function, and activity of RNAs. When reverse transcribed into cDNA during RNA sequencing library preparation, atypical (modified) ribonucleotides that affect Watson-Crick base pairing will interfere with reverse transcriptase (RT), resulting in cDNA products with mis-incorporated bases or prematurely terminated RNA products. These interactions with RT can therefore be inferred from mismatch patterns in the sequencing reads, and are distinguishable from simple base-calling errors, single-nucleotide polymorphisms (SNPs), or RNA editing sites. Here, we describe a computational protocol for the in silico identification of modified ribonucleotides from RT-based RNA-seq read-out using the *High-throughput Analysis of Modified Ribonucleotides* (HAMR) software. HAMR can identify these modifications transcriptome-wide with single nucleotide resolution, and also differentiate between different types of modifications to predict modification identity. Researchers can use HAMR to identify and characterize RNA modifications using RNA-seq data from a variety of common RT-based sequencing protocols such as Poly(A), total RNA-seq, and small RNA-seq.

Key words RNA modification, RNA posttranscriptional modification, RNA covalent modification, Small RNA, Small RNA sequencing, Messenger RNA, RNA sequencing, Machine learning, Classification

Abbreviations

3′	3-prime (3′)
5′	5-prime (5′)
bp	Base pair
cDNA	Complementary DNA
HAMR	High-throughput annotation of modified ribonucleotides
mRNA	Messenger RNA
mRNA-seq	messenger RNA sequencing
nt	Nucleotide
RNA-seq	RNA sequencing

Alexandra Lusser (ed.), *RNA Methylation: Methods and Protocols*, Methods in Molecular Biology, vol. 1562, DOI 10.1007/978-1-4939-6807-7_14, © Springer Science+Business Media LLC 2017

RT	Reverse transcriptase
smRNA	Small RNA
smRNA-seq	Small RNA sequencing
SNP	single nucleotide polymorphism
tRNA	Transfer RNA

1 Introduction

Covalent posttranscriptional modifications of specific nucleotide bases in RNA molecules are known to be highly prevalent and physiologically important [1–6]. RNA modifications play a role in maintaining the structure and stability of RNAs [7–10] and affect their maturation [3, 10], translation and cellular abundance [10, 15]. All known classes of RNA molecules harbor various levels of diverse modifications [1, 8, 9, 11–13]. However, the overall abundance and biological function of these modifications are incompletely understood [10, 14, 15].

Experimental methods for low-throughput detection of some types of RNA modifications are well established [16–19]. One such method is primer extension, which relies on the differential ability of reverse transcriptase to produce cDNAs with base-pair substitutions or premature termination at positions occupied by modified nucleotides [17]. However, because many RNA sequencing library preparation protocols require RNA to cDNA conversion by reverse transcriptase (RT), it is possible to use RNA sequencing data to identify sites of modified nucleotides across many classes of RNAs in a high-throughput and transcriptome-wide fashion by identifying positions within RNAs with significant mismatch rates after alignment. The protocol for in silico identification of RNA modifications presented here uses our previously developed HAMR software [20]. The protocol allows for fast and reliable identification of modified nucleotides at single-nucleotide resolution in all RNA classes transcriptome-wide through the analysis of nucleotide substitutions in high-throughput RNA sequencing datasets. This software provides an important tool for detecting and classifying modified RNA ribonucleotides that modulate RT incorporation, e.g., by affecting Watson-Crick base pairing. The examples of HAMR-detectable modifications include but are not limited to m1A, m3C, m5C, and pseudouridine [1, 8, 10, 13, 14, 20–22]. Modifications such as m6A [23], which do not significantly interfere with RT, will not be detected by HAMR.

In this chapter, we describe how to use HAMR [20] for in silico identification of RNA modifications using RNA sequencing data.

HAMR in a nutshell: The starting point of the HAMR analysis is the mapped sequencing reads in the standard BAM file format (https://samtools.github.io/hts-specs/SAMv1.pdf). The HAMR software is fully automated and the user can perform genome-wide analysis of RNA modifications in one command (*see* Subheading 3.5) using BAM files with the corresponding reference genome sequence. The user can also start from raw sequencing data and turn it into mapped data (BAM) as summarized in Subheading 3.4 for preparing raw sequencing data for HAMR analysis. Subheading 3.1 additionally describes suggestions for RNA library preparation.

Sample study: To illustrate RNA modification analysis using the HAMR software [20], we also provide a sample study of RNA modifications in human brain using RNA-seq data in Subheading 3.8.

Availability: HAMR is available as a stand-alone command-line pipeline (https://github.com/wanglab-upenn/HAMR/) and as a Web-based application (accessible at http://lisanwanglab.org/hamr/). The HAMR source code is freely available under the MIT license for academic and nonprofit use.

2 Materials

Always use the recommended version of all software dependencies. We also recommend performing a setup as described (*see* Subheading 3.3) before preprocessing the sequencing data from FASTQ to BAM file (*see* Subheading 3.4) and analyzing it using HAMR (*see* Subheadings 3.5 and 3.6).

2.1 Hardware Requirements

Desktop computer or server (multi-core server with 32GB RAM is recommended when starting from raw sequenced reads in human).

2.2 Software Requirements

1. Linux-based operating system (CentOS, Ubuntu, Debian, etc.).
2. C compiler (g++).
3. Standard POSIX programs (awk, grep, bash) (*see* **Note 1**).
4. Samtools suite [24] http://www.htslib.org/download/ (version 1.0 or later).
5. Python (version v2.7. x or above) (*see* **Note 2**).
6. R program https://www.r-project.org/(version 3.1.x or above).
7. Additionally, the following software is required for processing raw reads (*see* Subheading 3.4):
   (a) SRA toolkit (http://www.ncbi.nlm.nih.gov/Traces/sra/sra.cgi?cmd=show&f=software&m=software&s=software) when analyzing data from the Sequencing Read Archive.
   (b) Cutadapt [25] https://code.google.com/p/cutadapt/ (version 1.7 and above) to trim adapter sequences for sequenced reads.

(c) STAR [26] https://github.com/alexdobin/STAR/releases (version 2.4.x and above) for mapping reads to the genome.

The required external programs can be installed from Linux repositories (e.g., using package managers such as Apt-get in Ubuntu).

```
sudo apt-get install samtools
sudo apt-get install r-base
```

Alternatively, packages can be manually installed (*see* **Note 3**).

3 Methods

3.1 RNA Sequencing Data Generation

The protocol begins with extraction of the target population of RNAs such as mRNA or small RNA using a phenol-based extraction method from tissues or cells. The extracted material is then prepared as a sequencing library and sequenced using any standard RT-based Illumina sequencing protocol for RNA class of interest (e.g., mRNAs or small RNAs).

1. RNA extraction: Use a standard phenol-based RNA extraction to obtain a minimum of 4 μg total RNA using Illumina TruSeq® RNA Library Prep Kit or Illumina TruSeq® Stranded Total RNA Library Prep Kit; alternatively, obtain a minimum of 1 μg total RNA from tissues or cell lines using Illumina TruSeq® Small RNA Library Preparation.
2. Library preparation for RNA sequencing: Use Illumina TruSeq®RNA, Stranded total RNA, or TruSeq Small RNA library preparation kit protocol for standard RT-based sequencing. These include adaptor ligation, cDNA synthesis, and library amplification steps, all adapted from original Illumina library preparation kit protocol. The libraries should be sequenced on an Illumina Genome Analyzer machine (e.g., HiSeq2000, GAIIx, or above) with read length 50–100 bp.
3. Following sequencing is a series of preprocessing steps (not included in the HAMR software, summarized in Fig. 1) for turning the FASTQ file obtained from the Illumina sequencer to mapped (BAM) file format. These preprocessing steps ensure the input BAM file is compatible and ready for use with HAMR. The preprocessing steps include read trimming, read mapping with mismatches, read filtering, and handling of multi-mapped reads (optional for studying highly repetitive RNAs such as tRNAs, or other small RNAs). The HAMR software (*see* Fig. 2) can then be applied to the prepared BAM file. The HAMR software is generally applicable to any mRNA or small RNA populations [14].

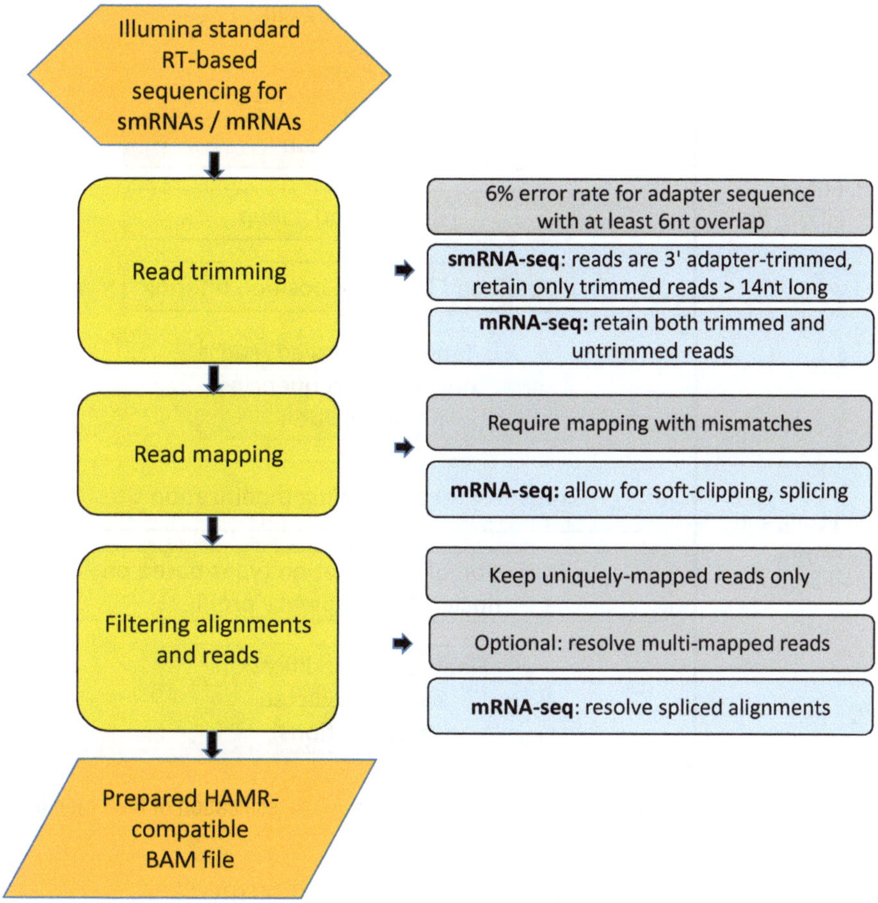

Fig. 1 Preparing the raw sequencing data (from FASTQ to BAM) for running HAMR software [20]

3.2 Downloading and Installing HAMR

All HAMR programs are packaged as compiled "binaries" (*see* **Note 4**) that can be run from the command line by typing in the program's file name. In general, HAMR programs will also require other external programs (*see* Subheading 2.2 on software requirements). To install HAMR:

1. Create a directory for HAMR analyses and enter this directory:
```
mkdir <HAMRdirectory>
cd <HAMRdirectory>
```
2. Download HAMR from https://github.com/wanglab-upenn/HAMR to the HAMR directory <HAMRdirectory>. For example, to download HAMR release v1.2 from the command line
```
wget https://github.com/wanglab-upenn/HAMR/archive/v1.2.tar.gz -O HAMR-v1.2.tar.gz
```
3. Extract HAMR source code from archive:
```
tar xzvf HAMR-v1.2.tar.gz
```

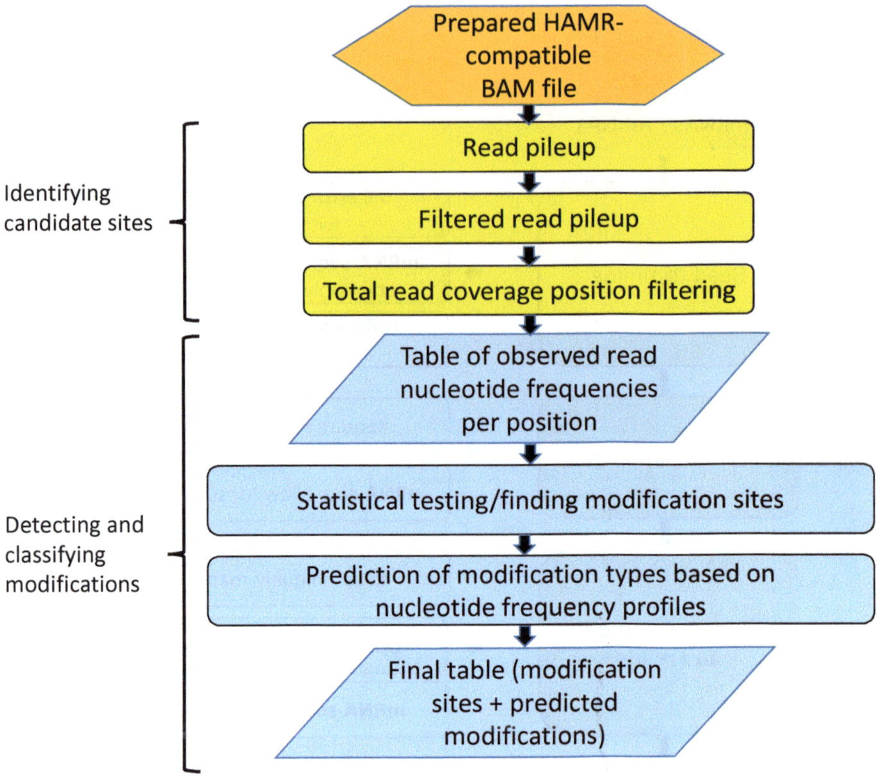

Fig. 2 Steps used by HAMR software [20] for in silico detection and classification of RNA modifications

4. Go to the HAMR program directory:
```
cd HAMR-1.2/
```
5. Compile HAMR source code (*see* **Note 4**):
```
Make clean
Make
```

3.3 Preparing the Reference Genome

1. Download a reference genomic sequence. Human genome data as well as genomes of other organisms can be obtained from the UCSC Genome Browser (http://hgdownload.soe.ucsc.edu/downloads.html). Users can also go to the HAMR website (http://lisanwanglab.org/hamr/genomes/) to get the genome FASTA files for human (hg19), mouse (mm9), *D. melanogaster* (dm3), *C. elegans* (ce6), yeast (sacCer3), and *A. thaliana* (TAIR10). Plant genomes, such as Arabidopsis, can be downloaded from the Ensembl Plant project (*see* **Note 5**). For instance, to download human reference genome:
```
cd <HAMRdirectory>
wget http://lisanwanglab.org/hamr/genomes/hg19_all_chr.fas
```
2. Index genome FASTA file (*see* **Note 6**):
```
samtools faidx hg19_all_chr.fas
```

3.4 Preparing RNA-Seq Data for HAMR

When starting from raw sequenced reads, the user needs to use a Linux-capable computer to execute the following steps (*see* Fig. 1) to prepare an input BAM file for HAMR analysis (*see* Subheading 3.8 for an example of HAMR analysis starting from raw sequenced data). All sorting is done with respect to chromosomal coordinates, not read name.

1. Read trimming

   Use Cutadapt software [25] to trim the adaptor sequences from sequencing reads (*see* Subheading 3.8 for an example of how to run cutadapt).

   (a) We suggest using the following trimming parameters: 6 % error rate for adapter sequence with at least 6 nt overlap, and the read length is at least 15 nt after trimming.

   (b) For smRNA-seq: The reads are 3′ adapter-trimmed, with at least 6 bp of adapter sequence having at most a 6 % mismatch rate is required. All untrimmed reads and trimmed reads shorter than 14 bp should be discarded (*see* **Note 7**). Only trimmed reads should be used.

   (c) For mRNA-seq: Use all reads (both trimmed and untrimmed) that are at least 15 nt.

2. Read mapping

   (a) Use STAR [26] for mapping reads to the genome (*see* Subheading 3.8 on an example of mapping reads using STAR aligner). The reads should be mapped allowing for mismatches between sequencing reads and the reference genomic sequence (*see* Subheading 3.8 on HAMR analysis of RNA modification in the human brain for an example of mapping with STAR).

   (b) For both smRNAs and mRNAs, map reads with mismatches. For shorter reads (e.g., smRNA-seq) X = 2 mismatches are suggested. For longer reads (e.g., mRNA-seq), setting the number of mismatches to <=5 % of mapped read length is suggested (*see* **Note 8**).

   (c) Allow for soft-clipping (*see* **Note 9**).

   (d) For mRNA only, filter out or resolve the spliced alignments (*see* **Note 10**).

   (e) Sort and index the resulting BAM file (*see* **Note 11**).

3. Handling multimapped reads (optional).

We strongly suggest using only uniquely mapped reads for general purposes and for analyzing longer RNAs such as mRNAs. For example, when using STAR for mapping, the multimapped reads can be filtered out with—outFilterMultimapNmax1 option (*see* Subheading 3.8). For analysis of highly repetitive classes of RNA (e.g., tRNA), additional steps need to be performed to assign multimapped reads to the repeat family before running HAMR. Resolving multimapped reads in case of tRNAs is illustrated below:

(a) Take the nuclear tRNA annotations from the "tRNAs" table in the UCSC genome browser (e.g., GRCh37) (*see* **Note 12**).

(b) Generate the annotations for mitochondrial tRNAs by running tRNAscan-SE (*see* **Note 13**). Set to organelle mode on the mitochondrial genome ("chrM" in GRCh37/hg19).

(c) Discard reads aligning to repeat regions or annotated RNAs other than tRNAs. This step will keep only the reads mapped exclusively to the target tRNAs. Discard reads mapped to more than one tRNA iso-acceptor family.

(d) Optional: filter out alignments with mismatches overlapping with single nucleotide polymorphisms (SNPs) based on dbSNP for human (*see* **Note 14**). This will prioritize alignments whose mismatches had no apparent explanation (i.e., likely caused by RNA modification) over mismatches caused by SNPs.

(e) Sort and index the output BAM file (*see* **Note 11**).

3.5 Running HAMR Pipeline

1. Description:

The HAMR program [20] takes as input a BAM file consisting of the sequencing reads aligned to the reference genome allowing for mismatches between reads and reference genomic sequences (*see* Subheading **3.4**). HAMR first produces a table of observed reference and non-reference (mismatched) read nucleotide frequencies per candidate genomic site. Only high-quality (e.g., Q score of at least 30) read positions are considered when tabulating mismatches. The resulting candidate set of genomic sites are strictly filtered for sequencing errors using a binomial test that assumes a sequencing error-rate tenfold higher than the error-rate corresponding to the minimum quality score. The sites that pass this filter are enriched for mismatches and consist of potential SNPs, RNA editing events, and RNA modification sites. To filter out sites corresponding to homozygous or heterozygous SNPs and RNA editing events, HAMR performs a series of binomial tests using an ensemble of biallelic null hypotheses corresponding to all possible genotypes (AA, AC, AG, AT, CC, etc.). Sites that cannot be explained by any of the homozygous and biallelic genotypes are then inferred to be modifications. These inferred modification sites will be classified as particular modification types based on machine learning models built from the observed mismatch patterns at known tRNA modification sites, as annotated in the MODOMICS database [20, 27, 28] (*see* **Note 15**).

Here, we describe how to use HAMR for processing mapped BAM files into predictions of RNA modification positions and identity (*see* Fig. 2) (if starting from raw sequencing reads, *see*

Subheading 3.4 on read preprocessing steps). The starting point for HAMR analysis is the mapped BAM. For the following steps, we assume that sequenced reads have been aligned, filtered and are stored in BAM format (*see* Subheading 3.4). All HAMR programs are operated from the command line. Using a command-line interface makes it easy to reproduce results and keep track of the analyses performed. Note that variable names within HAMR commands enclosed in <angle brackets> are placeholders for file names and analysis parameters and should be replaced with appropriate names without angle brackets.

HAMR analysis can be performed genome-wide (default) or for specific genomic regions (targeted mode). The only difference between the two modes is that user needs to prepare a BED file (BED6 format, also *see* **Note 16**) containing genomic regions of interest in BED format for the targeted mode.

Besides preparing a BAM file <prepared_sorted.bam> to run HAMR, the user needs to provide a reference genome file <genome.fa> in FASTA format (*see* Subheading 3.3 on how to download and prepare a reference genome sequence).

2 HAMR usage:

```
python hamr.py [-h] [--target_bed [TARGET_
BED]] [--paired_ends] [--filter_ends]
<bam> <genome_fas> <prediction_training_set>
<output_folder> <output_prefix> <min_qual>
<min_cov> <seq_err> <hypothesis> <max_p>
<max_fdr> <refpercent>
```

Required parameters:

(a) <bam> = prepared sorted BAM file (*see* Subheading 3.4 on preparing BAM file from raw sequenced read data)

(b) <genome_fas> = indexed <genome.fa>(FASTA file)

(c) <prediction_training_set> = data used for training HAMR

(d) <output_folder> = output directory

(e) <output_prefix> = output directory prefix name, contains the following outputs:

- <output_prefix>.mods.bed= BED file contains detected modification sites
- <output_prefix>.mods.txt = Text file contains detected modification sites
- <output_prefix>.raw.txt =full table contains the candidate sites for statistical test

(f) <min_qual> = Minimum base calling quality, suggested 30

(g) <min_cov> = Minimum number of non-reference nucleotide counts, suggested 10 (*see* **Note 17**)

(h) <seq_err> = expected sequencing error rate / mismatch rate, suggested 0.01–0.05 (*see* **Note 18**)

(i) <hypothesis> = Statistical test to be performed, H4 is recommended (*see* **Note 19**)

(j) <max_p> = Maximum p-value (0.01 is recommended)

(k) <max_fdr> = Maximum FDR with replacement

(l) <refpercent>= Minimum percentage of reads with reference nucleotide

<u>Optional parameters:</u>

(m) [-h] = prints help screen

(n) [-target_bed [TARGET_BED]] = targeted BED file containing genomic regions of interest (*see* **Note 16**)

(o) [--paired_ends] = set to 1 for paired-end sequencing data

(p) [--filter_ends] = set to 1 to filter out the read ends from read-pileups

3. Example for running HAMR genome-wide mode:

```
python hamr.py trial.human.bam hg19_all_chr.
fas models/euk_trna_mods.Rdata HAMR_out brain
30 10 0.05 H4 0.01 0.05 0.05 --filter_ends
Sample output HAMR_out/brain.mods.txt
```

```
chr    bp         strand refnuc A   C   G    T    nonref ref  h1.p      h1.padj   h4.p      h4.padj   sig  pred.mod
chr1   159111465  -      G      31  43  485  126  200    485  1.70e-93  1.55e-91  7.89e-10  8.22e-08  TRUE m1G
*chr1  161510088  +      A      29  1   16   54   71     29   1.21e-68  7.31e-67  9.43e-06  0.001     TRUE m1A|m1I|ms2i6A
chr1   222638403  +      A      78  1   114  418  533    78   0         0         2.56e-14  3.11e-12  TRUE m1A|m1I|ms2i6A
```

4. Example for running HAMR in targeted mode for specific genomic regions:

```
python hamr.py trial.human.bam hg19_all_chr.
fas models/euk_trna_mods.Rdata HAMR_out brain
30 10 0.05 H4 0.01 0.05 0.05 --filter_ends
```

3.6 Running HAMR Step-by-Step

Besides running HAMR in a genome-wide or target-bed batch mode, the HAMR pipeline (*see* Fig. 2) can also be run step-by-step.

3.6.1 HAMR-Step-1: RNA Pileup

1. Description: this step produces per-position read pileup for a given BAM file containing mapped reads.
2. Usage:

```
./rnapileup <align.bam> <genome.fa> >
<pileup.raw>
```

<align.bam>: sorted BAM file (*see* Subheading **3.4**) containing uniquely-mapped reads and continuous, un-spliced alignments

<genome.fa>: indexed reference genome (FASTA file)

<pileup.raw>: output file in pileup format

3. Example:
   `./rnapileup trial.human.bam hg19_all_chr.fas > trial.human.bam.pileup`
4. Sample RNA pileup output:

```
chr1    161510084   G   108   ..........................................................................................................^~.
chr1    161510085   T   109   ...................................................................................A.......................^~.
chr1    161510086   T   109   ............................................................................................................
chr1    161510087   C   109   ....................................................................................................A.......
chr1    161510088   G   109   ............................................................................................................
*chr1   161510089   A   109   .TTTTGT...TT..TTT.TTTT.TTTTGGGTT..TTTT.TTGTTG.T.TTTTTTG.GTTTGGT..GTTTTTTTTTGT.TGT..T..TTG.T..TGT.GTG......C..
chr1    161510090   T   109   ............................................................................................................
chr1    161510091   C   109   ..........A.................................................................................................
chr1    161510092   C   109   ............................................................................................................
chr1    161510093   C   109   .A..........................................................................................................
chr1    161510094   A   110   ............................................................................................................^~.
chr1    161510095   C   110   ..............................................................................................................
```

This output is in a pileup format (see http://samtools.sourceforge.net/pileup.shtml for details) and contains chromosome, physical position, reference nucleotide, the number of reads covering the site, read nucleotides, base qualities. In the read nucleotide column:

^~	indicates 5′ (start) of the read segment
.	indicates a match between the read nucleotide and reference nucleotide on the forward strand
,	indicates a match between the read nucleotide and reference nucleotide on the reverse strand
[ACGT]	indicates a mismatch (non-reference read nucleotide) on the forward strand
[acgt]	indicates a mismatch (non-reference nucleotide) on the reverse strand

3.6.2 HAMR-Step-2: Filter Pileup

1. Description: this step removes low-quality reads (i.e., reads with low base calling quality) from pileups and (optionally) filters reads whose 5′ or 3′ ends coincide with the predicted pileup position.
2. Usage:
   `./filter_pileup <pileup.raw> <minQ> <filter_ends=0|1> > <pileup.filtered>`
   This command reflects the following required parameters:

<pileup.raw>:	raw pileup file from **step 1**
<minQ>:	minimum read quality (30)
<filter_ends=0|1>:	read end filtering option

3. Example:
   `./filter_pileup trial.human.bam.pileup 30 1 > trial.human.bam.pileup.filtered.`
   Output of coverage <pileup.filtered> is of the same format as that of <pileup.raw>
4. Sample output after filtering RNA pileup:

```
chr1    161510082    T    97     .................................................................................................
chr1    161510083    G    99     ...................................................................................................
chr1    161510084    G    97     .................................................................................................
chr1    161510085    T    102    ......................................................................................................
chr1    161510086    T    104    ........................................................................................................
chr1    161510087    C    101    .....................................................................................................
chr1    161510088    G    104    ........................................................................................................
*chr1   161510089    A    100    .TTGT...TT.TTT.TTT.TTTTGGGTT..TTT.TTGTTG.T.TTTTTTG.GTTTGGT.GTTTTTTTTTGT.TGT..T..TTG.T..TGTGT.....C..
chr1    161510090    T    106    ..........................................................................................................
chr1    161510091    C    100    ....................................................................................................
chr1    161510092    C    102    ......................................................................................................
chr1    161510093    C    101    .....................................................................................................
chr1    161510094    A    99     ...................................................................................................
chr1    161510095    C    103    .......................................................................................................
```

3.6.3 HAMR-Step-3: Filter Out Pileups with Below Minimum Coverage

1. Description: this step filters out pileups with below minimum coverage.
2. Usage:
   `awk '$4>=<min_cov>' <pileup.filtered>><pileup.filtered.min_cov>`
   This command reflects the following required parameters:
   <min_cov>: minimum coverage
3. Example:
   `awk '$4>=10' trial.human.bam.pileup.filtered >trial.human.bam.pileup.filtered.min10`
4. Sample output of pileups filtered by read coverage:

```
chr1    161510084    G    97     .................................................................................................
chr1    161510085    T    102    ......................................................................................................
chr1    161510086    T    104    ........................................................................................................
chr1    161510087    C    101    .....................................................................................................
chr1    161510088    G    104    ........................................................................................................
*chr1   161510089    A    100    .TTGT...TT.TTT.TTT.TTTTGGGTT..TTT.TTGTTG.T.TTTTTTG.GTTTGGT.GTTTTTTTTTGT.TGT..T..TTG.T..TGTGT.....C..
chr1    161510090    T    106    ..........................................................................................................
chr1    161510091    C    100    ....................................................................................................
chr1    161510092    C    102    ......................................................................................................
chr1    161510093    C    101    .....................................................................................................
```

3.6.4 HAMR-Step-4: Convert Filtered Pileups to BED Format

1. Description: this step converts filtered pileups to BED format.
2. Usage:
   `./rnapileup2mismatchbed <filtered.min_cov> > <nuc2nuc.BED>`
3. Example:
   `./rnapileup2mismatchbed trial.human.bam.pileup.filtered.min10 > trial.human.bam.mismatch.bed`
4. Output is sorted by genomic coordinates.
5. The output BED file contains the number of observed reference nucleotides to read nucleotide transitions per candidate site.
6. Sample read nucleotide mismatch BED output

```
chr1    161510082    161510083    G>.    99;2:1,5:7,7:2,11:20,12:33,13:19,14:5,15:7,16:1,17:1,18:1,19:1,20:1    +
chr1    161510083    161510084    G>.    97;3:1,6:7,8:1,12:19,13:32,14:19,15:6,16:8,17:1,18:1,19:1,20:1 +
chr1    161510084    161510085    T>.    102;1:1,4:1,7:7,9:2,13:19,14:33,15:21,16:6,17:7,18:1,19:1,20:1,21:1,22:1    +
chr1    161510085    161510086    T>.    104;1:1,2:1,5:1,8:7,10:2,14:20,15:33,16:20,17:6,18:9,19:1,20:1,22:1,23:1    +
chr1    161510086    161510087    C>.    101;2:1,6:1,9:5,11:2,15:20,16:33,17:19,18:6,19:8,20:1,21:1,22:1,23:1,24:2    +
chr1    161510087    161510088    G>.    104;3:1,4:1,7:1,10:6,12:2,16:20,17:33,18:20,19:6,20:9,21:1,22:1,23:1,24:1,25:1 +
*chr1   161510088    161510089    A>.    29;4:1,5:1,11:5,17:8,18:4,19:4,20:1,21:4,26:1    +
*chr1   161510088    161510089    A>C    1;8:1   +
*chr1   161510088    161510089    A>G    16;13:1,17:3,18:7,19:4,22:1     +
*chr1   161510088    161510089    A>T    54;11:1,17:9,18:21,19:12,20:4,21:5,23:1,25:1    +
chr1    161510089    161510090    T>.    106;5:1,6:1,9:1,12:7,14:2,18:20,19:33,20:20,21:6,22:9,23:1,24:1,25:1,26:1,27:2 +
chr1    161510090    161510091    C>.    100;6:1,7:1,10:2,13:6,15:2,19:19,20:31,21:19,22:6,23:8,24:1,25:1,26:1,27:1,28:1 +
chr1    161510091    161510092    C>.    102;7:1,8:1,11:2,14:6,16:2,20:20,21:31,22:19,23:6,24:9,25:1,26:1,27:1,28:1,29:1 +
chr1    161510092    161510093    C>.    101;8:1,12:2,15:6,17:2,21:20,22:30,23:19,24:6,25:10,26:1,27:1,28:1,29:1,30:1    +
chr1    161510093    161510094    A>.    99;9:1,13:2,16:5,18:2,22:19,23:31,24:18,25:6,26:10,27:1,28:1,29:1,30:1,31:1    +
chr1    161510094    161510095    C>.    103;1:1,10:1,11:1,14:2,17:6,19:2,23:19,24:31,25:20,26:6,27:10,28:1,29:1,31:1,32:1    +
chr1    161510095    161510096    C>.    99;2:1,11:1,12:1,15:1,18:6,20:1,24:18,25:31,26:18,27:6,28:10,29:1,30:1,31:1,32:1,33:1    +
chr1    161510096    161510097    C>.    99;3:1,12:1,13:1,16:1,19:6,21:1,25:18,26:31,27:18,28:6,29:10,30:1,31:1,32:1,33:1,34:1    +
```

This output contains the following information:

Fourth column = type of mismatch (A>T, A>G, A>C, etc.) or match (A>., T>., C>., G>.) between the reference and read nucleotide

Fifth column = first number is the total number of reads with match or mismatch event, followed by the detailed list of positions within the reads where the event is observed

3.6.5 HAMR-Step-5: Convert BED to HAMR-FREQ- TABLE

1. Description: this step converts BED to HAMR-FREQ-TABLE while keeping only positions with >0 non-reference nucleotides, and filters the nucleotide frequency table by removing SITES (rows) with insufficient sequencing-based information.
2. Usage:
   ```
   mismatchbed2table <nuc2nuc.BED> > <HAMR-FREQ-TABLE>
   ```
3. Example:
   ```
   bash mismatchbed2table.sh trial.human.bam.mismatch.bed > trial.human.bam.freq.table
   ```
4. Sample output of read nucleotide frequencies:

chr	bp	strand	refnuc	A	C	G	T	nonref
chr1	161450359	+	A	36	1	0	0	1
chr1	161450360	+	G	0	0	180	17	17
chr1	161493634	-	T	0	3	0	7	3
chr1	161493635	-	C	0	9	0	1	1
chr1	161493641	-	A	14	1	0	0	1
chr1	161493651	-	A	5	0	0	7	7
chr1	161500129	-	T	0	9	0	2	9
chr1	161500130	-	A	2	8	0	0	8
chr1	161500146	-	A	15	1	0	0	1
chr1	161510080	+	G	1	0	98	0	1
*chr1	161510088	+	A	29	1	16	54	71
chr1	169076221	+	T	0	17	0	1	17
chr1	169101940	+	A	26	1	0	0	1
chr1	171570078	-	T	0	5	0	39	5
chr1	171966723	-	C	0	0	268	0	268

5. Additional filtering of sites based on observed read nucleotide counts is imposed, including:
   minimum number of reads with non-reference nucleotide [=10]
   minimum number of reference reads [=10]
   minimum ratio of non-reference/reference[=1 %]
6. Usage:
   ```
   awk '{cov=$5+$6+$7+$8;nonref=$9; ref.=cov-
   nonref; if (ref/cov>=+<refpercent>+)
   print;}''<HAMR-FREQ-TABLE>> <HAMR-FREQ-
   TABLE-FINAL>
   ```
   This command reflects the following required parameters:
   <refpercent> = Minimum percentage of reads with reference nucleotide
7. Example:
   ```
   awk '{cov=$5+$6+$7+$8;nonref=$9; ref=cov-
   nonref; if  (ref/cov>=0.05) print;}' trial.
   human.bam.freq.table > trial.human.bam.freq.
   final.table
   ```
8. Output table contains the HAMR frequency information on the FINALIZED set of candidate sites with enough sequencing support.

3.6.6 HAMR-Step-6: Finding Modification Sites Based on Statistical Testing

1. Description: this step finds modification sites based on significant deviations of the observed nucleotide frequencies from the expected homozygous or biallelic distributions.
2. Usage:
   ```
   Rscript detect_mods.R <HAMR-NUC-TABLE>
   <EXPECTED-SEQ-ERROR-RATE> <HYPOTHESIS-TYPE>
   <P-VALUE-THRESHOLD> <Q-VALUE-THRESHOLD> >
   <RAW-MOD-TABLE>
   ```
   This command reflects the following required parameters:
   <HAMR-NUC-TABLE> = HAMR frequency table from **step 5**
   <EXPECTED-SEQ-ERROR-RATE> = expected sequencing error rate / mismatch rate, suggested 0.01–0.05 (*see* **Note 18**)
   <HYPOTHESIS-TYPE> = Statistical test to be performed, H4 is recommended (*see* **Note 19**)
   <P-VALUE-THRESHOLD> = Maximum p-value (0.01 is recommended)
   <Q-VALUE-THRESHOLD> = Maximum FDR with replacement
3. Example:
   ```
   Rscript detect_mods.R trial.human.bam.freq.final.
   table 0.05 H4 0.01 0.05 > trial.human.bam.raw.txt
   ```

4. Sample output of HAMR predicted modification sites:

chr	bp	strand	refnuc	A	C	G	T	nonref	ref	h1.p	h1.padj	h4.p	h4.padj	sig
chr1	161500129	-	T	0	9	0	2	9	2	9.79e-11	2.38e-09	1	1	FALSE
chr1	161500130	-	A	2	8	0	0	8	2	1.60e-09	3.77e-08	1	1	FALSE
chr1	161500146	-	A	15	1	0	0	1	15	0.56	1	1	1	FALSE
chr1	161510080	+	G	1	0	98	0	1	98	0.99	1	1	1	FALSE
*chr1	161510088	+	A	29	1	16	54	71	29	1.21e-68	7.37e-67	9.43e-06	0.001	TRUE
chr1	169076221	+	T	0	17	0	1	17	1	1.31e-21	5.30e-20	1	1	FALSE
chr1	169101940	+	A	26	1	0	0	1	26	0.75	1	1	1	FALSE
chr1	171570078	-	T	0	5	0	39	5	39	0.07	1	1	1	FALSE
chr1	172108004	+	A	22	0	1	0	1	22	0.69	1	1	1	FALSE
chr1	172108013	+	A	23	0	1	0	1	23	0.71	1	1	1	FALSE

3.6.7 HAMR-Step-7—Predicting modification identity

1. Description: predict the identity of modifications for each of the modification sites (*see* **Note 20**).
2. Usage:
   ```
   Rscript classify_mods.R <TABLE-significant.sites.only> <training.set.Rdata> > <TABLE-with-predicted-mod-types>
   ```
   This command reflects the following required parameters:
   <TABLE-significant.sites.only> = Table contains detected modification sites that are statistically significant
   <training.set.Rdata> = prediction training set (e.g., the training set models/euk_trna_mods.Rdata provided in HAMR that consists of mismatch patterns observed in confirmed modification site in tRNAs)
3. Example:
   ```
   Rscript classify_mods.R trial.human.bam.raw.txt models/euk_trna_mods.Rdata > trial.human.bam.mods.txt and CHANGE font to match other command lines
   ```
4. Sample output of HAMR modification sites and predicted modification types:

chr	bp	strand	refnuc	A	C	G	T	nonref	ref	h1.p	h1.padj	h4.p	h4.padj	sig	pred.mod
chr1	159111465	-	G	31	43	485	126	200	485	1.70e-93	1.55e-91	7.89e-10	8.22e-08	TRUE	m1G
*chr1	161510088	+	A	29	1	16	54	71	29	1.21e-68	7.31e-67	9.43e-06	0.001	TRUE	m1A|m1I|ms2i6A
chr1	222638403	+	A	78	1	114	418	533	78	0	0	2.56e-14	3.11e-12	TRUE	m1A|m1I|ms2i6A

3.7 Using HAMR Webserver

HAMR is also available as a web-based application, which can be accessed at http://lisanwanglab.org/hamr/.

The web interface allows specification of a remote, indexed BAM file and BED file with targeted intervals for querying. The user may specify parameters for the preprocessing steps, such as minimum base call quality score, minimum read coverage at a site, assumed sequencing error rate, and significance level. Additionally, the user may use the webserver to predict the modification type based on mismatch patterns.

3.8 Application of HAMR to Existing Human Small RNA-Seq Data

1. Here, we present the application of HAMR to a Illumina small RNA-seq data generated from human brain tissue (GSE48552, http://www.ncbi.nlm.nih.gov/geo/query/acc.cgi?acc=GSE48552). With a few preprocessing steps prior to

HAMR, users can run a single command and all information about modification sites and types will be summarized in a table.

2. Download the SRA file from SRA database (`http://www.ncbi.nlm.nih.gov/geo/`):
```
wget ftp://ftp-trace.ncbi.nlm.nih.gov/sra/sra-instant/reads/ByStudy/sra/SRP/SRP026/SRP026562/SRR1103937/SRR1103937.sra -O brainpfc.sra
```
3. Build genomic index:
```
STAR--runMode genomeGenerate --runThreadN 8 --genomeDir genomes/star/hg19 --genomeFastaFiles hg19_all_chr.fas
```
4. Convert SRA to FASTQ file using SRAtoolkit:
```
fastq-dump --split-3 brainpfc.sra
```
5. Trim adapter sequences from reads (Illumina v1.5 small RNA 3′ adapter in this case):
```
cutadapt brainpfc.fastq -a ATCTCGTATGCCGTCTTCTGCTTG -e 0.06 -O 6 -m 14 --too-short-output=brainpfc.tooshort.fastq --untrimmed-output=brainpfc.untrimmed.fastq --length-tag="length=" -o brainpfc.trimmed.fastq
```
6. Map trimmed reads to the human reference genome hg19 using STAR:
```
mkdir -p STAR_brainpfc
STAR --genomeDir genomes/star/hg19 --genomeLoad LoadAndKeep --readFilesIn brainpfc.trimmed.fastq --runThreadN 4 --alignIntronMax 1 --outSAMattributes NH HI NM MD --outFilterMultimapNmax 1 --outReadsUnmapped Fastx --outFilterMismatchNmax 2 --outFilterMatchNmin 15 --outFileNamePrefix STAR_brainpfc/
```
7. Sort and index output BAM:
```
samtools sort -@ 8 -O bam -T Aligned.out.sort STAR_brainpfc/Aligned.out.sam > STAR_brainpfc/Aligned.out.sorted.bam
samtools index STAR_brainpfc/Aligned.out.sorted.bam
```
8. Perform modification analysis with HAMR:
```
python hamr.py STAR_brainpfc/Aligned.out.sorted.bam genomes/hg19.fa models/euk_trna_mods.Rdata HAMR_results brainpfc 30 10 0.05 H4 0.01 0.05 0.05 --filter_ends
```
HAMR_results/brainpfc.mods.txt will contain a list of RNA modifications found by HAMR.

4 Notes

1. These programs are readily available in popular Linux distributions such as Ubuntu, CentOS, or Debian.
2. You can check the version of python by typing python --version on the command line.
3. The simplest way to install programs from source is to build them in your own directory. Download the appropriate package from the source websites (*see* Subheading 2.2 on software requirements). Extract the files and check the readme files for installation instructions and a list of required dependencies.
4. HAMR is packaged as compiled "binaries" that can be run by typing in the program's filepath. On certain systems, however, the HAMR needs to be recompiled, to do so, please use make clean, then make.
5. Arabidopsis genome FASTA file can be downloaded here: ftp://ftp.ensemblgenomes.org/pub/release-25/plants/fasta/arabidopsis_thaliana/dna/Arabidopsis_thaliana.TAIR10.25.dna.genome.fa.gz
6. Make sure to index the FASTA file with genome index before running HAMR for faster execution.
7. Untrimmed reads are likely to correspond to larger RNA fragments that are outside the small RNA population.
8. If no mismatches are allowed, HAMR is not capable of detecting any modifications.
9. Soft clipping routinely allows for mapping reads with untrimmed partial adapters at 3′ end (e.g., shorter than the minimum overlap length), containing partial exon-exon junctions, etc.
10. The HAMR program handles continuous read alignments, and spliced reads require additional preprocessing.
11. BAM files need to be sorted by genomic chromosome and start positions, not by the read name.
12. The tRNAs table in the UCSC genome browser (hg19) can be downloaded here: http://genome.ucsc.edu/cgi-bin/hgTables?hgsid=482380859_VTSVW7eMgPDQJ0DdDzlKALAe02bM&clade=mammal&org=Human&db=hg19&hgta_group=genes&hgta_track=tRNAs&hgta_table=0&hgta_regionType=genome&position=chr1%3A1103243-1103332&hgta_outputType=primaryTable&hgta_outFileName=.
13. tRNAscan-SE can be downloaded and installed from http://gtrnadb.ucsc.edu/.

14. dbSNP can be downloaded from http://www.ncbi.nlm.nih.gov/projects/SNP/.
15. tRNA modification information from MODOMICS is chosen as this is by far the best documented list of RNA modification sites. Data from MODOMICS is downloadable at http://modomics.genesilico.pl/downloads/.
16. See https://genome.ucsc.edu/FAQ/FAQformat.html#format1 for a description of the BED file format.
17. Filter out genomic positions with insufficient number (e.g., less than 10) of non-reference nucleotide counts. Such positions do not provide sufficient evidence for HAMR to reliably establish mismatch pattern necessary to detect RNA modifications.
18. Filter out genomic positions with non-reference / reference read count ratio that less than, e.g., 1 %. Such positions are likely to be attributed to sequencing errors, not modification-related mismatches.
19. HAMR tests for homozygous, bi-allelic, and multi-allelic loci, *see* [20] for more information on statistical testing and hypotheses.
20. For a list of HAMR-detectable modification types, refer to Supplementary table 1 in [20].

Acknowledgments

This work is supported by the National Institute of General Medical Sciences [R01-GM099962 to P.P.K, Y.Y.L, B.D.G., and L.S.W], National Institute on Aging [U24-AG041689 to L.S.W.], National Science Foundation [CAREER Award MCB-1053846, MCB-1243947, and IOS-1444490 to B.D.G.]. We thank Alexandre Amlie-Wolf and other members of the Wang and Gregory labs for their comments and help with this work.

References

1. Dominissini D, Nachtergaele S, Moshitch-Moshkovitz S et al (2016) The dynamic N(1)-methyladenosine methylome in eukaryotic messenger RNA. Nature 530:441–446
2. Eigenbrod T, Keller P, Kaiser S et al (2015) Recognition of specified RNA modifications by the Innate immune system. Methods Enzymol 560:73–89
3. Li S, Mason CE (2014) The pivotal regulatory landscape of RNA modifications. Annu Rev Genomics Hum Genet 15:127–150
4. Lee M, Kim B, Kim VN (2014) Emerging roles of RNA modification: m(6)A and U-tail. Cell 158:980–987
5. Satterlee JS, Basanta-Sanchez M, Blanco S et al (2014) Novel RNA modifications in the nervous system: form and function. J Neurosci 34:15170–15177
6. Delatte B, Wang F, Ngoc LV et al (2016) Transcriptome-wide distribution and function of RNA hydroxymethylcytosine. Science 351:282–285

7. Sundaram M, Durant PC, Davis DR (2000) Hypermodified nucleosides in the anticodon of tRNALys stabilize a canonical Uturn structure. Biochemistry 39:12575–12584
8. Kierzek E, Malgowska M, Lisowiec J, Turner DH, Gdaniec Z, Kierzek R (2014) The contribution of pseudouridine to stabilities and structure of RNAs. Nucleic Acids Res 42:3492–3501
9. Schwartz S, Mumbach MR, Jovanovic M et al (2014) Perturbation of m6A writers reveals two distinct classes of mRNA methylation at internal and 5′ sites. Cell Rep 8:284–296
10. Meyer KD, Jaffrey SR (2014) The dynamic epitranscriptome:N6-methyladenosine and gene expression control. Nat Rev Mol Cell Biol 15:313–326
11. Karijolich J, Yu YT (2015) The new era of RNA modification. RNA 21:659–660
12. Sun WJ, Li JH, Liu S et al (2016) RMBase: a resource for decoding the landscape of RNA modifications from high-throughput sequencing data. Nucleic Acids Res 44:D259–D265
13. Schwartz S, Bernstein DA, Mumbach MR, Jovanovic M, Herbst RH, León-Ricardo BX, Engreitz JM, Guttman M, Satija R, Lander ES, Fink G, Regev A (2014b) Transcriptome-wide mapping reveals widespread dynamic-regulated pseudouridylation of ncRNA and mRNA. Cell 159:148–162
14. Vandivier LE, Campos R, Kuksa PP et al (2015) Chemical modifications mark alternatively spliced and uncapped messenger RNAs in arabidopsis. Plant Cell 27:3024–3037
15. Meyer KD, Saletore Y, Zumbo P (2012) Comprehensive analysis of mRNA methylation reveals enrichment in 3′ UTRs and near stop codons. Cell 149:1635–1646
16. Gupta RC, Randerath K (1977) Use of specific endonuclease cleavage in RNA sequencing. Nucleic Acids Res 4:1957–1978
17. Woodson SA, Muller JG, Burrows CJ et al (1993) A primer extension assay for modification of guanine by Ni(II) complexes. Nucleic Acids Res 21:5524–5525
18. Motorin Y, Muller S, Behm-Ansmant I, Branlant C (2007) Identification of modified residues in RNAs by reverse transcription-based methods. Methods Enzymol 425:21–53
19. Behm-Ansmant I, Helm M, Motorin Y (2011) Use of specific chemical reagents for detection of modified nucleotides in RNA. JNucleic Acids 2011:408053
20. Ryvkin P, Leung YY, Silverman IM et al (2013) HAMR: high-throughput annotation of modified ribonucleotides. RNA 19: 1684–1692
21. Dominissini D, Moshitch-Moshkovitz S, Schwartz S et al (2012) Topology of the human and mouse m6A RNA methylomes revealed by m6A-seq. Nature 485:201–206
22. Squires JE, Patel HR, Nousch M et al (2012) Widespread occurrence of 5-methylcytosine in human coding and non-coding RNA. Nucleic Acids Res 40:5023–5033
23. Horowitz S, Horowitz A, Nilsen TW et al (1984) Mapping of N6-methyladenosine residues in bovine prolactin mRNA. Proc Natl Acad Sci U S A 81:5667–5671
24. Li H, Handsaker B, Wysoker A et al (2009) The sequence alignment/map format and SAMtools. Bioinformatics 25:2078–2079
25. Martin M (2011) Cutadapt removes adapter sequences from high-throughput sequencing reads. EMBnetJ 17(1):10–12
26. Dobin A, Davis CA, Schlesinger F et al (2013) STAR: ultrafast universal RNA-seq aligner. Bioinformatics 29:15–21
27. Machnicka MA, Milanowska K, Osman OO et al (2013) MODOMICS: a database of RNA modification pathways: 2012 update. Nucleic Acids Res 41:D262–D267
28. Leung YY, Ryvkin P, Ungar LH et al (2013) CoRAL: predicting non-coding RNAs from small RNA-sequencing data. Nucleic Acids Res 41:e137

Chapter 15

High-Throughput Small RNA Sequencing Enhanced by AlkB-Facilitated RNA de-Methylation (ARM-Seq)

Eva Hrabeta-Robinson*, Erin Marcus*, Aaron E. Cozen, Eric M. Phizicky, and Todd M. Lowe

Abstract

N[1]-methyladenosine (m^1A), *N*[3]-methylcytidine (m^3C), and *N*[1]-methylguanosine (m^1G) are common in transfer RNA (tRNA) and tRNA-derived fragments. These modifications alter Watson-Crick base-pairing, and cause pauses or stops during reverse transcription required for most high-throughput RNA sequencing protocols, resulting in inefficient detection of methyl-modified RNAs. Here, we describe a procedure to demethylate RNAs containing m^1A, m^3C, or m^1G using the *Escherichia coli* dealkylating enzyme AlkB, along with instructions for subsequent processing with widely used protocols for small RNA sequencing.

Key words RNA Sequencing, Transfer RNA (tRNA), AlkB, RNA demethylation, *N*[1]-methyladenosine (m^1A), *N*[3]-methylcytidine (m^3C), *N*[1]-methylguanosine (m^1G)

1 Introduction

High-throughput sequencing has become commonplace for characterizing small RNA populations, which can include microRNAs (miRNAs), small nucleolar RNAs (snoRNAs), and transfer RNAs (tRNAs), as well as partial fragments of tRNAs and ribosomal RNAs (rRNAs). tRNAs in particular are highly modified, often containing multiple nucleoside modifications that interfere with Watson-Crick base-pairing and are known to block reverse transcription during cDNA synthesis [1, 2]. Because most small RNA sequencing protocols are designed specifically to clone only full-length cDNAs rather than partial cDNAs derived from segments of larger RNAs, RNAs containing modifications that pause or block reverse transcription are typically underrepresented or absent in small RNA sequencing results.

---

* These authors contributed equally to this work.

Alexandra Lusser (ed.), *RNA Methylation: Methods and Protocols*, Methods in Molecular Biology, vol. 1562, DOI 10.1007/978-1-4939-6807-7_15, © Springer Science+Business Media LLC 2017

One approach to facilitate sequencing of modified RNAs is to enzymatically remove problematic modifications prior to sequencing library preparation. The *Escherichia coli* AlkB protein is a dealkylating enzyme that has been shown to remove methyl groups from m^1A, m^3C, and m^1G residues in RNA [3–5]. Purified AlkB enzyme was used previously to identify AlkB-sensitive m^3C residues in tRNAs in primer extension assays [6]. We subsequently adapted the approach for demethylation pretreatment to facilitate high-throughput sequencing of m^1A, m^3C, or m^1G modified small RNAs, revealing an abundance of methyl-modified tRNA-derived small RNAs [7]. Here, we describe procedures for overexpression and purification of the AlkB enzyme, small RNA isolation, and sequencing library preparation. The small RNA library preparation kit used here is NEBNext, produced by New England Biolabs. Similar results have been obtained using TruSeq small RNA kits from Illumina. Specific RNAs that are methylated can be subsequently identified by comparing sequencing results from control (buffer-treated) versus demethylated RNA using statistical methods for differential abundance analysis such as DESeq2 [8]. Our lab has implemented specialized tRNA comparative read analysis and visualization in a freely available software package (http://trna.ucsc.edu/ARM-seq/).

2 Materials

Use nuclease-free reagents for all buffers and decontaminate all equipment using RNA-Zap and 70 % ethanol prior to starting any step. Use nuclease-free low-binding microcentrifuge tubes to maximize yield at each step. Cloning of the *E. coli* AlkB gene into the AVA421 plasmid was described by Quartley et al. [9]. *E. coli* BL21(DE3)pLysS competent cells (Promega) were transformed with AVA421 plasmid according to manufacturer's protocol (AVA421 contains ampicillin resistance, so transformed cells should be grown in media at standard 100 ug/mL ampicillin to retain the vector; any equivalent *E. coli* competent cells commonly used for protein expression may be used). Cells were flash frozen on dry ice and stored at −80 °C.

2.1 Purification of Recombinant AlkB

1. Frozen cell pellet from 12 L culture of AlkB-AVA421-transformed cells grown at 37 °C to OD^{600}=0.5 and induced with IPTG for 2 h.
2. Sonication buffer: 20 mM Hepes-KOH pH 7.5, 5 % glycerol, 1 M NaCl, 2 mM β-mercaptoethanol (BME), 2 μg/mL Leupeptin, 1 μg/mL Pepstatin, 1 mM Pefabloc.
3. TALON Metal Affinity Resin (Clontech Laboratories, Inc), 2 mL per liter of bacterial culture.

4. 0.5 M NaCl Wash Buffer: 20 mM Tris–HCl pH 8.0, 5 % glycerol, 2 mM BME, 0.5 M NaCl.
5. 0.8 M NaCl Wash Buffer: 20 mM Tris–HCl pH 8.0, 5 % glycerol, 2 mM BME, 800 mM NaCl.
6. No Salt Buffer: 20 mM Tris–HCl pH 8.0, 5 % glycerol, 2 mM BME.
7. E1 buffer: 5 mM imidazole dissolved in 0.5 M NaCl Wash Buffer.
8. E2 buffer: 10 mM imidazole dissolved in 0.5 M NaCl Wash Buffer.
9. E3 buffer: 250 mM imidazole.
10. Bradford assay components (Bio-Rad).
11. HRV 3C protease (Clontech), or equivalent 3C protease.
12. Dialysis Buffer: 20 mM Tris–HCl pH 8, 5 % glycerol, 200 mM NaCl, 2 mM BME.
13. HiLoad 16/60 Superdex 200 gel filtration column.
14. HiLoad Buffer: 20 mM Tris–HCl pH 8, 5 % glycerol, 500 mM NaCl, 2 mM DTT, 0.025 % NaN_3.
15. Dialysis cassette 10,000 MWCO (Thermo Scientific Pierce Slide-A-Lyzer cassette).
16. Storage buffer: 20 mM Tris–HCl pH 8.0, 50 % glycerol, 200 mM NaCl, 2 mM DTT.
17. 8–16 % SDS polyacrylamide gel (SDS-PAGE), Tris–HCl buffered, 18 well.
18. Glass chromatography column, 1.5 × 10 cm, max volume 18 mL (e.g., BioRad Econo-Columns, 1.5 × 10 cm).

2.2 RNA Isolation

1. TRI-Reagent (Ambion), used to harvest and freeze down mammalian cells for storage in −80 °C until RNA isolation.
2. Direct-Zol RNA MiniPrep Kit (Zymo Research) for RNA extraction, with Pre-wash buffer.
3. mirVana miRNA Isolation kit without phenol (Ambion) for selection of RNA <200 nt.
4. BioAnalyzer Small RNA kit (Agilent Technologies) for assessment of quality and sizes of RNA input into library synthesis.

2.3 DNase Treatment

1. DNaseI (NEB): 2000 U/mL, supplied with 10× reaction buffer.
2. 0.5 M EDTA, pH 8.0.
3. RNA Clean and Concentrator (Zymo Research) for RNA cleanup.

2.4 AlkB Demethylation

1. 2× common mix: 50 μL made directly before use by adding 10 μL of each of the following stock solutions on ice:

   500 mM Hepes-KOH, pH 8.0.

   750 mM Ferrous Ammonium Sulfate pH 5.0.

10 mM Alpha-ketoglutarate in 50 mM Hepes-KOH pH 8.0.

20 mM Na-Ascorbate (made fresh, and shielded from light).

500 μg/mL BSA.

2.5 cDNA Library Preparation, Size Selection, and Analysis

1. NEBNext Multiplex Small RNA Library Prep Set for Illumina (New England BioLabs) for preparation and indexing of cDNA libraries.
2. E-Gel SizeSelect™ Agarose Gels, 2 % (Invitrogen) for size selection of libraries.
3. E-Gel 50 bp DNA Ladder (Invitrogen) for guiding the size selection of libraries.
4. TE buffer: 10 mM Tris–HCL pH 8.0, 1 mM EDTA.
5. DNA Clean and Concentrator-5 kit (Zymo Research) for clean-up steps.
6. Agilent High Sensitivity DNA Kit (Agilent Technologies).

3 Methods

3.1 Purification of Recombinant AlkB

Carry out all parts of this protocol at 4 °C in a cold room. All centrifugation steps should be performed with a swing bucket rotor, except as noted in **step 1**. Equivalent RCF x $\mathcal{g}r_{av}$ values given.

1. Thaw cell pellet from cells expressing AlkB, resuspend in 240 mL sonication buffer (12 L cell culture × 20 mL buffer per liter of culture), sonicate eight times for 10 s with a probe sonicator (each time followed by 1 min on ice), and remove debris by centrifugation at 10,100 RCFr_{av} (e.g., we use 10,080 RCF in a Beckman Coulter JLA-16.250 fixed angle rotor). Carefully remove clarified supernatant crude extract ("CE") to a clean tube. It can be used immediately or flash frozen on dry ice for storage at −80 °C until needed. Discard cell debris pellet.
2. Prepare 24 mL TALON resin (2 mL resin per L of cell culture) by washing three times with 0.5 M NaCl Wash Buffer. Use 25 mL Wash Buffer per 2 mL resin, centrifuge at 700 RCF for 4 min (e.g., 2000 rpm in a Beckman Coulter JS-5.3 swing bucket rotor). Remove and discard supernatant after each centrifugation.
3. Add 240 mL of No Salt buffer to 240 mL of CE. Save 10 μL of diluted CE for SDS-PAGE.
4. Divide the diluted CE into 12 × 50 mL conical tubes (40 mL per tube), each containing 2 mL of prewashed resin.
5. Mix slowly for 1 h (e.g., on an orbital shaker) to allow his-tagged protein to bind resin.
6. Spin at 2787 RCF for 5 min (e.g., 4000 rpm in JS-5.3 swing bucket rotor.

7. Carefully remove as much supernatant as possible without disturbing resin pellet. Keep resin (with bound protein), but save 10 μL of flow-through ("FT") supernatant for SDS-PAGE. (Optionally, keep all of supernatant, temporarily, until protein recovery is confirmed at the end of isolation.)
8. Consolidate 12 tubes of resin into two tubes by adding 25 mL 0.5 M NaCl Wash Buffer in each of two tubes, and combing resins across six total tubes (leaving two final tubes with approximately 12 mL of combined resin and 25 mL of wash buffer per tube).
9. Mix gently for 10 min.
10. Spin at 2787 RCF for 5 min.
11. Remove and discard supernatant.
12. Repeat wash procedure three times: add 25 mL 0.5 M NaCl Wash Buffer to each of two tubes, mix for 10 min, spin at 2787 RCF for 5 min, and discard supernatant.
13. Add 25 mL of E1 buffer to the resin.
14. Mix gently for 10 min.
15. Spin at 2787 RCF for 5 min.
16. Remove and discard supernatant as Eluent 1 ("E1"). Optionally, keep 10 μL for SDS-PAGE analysis.
17. Add 15 mL of E2 buffer to each of 2 tubes, and mix gently.
18. Combine all of sample (E2 buffer and all resin from both tubes) by pouring slowly into one glass chromatography column.
19. Collect drip flowthrough into a 50 mL beaker (don't let the resin dry out) as Eluent 2 ("E2").
20. Add 10 mL more of E2 buffer and let drip into the same 50 mL beaker. Protein should still be bound to resin in column. (Optionally, keep all of E2 in beaker, temporarily, until protein recovery is confirmed at the end of isolation.)
21. To elute the bound protein, add 30 mL E3 buffer to the column (add buffer slowly, only 10 mL at a time).
22. Collect the first eight 3 mL fractions ("E3 fractions") in individual tubes.
23. Do a Bradford assay on all eight collected E3 fractions, and combine only fractions with most protein (generally two or three fractions, so final volume will be 6–9 mL of eluted protein).
24. Quick freeze samples collected for SDS-PAGE: 10 μL diluted CE, 10 μL FT, and 4 μL E3 (and 10 μL optional samples of E1 and E2) and store at −70°.
25. Calculate the amount of HRV 3C protease needed and add to pooled E3 protein fractions from **step 23** (1:75 by mass, e.g., 1 mg protease/75 mg protein).

26. Dialyze in dialysis cassette into 2 L of Dialysis Buffer overnight (*see* **Note 1**).
27. Next Day: harvest dialyzed protein supernatant ("DP") from dialysis cassette. If supernatant is cloudy from a small amount of protein that has precipitated during dialysis, remove precipitate by dividing DP sample into 2 mL microfuge tubes and spin for 1 min at 2787 RCF, and transfer supernatant (with protein) to a new tube, discarding pellet.
28. Take a sample of DP (4 μL) for SDS-PAGE analysis and determine protein concentration by Bradford assay.
29. Prepare fresh TALON resin by combining 1.5 mL resin slurry and 10 mL of 0.5 M NaCl Wash buffer in a 15 mL conical tube. Gently mix and spin at 2787 RCF for 5 min to pellet resin. Discard the supernatant and repeat the wash twice.
30. Dilute DP (should have 5–10 mL from **step 27**) with an equal volume of 0.8 M NaCl Wash Buffer (to bring NaCl back up to 0.5 M for TALON binding), and move to 50 mL conical tube.
31. Add 0.75 mL washed TALON resin to diluted DP (this step binds HRV 3C protease, and any other non-AlkB contaminant proteins to TALON resin).
32. Mix gently for 1 h.
33. Spin down the resin at 2787 RCF for 5 min.
34. Remove and save the protein supernatant ("PS"), discard resin pellet.
35. Do Bradford assay on PS, and keep 8 μL for SDS-PAGE.
36. Run SDS-PAGE on all samples previously saved for gel analysis (CE, FT, E3, DP, PS) for 1 h at 200 V and then stain with Coomassie blue.
37. Concentrate down the entire PS from **step 33** to 5 mL or less for sizing column. If the sample is cloudy, spin the sample for 5 min at 2787 RCF, and discard pellet, keeping supernatant, as in **step 27**. Perform Bradford assay to determine protein concentration. Save 20 μL of the concentrated sample ("CS") for SDS-PAGE analysis.
38. Load all of CS on the HiLoad 16/60 Superdex 200, 120 mL column in HiLoad Buffer (pre-equilibrated with 5 column volumes of HiLoad buffer), and collect fractions of 3 mL overnight.
39. Run SDS-PAGE analysis, as in **step 35**, of 20 μL aliquots of all sizing column fractions, combine peak fractions (about 3–5 fractions), and perform Bradford assay of the pooled sample.
40. Concentrate down to appropriate volume (0.5–3 mL) for dialysis cassette. Save 10 μL sample for SDS-PAGE.
41. Dialyze sample overnight into 2 L Storage Buffer.

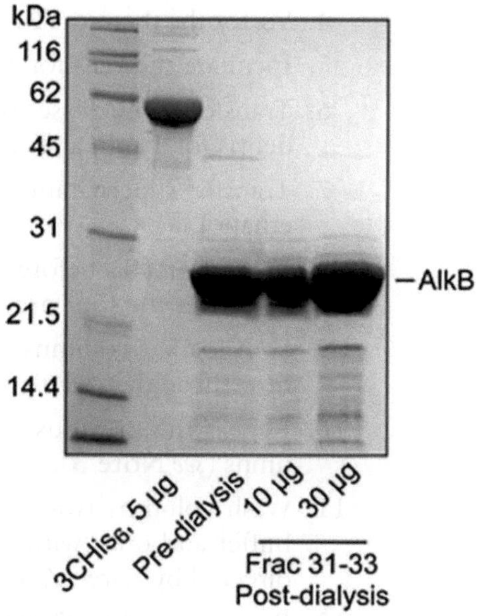

Fig. 1 Typical SDS-PAGE analysis of purified AlkB protein

42. The next day, collect protein from the dialysis cassette. If the solution is cloudy, spin down, discard pellet as in **step 27**. Keep supernatant and transfer to a new tube (*see* **Note 2**).
43. Determine protein concentration by Bradford assay and take a 2–20 µL sample (20 µg) for SDS-PAGE.
44. Run SDS-PAGE analysis, as in **step 35**, with different concentrations and size standards on gel to assess purity (see example SDS-PAGE gel in Fig. 1; should see a single clear band of ~26 kDa, with only minor contaminants visible in the 10 µg sample).
45. Aliquot ¾ of the final dialyzed protein into small aliquots, freeze on dry ice and store at −80 °C.
46. Store remaining ¼ of protein and 5 aliquots of 1 mL of storage buffer (to use as enzyme dilution buffer later, or in control "no enzyme" ARM-seq experiments) at −20 °C for use within 60 days (AlkB activity is retained for at least 60 days at −20 °C).

3.2 Total RNA Isolation from Mammalian Cells Frozen in TRI-Reagent

1. Remove two cryovials of 10^7 cells each frozen down from an 80–90 % confluent culture in 1 mL of TRI-Reagent and place on ice. Work with TRI-Reagent should be done in the fume-hood and waste should be disposed of appropriately, to avoid phenol and guanidinium thiocyanate exposure and contamination.
2. While thawing on ice, pipet up and down to quickly expose cell contents to TRI-Reagent to limit RNA degradation.
3. Prepare two columns from the DirectZol kit for each cryovial (four total).

4. Vortex the thawed TRI-Reagent cell extracts vigorously for 10 s.
5. Incubate the vials for 5 min at room temperature.
6. Transfer to nuclease-free microcentrifuge tubes and spin out debris for 1 min at 12,000 RCF.
7. Transfer supernatant to clean tubes and add 1 mL 100 % ethanol.
8. Vortex for 10 s before distributing 700 μL of lysate from each tube over the four prepared columns.
9. Spin loaded columns for 1 min at 12,000 RCF and discard flow through.
10. Repeat the previous step until all material is loaded on columns (*see* **Note 3**).
11. Wash columns twice with 400 μL of DirectZol kit Prewash buffer and once with 700 μL of DirectZol kit Wash buffer as directed by DirectZol miniprep kit instructions.
12. Remove residual DirectZol kit Wash buffer with a final 2 min spin at 12,000 RCF and discard flow through.
13. Elute RNA from each column into a clean tube with 25 μL of nuclease-free water.
14. Remove 1 μL for 1:10 dilution in nuclease-free water for concentration determination using a NanoDrop spectrophotometer before storing purified total RNA at −80 °C.
15. Expected yield from each cryovial of 10^7 cells is 30 μg (combined total of 60 μg total RNA available at this step).

3.3 Size Selection for <200 nt RNA Using MirVana Kit (See Note 4*)*

1. Raise volume of total RNA solution from **step 13** of Subheading 3.2 to 150 μL by adding nuclease-free water.
2. Prepare two columns from MirVana miRNA Isolation Kit following the manufacturer's instructions.
3. Preheat Elution solution to 95 °C.
4. In a fume hood, add 5× volume of Lysis/Binding Buffer containing the denaturant guanidinium thiocyanate, a potential hazard (750 μL to 150 μL of RNA suspension) and briefly vortex.
5. Add 1/10th volume of miRNA Homogenate Additive (90–900 μL from the above step) and briefly vortex.
6. Place on ice for 10 min.
7. Add 1/3 volume of 100 % ethanol (330 μL to 990 μL from the above step) and vortex briefly.
8. Filter the entire sample from the step above through one column by adding 700 μL at a time and spinning 5000 RCF for 1 min. Repeat until the entire volume has been applied to the column. The flow-through contains the <200 nt fraction (*see* **Note 5**).

9. Combine flow-through (~1220 μL total) and split into two nuclease-free microfuge tubes (~610 μL each).
10. Raise the ethanol concentration of the flow through to promote column binding of the smaller RNA by adding 2/3 volume of 100 % ethanol (~406.6 μL) to each tube and vortex briefly.
11. Load the contents of both tubes on a single fresh column by repeatedly adding 700 μL at a time and spinning for 1 min as above. The flow through can be discarded each time because the <200 nt RNA is bound to the column by this step.
12. Wash the column with 700 μL of Wash 1 followed by two washes of 500 μL Wash 2/3 each and spinning as above according to kit instructions.
13. To dry the column, spin empty at 10,000 RCF for 1 min.
14. Transfer column to clean, low-binding, nuclease-free tube and add 50 μL warmed Elution Solution provided in the kit.
15. Incubate for 2 min at room temperature before spinning as above to elute the RNA.
16. Repeat **steps 14–15**. Final volume of <200 nt RNA is 100 μL.
17. Expected yield at this step is 20–30 % of the amount of input RNA (expect 12–15 μg of <200 nt RNA).

3.4 DNAse I Treatment and Cleanup of <200 nt RNA (see Note 6)

1. Combine <200 nt RNA from above and place 90 μL into each of two thin-walled 0.2 mL PCR tubes.
2. Add 10 μL of 10×DNase I reaction buffer.
3. Add 1 μL of DNase I.
4. Incubate in a thermocycler at 37 °C for 10 min.
5. Add 1 μL of 0.5 M EDTA and incubate at 75 °C for 10 min.
6. Combine DNase I reactions and clean and concentrate RNA by following directions in the RNA Clean and Concentrator-5 kit.
7. Elute RNA from one column using 17 μL of nuclease-free water.
8. Expected yield at this step is 85 % of input RNA (expect 10–12 μg of <200 nt DNAse-free RNA at approximately 0.6–0.7 μg/μL available at this step).

3.5 Demethylation of RNA by AlkB Treatment

Equal amounts (w/w) of AlkB enzyme and RNA are used in the demethylation reaction. For use in the protocol below adjust purified AlkB to a concentration of 15.4 μg/μL. Control samples used for differential abundance analysis should be treated with AlkB Storage buffer only, from **step 45** of Subheading 3.1 (no enzyme addition, *see* **Note** 7).

1. Immediately before use, prepare 50 μL of a 2× common mix on ice:

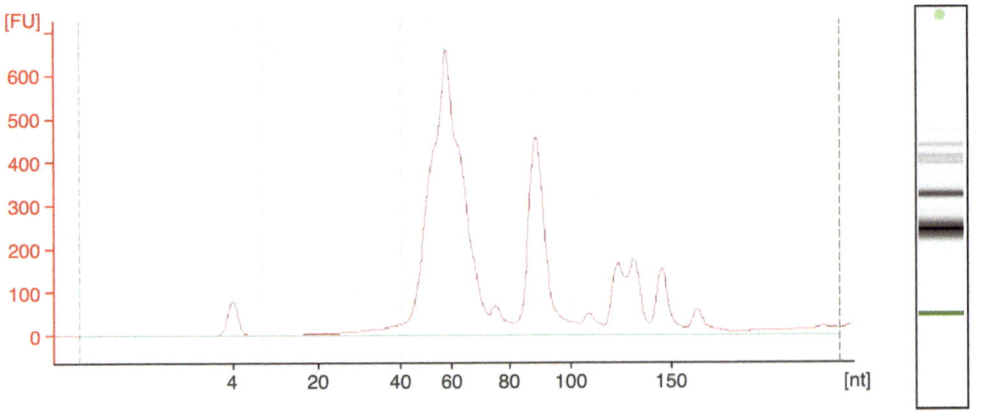

Fig. 2 Typical BioAnalyzer Trace of human small RNA preparation after DNase and AlkB treatment, 1:20 dilution

2. In a thin-walled 0.2 mL PCR tube, add the following:

   15 μL of <200 nt RNA.

   0.65 μL of 15.4 μg/μL AlkB (or AlkB buffer only for control samples).

   4.35 μL of water.

   20 μL of 2× common mix.

3. In a thermocycler, incubate the reaction at 37 °C for 100 min.
4. Stop and clean up reaction using Zymo RNA Clean and Concentrator-5 cleanup, eluting with 10 μL of ddH_2O.
5. Measure RNA concentration using a NanoDrop spectrophotometer. Check the quality of the RNA using Agilent Small RNA kit for the BioAnalyzer and 1:20 dilutions of the RNA samples (Fig. 2). The typical yield at this step is 4–6 μg of <200 nt demethylated DNA-free RNA in 8 μL of water.

3.6 Small RNA Library Preparation and Size Selection of Libraries

(For guidance on sequencing platforms and read processing see **Notes 8** and **9**).

1. Typically, 1 μg RNA from the above is used as input into the NEBNext (or comparable) small RNA library construction kit, and instructions are followed with the exception of 60 °C temperature used for the reverse transcription.
2. After 15 cycles of PCR, clean samples and concentrate using the DNA Clean and Concentrator-5, eluting with 27 μL of Elution buffer provided with the kit. At this point, samples can be stored at−20 °C until size selection of the libraries.
3. To size select the adapted libraries from adaptor-adaptor ligation products, load samples on prerinsed 2 % Agarose SizeSelect E-gels with a 50 bp E-gel ladder and run for 10 min. Rinse the

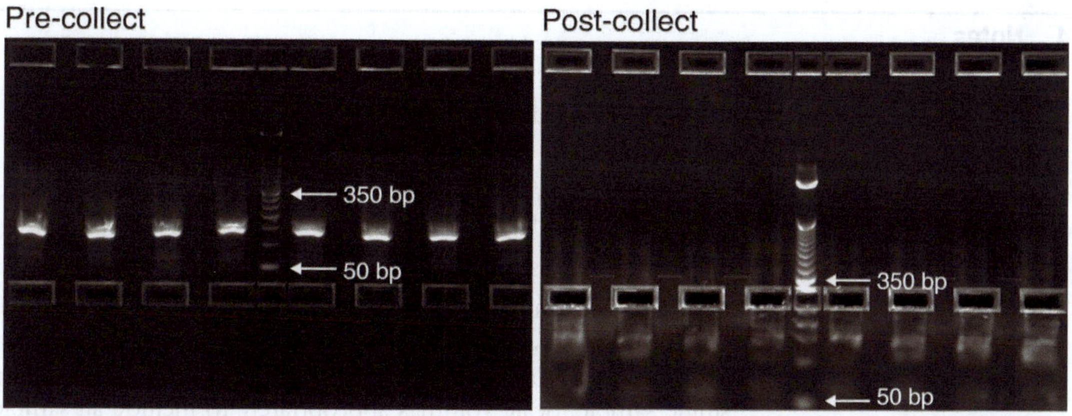

Fig. 3 Typical size selection of cDNA libraries adapted for Illumina sequencing using 2 % Agarose SizeSelect E-gel. Photographs are taken 10 min after starting the run (**a**), and after the collection of the desired fraction (**b**). A 50 bp E-gel ladder was used as a guide for fraction collection

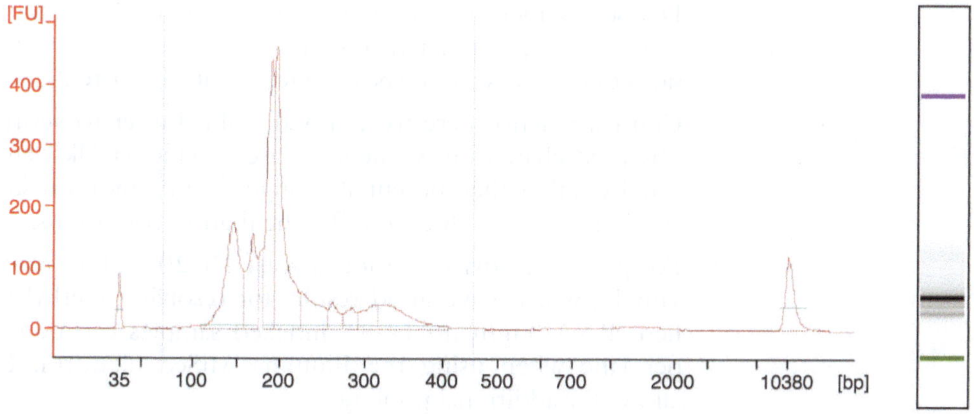

Fig. 4 Typical BioAnalyzer trace of 1:10 dilution of cDNA library used as input for Illumina sequencing in ARM-seq

lower wells and image the gel. Then continue electrophoresis and withdraw 20 μL every 10–12 s and replace with fresh TE starting once the 100 bp band of the ladder passes from the bottom well and finishing as the 250 bp band clears the well. Typically, 350 μL is collected per sample (Fig. 3).

4. Purify the size-selected libraries using the DNA clean and concentrator-5 and elute with 18–20 μL TE.
5. BioAnalyzer traces using the Agilent DNA High Sensitivity kit are done on 1:10 dilutions of the libraries to assess the amount of adaptor-adaptor product at 120 bp and for concentration determinations (Fig. 4).
6. Samples are ready for indexing, pooling, and sequencing, following library kit instructions.

4 Notes

1. 200 mM NaCl dialysis buffer enhances efficiency of 3C protease cleavage.
2. In-column DNaseI treatment using DirectZol kit-provided components and instructions can be performed at this step.
3. AlkB protein has "peachy" color to it.
4. To facilitate calculation of reagent volumes needed for processing of multiple samples, RNA size selection and all subsequent processing steps as described are scaled for the preparation of a single sample. Scale volumes appropriately to include all sample replicates and untreated controls. Be sure to follow manufacturer's instructions and dispose of kit components appropriately.
5. This step binds large RNAs to the column, allowing the <200 nt RNA to flow through.
6. DNase treatment and cleanup can now be performed at the RNA extraction step with an in-column protocol sold with newer versions of the Direct-Zol RNA extraction kit (*see* **Note 2**).
7. Control reactions are treated with AlkB buffer to verify that observed changes in sequencing are specific to AlkB enzyme activity rather than potential changes in modification state or cloning efficiency due to buffer incubation conditions.
8. For good sensitivity aim for at least 10–20 million reads per sample, which gave good results for resolving methyl-modified tRNA fragments.12–18 indexed samples can be pooled per lane when using the Illumina MiSeq platform. HiSeq allows for additional pooling.
9. Sequencing reads derived from tRNAs should be processed with specialized software to appropriately map transcripts and fragments that contain posttranscriptional modifications common in mature tRNAs (e.g., 3′-CCA ends). Software designed for this purpose is freely available (*see* http://trna.ucsc.edu/ARM-seq/). Examples of ARM-seq sequencing results mapped to individual tRNAs can be viewed in the Genomic tRNA Database (http://gtrnadb.ucsc.edu) by navigating to any individual tRNA gene entry for *Homo sapiens* (GRCh37/hg19) or *Saccharomyces cerevisiae*S288c, and scrolling to the bottom of the page to view "Expression Profiles."

Acknowledgments

This work was supported by the US National Institutes of Health (NIH) NHGRI grant 5R01HG006753 to T.M.L. and by NIH grant GM052347 to E.M.P. Thanks are given to Patricia Chan for the assistance in the preparation of the figures.

References

1. Machnicka MA, Milanowska K, Osman Oglou O, Purta E, Kurkowska M, Olchowik A, Januszewski W, Kalinowski S, Dunin-Horkawicz S, Rother KM, Helm M, Bujnicki JM, Grosjean H (2013) MODOMICS: a database of RNA modification pathways--2013 update. Nucleic Acids Res 41(Database issue):D262–D267. doi:10.1093/nar/gks1007
2. Motorin Y, Muller S, Behm-Ansmant I, Branlant C (2007) Identification of modified residues in RNAs by reverse transcription-based methods. Methods Enzymol 425:21–53. doi:10.1016/S0076-6879(07)25002-5
3. Delaney JC, Essigmann JM (2004) Mutagenesis, genotoxicity, and repair of 1-methyladenine, 3-alkylcytosines, 1-methylguanine, and 3-methylthymine in alkB *Escherichia coli*. Proc Natl Acad Sci U S A 101(39):14051–14056. doi:10.1073/pnas.0403489101
4. Aas PA, Otterlei M, Falnes PO, Vagbo CB, Skorpen F, Akbari M, Sundheim O, Bjoras M, Slupphaug G, Seeberg E, Krokan HE (2003) Human and bacterial oxidative demethylases repair alkylation damage in both RNA and DNA. Nature 421(6925):859–863. doi: 10.1038/nature01363
5. Falnes PO (2004) Repair of 3-methylthymine and 1-methylguanine lesions by bacterial and human AlkB proteins. Nucleic Acids Res 32(21):6260–6267. doi:10.1093/nar/gkh964
6. D'Silva S, Haider SJ, Phizicky EM (2011) A domain of the actin binding protein Abp140 is the yeast methyltransferase responsible for 3-methylcytidine modification in the tRNA anti-codon loop. RNA 17(6):1100–1110. doi:10.1261/rna.2652611
7. Cozen AE, Quartley E, Holmes AD, Hrabeta-Robinson E, Phizicky EM, Lowe TM (2015) ARM-seq: AlkB-facilitated RNA methylation sequencing reveals a complex landscape of modified tRNA fragments. Nat Methods 12(9):879–884. doi:10.1038/nmeth.3508
8. Love MI, Huber W, Anders S (2014) Moderated estimation of fold change and dispersion for RNA-seq data with DESeq2. Genome Biol 15(12):550. doi:10.1186/s13059-014-0550-8
9. Quartley E, Alexandrov A, Mikucki M, Buckner FS, Hol WG, DeTitta GT, Phizicky EM, Grayhack EJ (2009) Heterologous expression of *L. major* proteins in *S. cerevisiae*: a test of solubility, purity, and gene recoding. J Struct Funct Genomics 10(3):233–247. doi:10.1007/s10969-009-9068-9

Chapter 16

Transcriptome-Wide Mapping of N^1-Methyladenosine Methylome

Xiaoyu Li*, Jinying Peng*, and Chengqi Yi

Abstract

N^1-Methyladenosine (m^1A) is a prevalent posttranscriptional RNA modification and commonly found in tRNA and rRNA. Very recent works have also demonstrated the prevalence of m^1A in mammalian mRNA. Hence, high-throughput methods that allow transcriptome-wide mapping of m^1A will be important for further functional investigations. Here, we describe a technique called "m^1A-ID-Seq", which is based on m^1A immunoprecipitation and the inherent ability of m^1A to stall reverse transcription, to map m^1A in the transcriptome. Utilizing this technique, highly confident m^1A peaks can be obtained on a transcriptome-wide level.

Key words N^1-Methyladenosine (m^1A), Transcriptome-wide profiling, AlkB, Demethylase, m^1A-ID-Seq, Epitranscriptome, RNA epigenetics

1 Introduction

Until now, more than 100 types of posttranscriptional modifications to RNA molecules have been characterized [1]. Most of them are distributed in noncoding RNAs (ncRNA) including rRNA, tRNA, and snRNA. In contrast to the extensive knowledge of modifications in ncRNA, progress in mRNA modifications has been lagging and only a few internal mRNA modifications have been reported to date, including N^6-methyladenosine (m^6A), inosine (I), 5-methylcytosine (m^5C), and pseudouridine (Ψ) [2, 3]. Very recently, our group as well as the labs of Chuan He and Gideon Rechavi independently developed two different methods for the transcriptome-wide detection of m^1A [4, 5]. N^1-methyladenosine (m^1A) has long been known to be present at position 58 in tRNA, playing an important role in stabilizing its tertiary structure [6]; it is also present in rRNA and can affect

* These authors contributed equally to this work.

Alexandra Lusser (ed.), *RNA Methylation: Methods and Protocols*, Methods in Molecular Biology, vol. 1562, DOI 10.1007/978-1-4939-6807-7_16, © Springer Science+Business Media LLC 2017

rRNA biogenesis [7]. In the two recent studies, m^{1}A was shown to also be present in mRNA and, interestingly, to be enriched in the 5′ UTR of mRNA [4, 5]. While biological consequences of m^{1}A in mRNA are still unclear, it is evident that the existing sequencing tools will be important for future functional studies of m^{1}A. In this chapter, we provide a detailed description of "m^{1}A-ID-Seq," the transcriptome-wide m^{1}A mapping technology that we recently developed. In m^{1}A-ID-Seq, we first enriched m^{1}A-containing RNA by m^{1}A-specific immunoprecipitation. As m^{1}A stalls reverse transcription and results in truncated RT products [8–11], we take advantage of the biochemical activity of the *Escherichia coli* AlkB enzyme to remove m^{1}A from the immunoprecipitated m^{1}A-containing RNA [12, 13]. Therefore, by comparing the sequencing profiles of the (+) and (−) demethylase samples, we are able to identify high-confidence m^{1}A peaks (Fig. 1). In summary, m^{1}A-ID-Seq allows m^{1}A mapping on a transcriptome-wide level and provides a tool for functional studies of m^{1}A methylation.

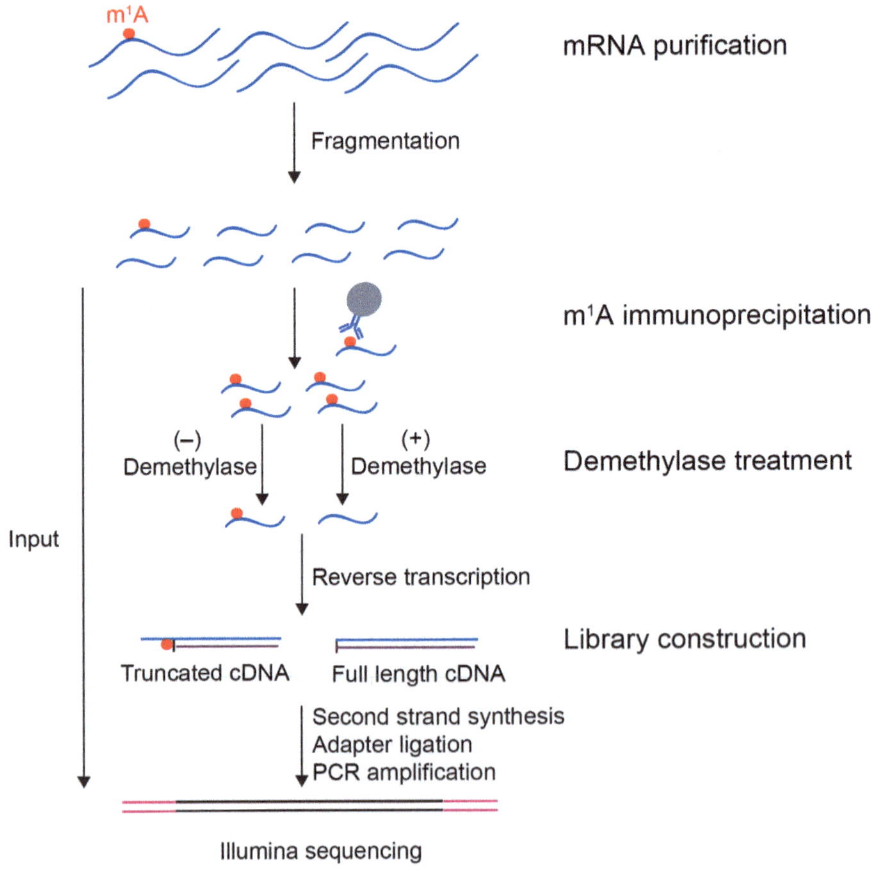

Fig. 1 Flowchart of m^{1}A-ID-Seq. Purified mRNA is fragmented ("Input" sample) and subject to m^{1}A immunoprecipitation ("(−) demethylase" sample). One part of the enriched m^{1}A-containing RNA is treated with demethylase ("(+) demethylase" sample). The three samples are then subject to library construction and Illumina sequencing

2 Materials

2.1 Total RNA Isolation

1. Cold PBS.
2. TRIzol.
3. Chloroform.
4. Isopropanol.
5. 75 % Ethanol in DEPC-treated water.
6. RNase-free water.

2.2 Small RNA Removal

1. MEGAclear™ Transcription Clean-Up Kit (Ambion) (*see* **Note 1**).
2. RNase-free water.

2.3 mRNA Purification

1. Ethanol.
2. Oligo(dT)25 dynabeads.
3. Binding buffer: 20 mM Tris–HCl, pH 7.5, 1.0 M LiCl, 2 mM EDTA.
4. Washing buffer: 10 mM Tris–HCl, pH 7.5, 0.15 M LiCl, 1 mM EDTA.
5. Elution buffer: 10 mM Tris–HCl, pH 7.5 (*see* **Note 2**).
6. Sodium acetate: 3 M in water, pH 5.2.
7. Glycogen: 5 mg/mL in water.
8. 75 % Ethanol in DEPC-treated water.
9. RNase-free water.

2.4 m¹A Immunoprecipitation

1. RNase-free water.
2. Fragmentation Reaction buffer: 0.4 M Tris-OAc,1 M KOAc, 0.3 M $Mg(OAc)_2$, pH 8.3.
3. Fragmentation Stop solution: 0.5 M EDTA in water, pH 8.0.
4. Ethanol.
5. Sodium acetate: 3 M in water, pH 5.2.
6. Glycogen: 5 mg/mL in water.
7. 75 % ethanol in DEPC-treated water.
8. RNase Inhibitor: 40 U/μL.
9. Anti-m^1A antibody.
10. Protein A/G UltraLink Resin.
11. 10× IPP buffer: 1.5 M NaCl, 1 % NP-40, 100 mM Tris–HCl, pH 7.4.
12. 1× IPP buffer: 150 mM NaCl, 0.1 % NP-40, 10 mM Tris-HCl, pH 7.4.
13. Low salt buffer: 75 mM NaCl, 0.1 % NP-40, 10 mM Tris, pH 7.4.

14. High salt buffer: 200 mM NaCl, 0.1 % NP-40, 10 mM Tris, pH 7.4.
15. TEN buffer: 10 mM Tris, pH 7.4,1 mM EDTA, pH 8.0, 0.05 % NP-40.
16. N^1-methyladenosine buffer: 3 mg/mL N^1-methyladenosine in 1× IPP buffer.
17. Phenol:chloroform:isoamyl alcohol (25:24:1) (pH <5.0).
18. Chloroform:isoamyl alcohol (24:1).

2.5 Demethylase Purification

1. *E. coli* BL21(DE3).
2. LB medium: Dissolve 10 g tryptone, 5 g yeast extract, and 10 g NaCl in 950 mL water, and autoclave at 121 °C for 15 min.
3. IPTG solution: 1 M IPTG in water.
4. Lysis buffer: 10 mM Tris–HCl, pH 7.4, 1 mM DTT, 300 mM NaCl, and 1 mM PMSF.

2.6 Demethylase Treatment

1. MES buffer: 0.5 M MES in water, pH 6.5.
2. $(NH_4)_2Fe(SO_4)_2$ buffer: 2.83 mM of $(NH_4)_2Fe(SO_4)_2{\cdot}6H_2O$ in water.
3. L-ascorbic acid buffer: 20 mM L-ascorbic acid in water.
4. 2-ketoglutarate buffer: 3 mM 2-ketoglutarate in water (*see* **Note 3**).
5. RNase Inhibitor: 40 U/μL.
6. RNase-free water.
7. Phenol: chloroform: isoamyl alcohol (25:24:1) (pH <5.0).
8. Chloroform: isoamyl alcohol (24:1).
9. Ethanol.
10. Sodium acetate: 3 M in water, pH 5.2.
11. Glycogen: 5 mg/mL in water.
12. 75 % ethanol in DEPC-treated water.
13. EDTA solution: 50 mM EDTA in water, pH 8.0.

2.7 Library Construction

1. Reverse transcription buffer: 500 mM Tris, pH 7.0, 500 mM KCl, 100 mM DTT.
2. dNTPs solution: 2 mM each in Tris buffer.
3. $MgCl_2$ buffer: 100 mM in water.
4. Random Hexamer Primers: 0.2 μg/μL in water.
5. AMV Reverse Transcriptase.
6. NEBNext® mRNA Second Strand Synthesis Module (NEB).
7. NEBNext® Ultra™ DNA Library Prep Kit for Illumina® (NEB) (*see* **Note 4**).

8. AMPure XP beads.
9. 80 % freshly prepared ethanol.
10. Nuclease-free water.

3 Methods

3.1 Total RNA Isolation (See Note 5)

1. Harvest three to four 152 cm^2 culture dishes of adherent cells with ~80 % confluence. Remove the growth media from culture dishes and carefully add 10 mL cold PBS per dish. Remove the cold PBS and then add 6–8 mL TRIzol directly to each dish. Incubate at room temperature for at least 5 min and then pipet up and down several times to allow for thorough homogenization. Transfer 1 mL aliquots of homogenized sample to fresh, RNase-free 1.5 mL tubes.
2. Add 0.2 mL of chloroform per 1 mL of TRIzol. Shake tubes vigorously by hand for 15 seconds and incubate for 2–3 min at room temperature.
3. Centrifuge the sample at 12,000 × *g* for 15 min at 4 °C. Carefully transfer the aqueous layer to fresh, RNase-free 1.5 mL tubes (*see* **Note 6**).
4. Add an equal volume of 100 % cold isopropanol to the aqueous phase and mix thoroughly by gently inverting tubes 4–6 times. Incubate at room temperature for 10 min.
5. Centrifuge at 12,000 × *g* for 10 min at 4 °C. Remove the supernatant and wash the pellet with 1 mL of 75 % ethanol. Centrifuge the tube at 12,000 × *g* for 5 min at 4 °C.
6. Air dry the pellets for 10 min. Dissolve the RNA pellets in 20 μL RNase-free water completely (*see* **Note 7**).
7. Measure the RNA concentration by Nanodrop photometer and analyze the RNA quality using agarose gel electrophoresis (*see* **Note 8**).

3.2 Small RNA Removal (See Note 9)

1. As m^1A is very abundant in tRNA, small RNA (size < 200 nt) needs to be removed before mRNA purification to minimize tRNA contamination (*see* **Note 10**).
2. Adjust the total RNA concentration to ~3 μg/μL with RNase-free water and then divide the RNA solution into 100 μL aliquots. Add 350 μL of Binding Solution per 100 μL RNA solution. Mix gently by pipetting. Then add 250 μL 100 % ethanol. Mix gently by pipetting and apply 700 μL sample per filter cartridge. Centrifuge at 10,000–15,000 × *g* for 15 seconds to 1 min (*see* **Note 11**).
3. Add 500 μL Wash Solution per sample and then centrifuge at 15,000 × *g* for 15 s to 1 min. Repeat this wash step once.

Centrifuge at 15,000 × *g* for 3 min to ensure complete removal of Wash Solution.

4. Preheat 210 μL of Elution Solution per sample to 95 °C. Add 100 μL preheated Elution Solution to the center of the filter cartridge and incubate at room temperature for 1 min and then centrifuge at 15,000 × *g* for 1 min. Repeat this step once to get maximum RNA recovery. Pool the eluates into the same tube (*see* **Note 12**).
5. Adjust the total RNA concentration to ~0.8–1 μg/μL with RNase-free water and then divide the RNA solution into 500 μL aliquots for polyA selection.

3.3 mRNA Purification (See Note 13)

1. Heat RNA solution at 65 °C for 3 min and chill on ice immediately to disrupt secondary structures.
2. Transfer 1 mL well resuspended oligo(dT)25 dynabeads to a fresh, RNase-free microcentrifuge tube. Place the tube on the magnet for 30 seconds. Remove the supernatant and wash beads by adding 1 mL Binding Buffer and pipetting up and down ~10 times. Put the tube back on the magnet and then remove the supernatant. Resuspend the dynabeads with 500 μL Binding buffer. Place the resuspended dynabeads on ice.
3. Per 500 μL RNA solution, add 500 μL resuspended oligo(dT)25 dynabeads. Mix thoroughly by pipetting up and down. Rotate end-over-end for 5–10 min at room temperature to allow mRNA to anneal to the oligo(dT)25 probe.
4. Place the tube on the magnet until the solution is clear. Remove the supernatant. Wash the mRNA-bead complex by adding 1 mL Washing Buffer and pipetting up and down ~10 times. Put the tube back on the magnet and then remove the supernatant. Repeat this wash step twice.
5. Add 500 μL Elution buffer to the mRNA-bead complex. Incubate at 70 °C for 2 min shaking at 1000 rpm and place the tube immediately on the magnet. Transfer the eluted mRNA to a new RNase-free tube and put on ice.
6. Add 1 mL Elution buffer to the resuspended beads again, and then incubate at 80 °C for 3 min shaking at 1200 rpm, place the tube immediately on the magnet, and discard the supernatant.
7. Repeat **step 1–4** once.
8. Elute the RNA with 200 μL Elution buffer as in **step 5**. Add 20 μL sodium acetate, 4 μL glycogen, and 0.5 mL 100 % ethanol to the eluted RNA. Mix well and incubate at −20 °C for at least 1 h.
9. Centrifuge at > 12,000 × *g* for 30 min at 4 °C. Remove the supernatant and wash the pellet with 1 mL of 75 % ethanol. Centrifuge the tube at 12,000 × *g* for 5 min at 4 °C. Repeat the wash step once.

10. Air dry the pellets for 10 min. Dissolve the RNA pellets in 18 μL RNase-free water completely.
11. Measure the RNA concentration by Nanodrop photometer. Analyze the RNA quality and rRNA contamination using RNA 6000 Pico Chips on an Agilent 2100 Bioanalyzer (*see* **Note 14**).

3.4 m¹A Immunoprecipitation (See Note 15)

1. Adjust the volume of ~45 μg polyA⁺ RNA to 72 μL with RNase-free water. Add 8 μL RNA Fragmentation Reaction Buffer and mix well by pipetting up and down. Divide the RNA mix into 20 μL aliquots. Incubate in a preheated thermal cycler for 5 min at 94 °C and then chill on ice immediately (*see* **Note 16**).
2. Add 2 μL RNA Fragmentation Stop Solution per 20 μL aliquot and then mix well by pipetting up and down. Combine all samples into one tube and adjust to 200 μL with RNase-free water. Add 20 μL sodium acetate, 4 μL glycogen, and 0.5 mL 100 % ethanol to the fragmented RNA. Mix well and incubate at −20 °C for at least 1 h. Centrifuge at > 12,000 × *g* for 30 min at 4 °C. Remove the supernatant and wash the pellet with 1 mL of 75 % ethanol. Centrifuge the tube at 12,000 × *g* for 5 min at 4 °C. Repeat the wash step once. Air dry the pellets for 10 min. Dissolve the RNA pellets with 110 μL RNase-free water completely.
3. Save 10 μL of fragmented RNA as input control in m¹A-ID-seq. Heat 100 μL of fragmented RNA at 65 °C for 5 min, followed by chilling on ice to disrupt secondary structures. Add 40 μL 10× IPP buffer, 10 μL RNase Inhibitor, 5 μg anti-m¹A antibody and adjust the volume to 400 μL with RNase-free water. Rotate at 4 °C overnight.
4. Add 40 μL Protein A/G Ultra Link Resin into 1 mL 1× IPP buffer, centrifuge for 2 min at 1000 × *g*, and discard supernatant. Repeat this wash step twice and resuspend in 400 μL 1× IPP buffer. Add the 400 μL resuspended Protein A/G UltraLink Resin to the RNA antibody and rotate at 4 °C for additional 2 h.
5. Centrifuge for 2 min at 1000 × *g* and discard supernatant. Add 1 mL 1× IPP buffer and rotate at 4 °C for 5 min. Centrifuge for 2 min at 1000 × *g* and discard supernatant. Repeat this step once.
6. Add 1 mL low salt buffer and rotate at 4 °C for 5 min. Centrifuge for 2 min at 1000 × *g* and discard supernatant.
7. Add 1 mL high salt buffer and rotate at 4 °C for 5 min. Centrifuge for 2 min at 1000 × *g* and discard supernatant.
8. Add 1 mL TEN buffer and rotate at 4 °C for 5 min. Centrifuge for 2 min at 1000 × *g* and discard supernatant. Repeat this step once.

9. Add 200 μL of N^1-methyladenosine buffer and rotate at room temperature for 1 h. Centrifuge for 2 min at 1000 × *g* and collect the supernatant. Repeat this step once and pool the supernatants as immunoprecipitated sample (IP).
10. Add 400 μL phenol:chloroform:isoamyl alcohol and mix well. Centrifuge for 5 min at 16,000 × *g* and transfer the aqueous upper layer to a new tube. Add an equal volume of chloroform: isoamyl alcohol and centrifuge for 10 min at 16,000 × *g*.
11. Transfer the upper aqueous layer to a new tube and add 1/10 volume sodium acetate, 4 μL glycogen, and 2.5 volumes of 100 % ethanol to the RNA. Mix well and incubate at −20 °C for at least 1 h. Centrifuge at > 12,000 × *g* for 30 min at 4 °C. Remove the supernatant and wash the pellet with 1 mL of 75 % ethanol. Centrifuge the tube at 12,000 × *g* for 5 min at 4 °C. Repeat the wash step once. Air dry the pellets for 10 min. Dissolve the RNA pellets in 20 μL RNase-free water completely. Measure the RNA concentration by Qubit (*see* **Note 17**).

3.5 Demethylase Purification (See Note 18)

1. Wild-type AlkB is used as the demethylase to remove m^1A in immunoprecipitated RNA. Clone a truncated AlkB with the deletion of the N-terminal 11 amino acids into pET30a and transform the plasmids to *E. coli* BL21(DE3).
2. Grow in LB medium at 37 °C until the OD600 reaches 0.6–0.8. Add 1 mL IPTG per 1 L LB medium and incubate at 30 °C for additional 4 h to express wild-type Alkb protein.
3. Collet *E. coli* by centrifugation, remove the medium, and resuspend *E. coli* in 40 mL cold lysis buffer. Make cell lysates by sonication.
4. Centrifuge for 30 min at 15,000 × *g* and collect the supernatant.
5. Purify AlkB protein from the supernatant using Ni-NTA chromatography (GE Healthcare), gel-filtration chromatography (Superdex 200), and Mono-Q anion exchange chromatography successively.

3.6 Demethylase Treatment (See Note 19)

1. Heat 10 μL immunoprecipitated RNA from Subheading 3.4, **step 11**, at 65 °C for 5 min and then chill on ice immediately.
2. Prepare a demethylation mix as follows. Add 2 μL MES buffer, 2 μL$(NH_4)_2Fe(SO_4)_2$ buffer, 2 μL 2-ketoglutarate buffer, 2 μL L-ascorbic acid buffer, 0.2 μL RNase Inhibitor, 0.4 pmol purified AlkB and adjust the volume to 10 μL with RNase-free water.
3. Add the demethylation mix to the immunoprecipitated RNA and mix well. Incubate at 37 °C for 2 h. Quench the reaction by the addition of 2 μL EDTA solution.

4. Adjust the reaction volume to 200 μL with RNase-free water. Add 200 μL phenol: chloroform: isoamyl alcohol and mix well. Centrifuge for 5 min at 16,000 × *g* and transfer the upper aqueous layer to a new tube.
5. Add an equal volume of chloroform:isoamyl alcohol and centrifuge for 10 min at 16,000 × *g*. Transfer the upper aqueous layer to a new tube and add 1/10 volume sodium acetate, 4 μL glycogen, and 2.5 volume 100 % ethanol to the fragmented RNA.
6. Mix well and incubate at −20 °C for at least 1 h. Centrifuge at > 12,000 × *g* for 30 min at 4 °C. Remove the supernatant and wash the pellet with 1 mL of 75 % ethanol. Centrifuge the tube at 12,000 × *g* for 5 min at 4 °C. Repeat the wash step once.
7. Air dry the pellets for 10 min. Dissolve the RNA pellets in 10 μL RNase-free water completely. These are the"(+) demethylase" sample.

3.7 Library Construction (See Note 20)

1. Subject the input, IP, and "(+) demethylase" samples to library construction.
2. Add 1 μL Random Hexamer Primers to 10 μL RNA sample and adjust the volume to 12 μL with RNase-free water. Heat at 65 °C for 5 min and then chill on ice immediately. Add 2 μL reverse transcription buffer, 2 μL dNTPs solution, 2 μL $MgCl_2$ buffer, 1 μL AMV Reverse Transcriptase and 1 μL RNase Inhibitor. Mix well and incubate at 42 °C for 1 h.
3. Perform second strand synthesis with NEBNext® mRNA Second Strand Synthesis Module according to manufacturer's instructions. Purify the double-stranded cDNA using144 μL AMPure XP Beads.
4. Perform end repair and adaptor ligation with NEBNext® Ultra™ DNA Library Prep Kit according to manufacturer's instructions. Purify the adaptor ligated DNA using 100 μL AMPure XP Beads.
5. Perform PCR amplification with NEBNext® Ultra™ DNA Library Prep Kit according to manufacturer's instructions. Purify the PCR reaction using 45 μL AMPure XP Beads.
6. Subject libraries to next-generation sequencing on the Illumina HiSeq 2500 with paired-end 2 × 100 bp read length (*see* **Note 21**).

4 Notes

1. We use MEGAclear™ Transcription Clean-Up Kit (Ambion) and other size selection kits can also be used in this step.
2. DEPC cannot be used directly to treat Tris buffer as it can react with primary amines. To prepare Tris buffer, treat water

with DEPC first. Dissolve Tris in DEPC-treated water and adjust the pH with HCl.

3. Dissolve $(NH_4)_2Fe(SO_4)_2 \cdot 6H_2O$, L-ascorbic acid, 2-ketoglutarate with RNase-free water freshly just before using. Stocks are not recommended.
4. In this step, the library construction kits we used are bought from NEB and the library construction kits from other suppliers can also be used in this step.
5. The duration of this step is about 2 h.
6. Carefully transfer the upper aqueous layer using a pipette with 200 μL tips. Do not use 1 mL tips that would disturb the phase separation. Do not attempt to draw off the entire aqueous layer after phase separation to avoid contamination with interphase/organic phase.
7. Do not allow the RNA to dry completely or the pellet will lose solubility.
8. Generally a total amount of 2–3 mg total RNA can be obtained from 4 × 152 cm^2 dishes of HEK293T cells. The integrity of total RNA is very important. The degradation of RNA would lead to low yield of mRNA purification and affect sequencing reads distribution.
9. The duration of this step is about 1 h.
10. Solutions in the kit should be stored at 4 °C and brought to room temperature before use. Perform all the steps of this procedure and all the steps of centrifugation at room temperature.
11. For one filter cartridge, the capacity is about 350 μg RNA. Thus, do not apply too much RNA for one column; otherwise, it will lead to low yield.
12. Generally, the recovery after small RNA depletion is about 80 %.
13. The duration of this step is about 3 h.
14. Generally, the yield of polyA selection is about 2–2.5 % of the total RNA amount.
15. The duration of this step is about 1 day.
16. In this step, PCR tubes need to be put on ice immediately after heating, and the cooling step in the thermal cycler is not recommended.
17. Generally, the yield of m^1A immunoprecipitation is about 0.05 % of the input RNA amount and about 20 ng RNA can be immunoprecipitated from 40 μg fragmented RNA.
18. The duration of this step is about 1 day.
19. The duration of this step is about 6 h.
20. The duration of this step is about 1 day.

21. Raw sequencing data are subjected to Trimgalore for quality control. Peak calling is performed using "(−) demethylase" samples and "input" samples. Demethylase-Sensitivity score (DS score) is calculated to evaluate the increase of normalized sequencing depth between the (+) and (−) demethylase samples. Considering m^1A could locate to the 5′ of the peaks due to RT stops, m^1A peaks, and the corresponding 50 nucleotides 5′ to each peak are used for DS score calculations. Then the DS scores can be used to identify the demethylase-sensitive regions.

References

1. Machnicka MA, Milanowska K, Osman Oglou O, Purta E, Kurkowska M, Olchowik A, Januszewski W, Kalinowski S, Dunin-Horkawicz S, Rother KM, Helm M, Bujnicki JM, Grosjean H (2013) MODOMICS: a database of RNA modification pathways--2013 update. Nucleic Acids Res 41(Database issue): D262–D267
2. Liu N, Pan T (2015) RNA epigenetics. Trans Res 165(1):28–35
3. Roundtree IA, He C (2016) RNA epigenetics--chemical messages for posttranscriptional gene regulation. Curr Opin Chem Biol 30: 46–51
4. Dominissini D, Nachtergaele S, Moshitch-Moshkovitz S, Peer E, Kol N, Ben-Haim MS, Dai Q, Di Segni A, Salmon-Divon M, Clark WC, Zheng G, Pan T, Solomon O, Eyal E, Hershkovitz V, Han D, Dore LC, Amariglio N, Rechavi G, He C (2016) The dynamic N(1)-methyladenosine methylome in eukaryotic messenger RNA. Nature 530(7591):441–446
5. Li X, Xiong X, Wang K, Wang L, Shu X, Ma S, Yi C (2016) Transcriptome-wide mapping reveals reversible and dynamic N(1)-methyladenosine methylome. Nat Chem Biol 12(5):311–316
6. Schevitz RW, Podjarny AD, Krishnamachari N, Hughes JJ, Sigler PB, Sussman JL (1979) Crystal structure of a eukaryotic initiator tRNA. Nature 278(5700):188–190
7. Peifer C, Sharma S, Watzinger P, Lamberth S, Kotter P, Entian KD (2013) Yeast Rrp8p, a novel methyltransferase responsible for m1A 645 base modification of 25S rRNA. Nucleic Acids Res 41(2):1151–1163
8. Motorin Y, Muller S, Behm-Ansmant I, Branlant C (2007) Identification of modified residues in RNAs by reverse transcription-based methods. Methods Enzymol 425: 21–53
9. Behm-Ansmant I, Helm M, Motorin Y (2011) Use of specific chemical reagents for detection of modified nucleotides in RNA. J Nucleic Acids 2011:408053
10. Hodgkinson A, Idaghdour Y, Gbeha E, Grenier JC, Hip-Ki E, Bruat V, Goulet JP, de Malliard T, Awadalla P (2014) High-resolution genomic analysis of human mitochondrial RNA sequence variation. Science 344(6182): 413–415
11. Hauenschild R, Tserovski L, Schmid K, Thuring K, Winz ML, Sharma S, Entian KD, Wacheul L, Lafontaine DL, Anderson J, Alfonzo J, Hildebrandt A, Jaschke A, Motorin Y, Helm M (2015) The reverse transcription signature of N-1-methyladenosine in RNA-Seq is sequence dependent. Nucleic Acids Res 43(20):9950–9964
12. Falnes PO, Johansen RF, Seeberg E (2002) AlkB-mediated oxidative demethylation reverses DNA damage in *Escherichia coli.* Nature 419(6903):178–182
13. Trewick SC, Henshaw TF, Hausinger RP, Lindahl T, Sedgwick B (2002) Oxidative demethylation by *Escherichia coli* AlkB directly reverts DNA base damage. Nature 419(6903): 174–178

Part V

Functional Assays for Modified and Unmodified RNA

Chapter 17

In Vitro Assays for RNA Methyltransferase Activity

Sara Haag, Katherine E. Sloan, Claudia Höbartner, and Markus T. Bohnsack

Abstract

RNA methyltransferases (MTases) are responsible for co- and posttranscriptional methylation of nucleotides in a wide variety of RNA substrates. Examination of the target specificity, catalytic activity, and function of these enzymes requires in vitro methylation assays. Here, we provide a detailed protocol for the methylation of in vitro transcripts, synthetic RNAs, and total cellular RNA using recombinant RNA methyltransferases and S-adenosylmethionine (SAM) as a methyl group donor. We describe how this method can be coupled to fluorographic detection of RNA methylation if ^{3}H–labeled SAM is used, and discuss alternative chromatography-based methods for the detection of methylated nucleotides, focusing on reversed-phase high-performance liquid chromatography (RP-HPLC). In both cases, mutagenesis of the methyltransferase or the RNA substrate can be employed to elucidate the catalytic mechanisms and target specificity of the enzymes. Together these approaches provide valuable insight into the action of RNA methyltransferases on the molecular level and serve as a basis for further functional characterization of RNA methyltransferases in vivo.

Key words Methyltransferase, RNA modification, Methylation, S-Adenosyl methionine, Epitranscriptome, Fluorography, RP-HPLC

1 Introduction

Co- and posttranscriptional modifications of RNA molecules occur in all kingdoms of life and the majority of these modifications include a methylation step [1]. Transfer of methyl groups to RNA nucleosides is catalyzed by RNA methyltransferases (RNA MTases), which almost exclusively employ S-adenosylmethionine (SAM) as a methyl group donor (*see,* for example, Fig. 1a, for m^{5}C modification) [2].

In vitro methylation analysis is a powerful tool to gain insight into the catalytic activity and specificity of RNA MTases. The method utilizes recombinant proteins and (often radioactively-labeled) SAM to monitor methylation activity, and can be applied to in vitro transcripts, synthetic RNAs, or RNA isolated from cells [3–5]. Here, we describe an in vitro methylation protocol that can be used in combination with denaturing polyacrylamide gel

Alexandra Lusser (ed.), *RNA Methylation: Methods and Protocols*, Methods in Molecular Biology, vol. 1562, DOI 10.1007/978-1-4939-6807-7_17,

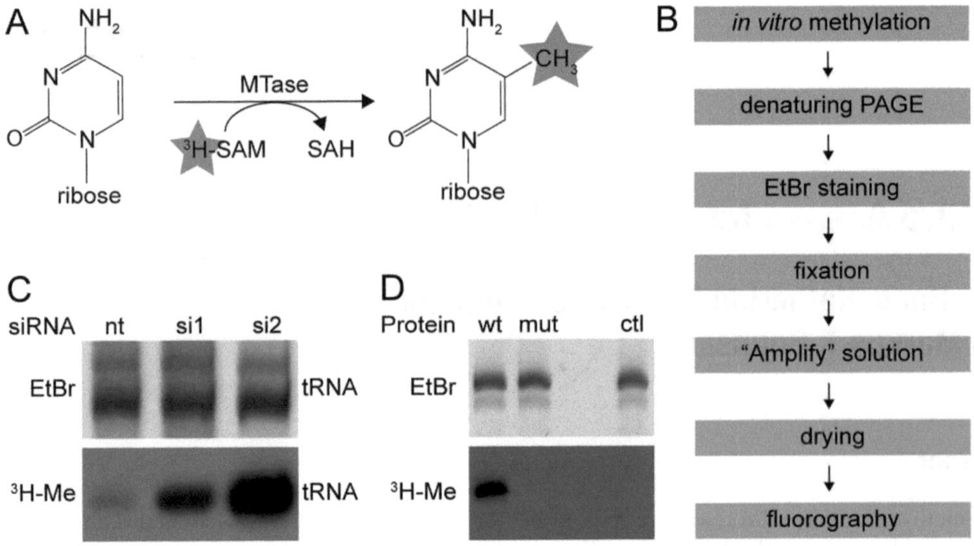

Fig. 1 RNA in vitro methylation assay using recombinant RNA methyltransferases (MTases) and S-adenosyl-methionine (SAM). (**a**) Scheme of the in vitro methylation reaction using radioactively labeled SAM (^{3}H–SAM) exemplified by cytosine methylation. (**b**) Flow diagram showing the major steps of the in vitro methylation assay coupled with denaturing gel electrophoresis and fluorography as detection method (PAGE—polyacrylamide gel electrophoresis, EtBr—ethidium bromide). (**c**) In vitro methylation of total RNA from wildtype and knockdown HeLa cells using two different siRNAs against the methyltransferase. The increase of the in vitro methylation signal on the RNA isolated from cells after RNAi indicates the loss of methylation in vivo after knockdown of the RNA methyltransferase. (**d**) Radioactive in vitro methylation of a substrate RNA using wildtype MTase (wt), catalytic MTase mutant (mut), and control protein (ctl)

electrophoresis and fluorography (Fig. 1b) or together with chromatographic separation of nucleosides/nucleotides after enzymatic digestion of the methylated RNA followed by UV absorbance detection (Fig. 2). These methods are generally applicable for the analysis of SAM-dependent MTases, irrespective of the target nucleotide and methylation position. Alternative methods that monitor consumption of SAM by enzyme-coupled assays have also been described [6]. Although fluorescent or colorimetric readouts allow for parallelization and high-throughput analyses, the formation of methylated products needs to be confirmed in separate experiments [7].

The in vitro methylation assays described in this chapter can be used for the validation of putative substrates and protein-RNA binding sites identified by genome-wide methods such as crosslinking and analysis of cDNA (CRAC) [5, 8, 9]. However, they additionally enable the identification/verification of the target nucleotide and the delineation of additional nucleotides or secondary structural elements that are important for target recognition, as site-directed mutagenesis can be used to introduce point mutations into in vitro transcribed substrates or mutations can be integrated in synthetic RNAs. If defined ends of the in vitro transcripts are essential

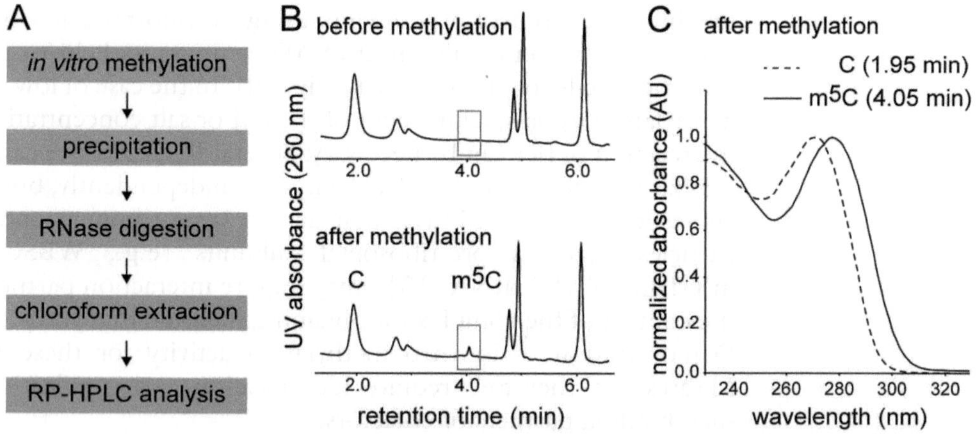

Fig. 2 Analysis of in vitro RNA methylation by RP-HPLC. (**a**) Flow diagram showing the major steps of the in vitro methylation assay coupled with RP-HPLC analysis and detection by UV absorbance. (**b**) RP-HPLC traces of digested synthetic RNA (20 nt) monitored at 260 nm before and after methylation of cytosine. The peak for m^5C is only visible in the bottom trace. (**c**) Normalized extracted UV spectra of C and m^5C in methylated RNA

for target recognition, the restriction enzyme that is used for plasmid linearization prior to in vitro transcription has to be chosen carefully. Alternatively, the in vitro transcript can be processed by nucleic acid enzymes to achieve defined RNA ends. Examples include fusing the transcript to self-cleaving hammerhead or hepatitis delta virus ribozymes, or site-specific intermolecular cleavage by deoxyribozymes [10, 11]. Solid-phase synthesis provides defined RNA substrates at a length of up to 50 nucleotides, and enables the site-specific incorporation of modified nucleotides [12].

In vitro methylation assays can also give insight into whether the activity of an RNA MTase requires pre-existing modifications in the target RNA using a comparative approach with unmodified and premodified synthetic RNA substrates. Furthermore, these assays can be used to provide indirect evidence of methylation activity in vivo; in vitro methylation signals on RNA isolated from wildtype and knockdown or knockout cell lines can be compared with stronger methylation in vitro suggesting a previous loss of methylation in vivo (Fig. 1c) [5]. Moreover, blocking methylation of the main substrate in total RNA can be achieved by hybridization of a DNA oligonucleotide covering the target site prior to methylation and thus allowing the detection of additional/alternative substrates [4].

One of the challenges for in vitro methylation is to obtain recombinant proteins of sufficiently high purity since many RNA MTases can multimerize and therefore often copurify with other MTases from *E. coli*, leading to unspecific methylation signals in the in vitro assay. The use of a bipartite tag for tandem affinity purification of the RNA MTases combined with stringent washing conditions during protein purification is therefore recommended.

Furthermore, control experiments using nonsubstrate RNAs and generation of catalytically inactive MTase mutants help to prove catalytic specificity of the protein (Fig. 1d). In the case of low or no methylation signals, adjustment of the pH or salt concentration of the reaction buffer can be necessary. Several RNA MTases can recognize and methylate their RNA target(s) independently, but others, especially RNA MTases that act on large ribonucleoprotein particles such as pre-ribosomal subunits (e.g., WBSCR22; m^7G1639 of 18S rRNA [13]), may require interaction partners or the context of the complex for obtaining their full catalytic activity. Demonstration of in vitro methylation activity for these RNA MTases can therefore require identification and purification of such binding partners or cofactors.

Fluorographic detection of ^{3}H–labelled methylation products after separation by denaturing gel electrophoresis can be carried out semiquantitatively with X-ray films or quantitatively using fluoroimaging screens. Both variants have several advantages over scintillation counting [14], since separation of the reaction products by gelelectrophoresis allows size determination of the methylated RNA, which can provide an indication for the type of RNA methylated if total cellular RNA was used as a substrate. Secondly, the gel can be stained with ethidium bromide to monitor the integrity and quantity of the RNA after performing the methylation reaction. This is especially useful for control samples where no methylation is expected.

Another method for the detection of modified residues in RNA after in vitro methylation reactions is based on chromatographic separation of individual nucleosides or nucleotides. After nonradioactive methylation in vitro, the RNA is enzymatically digested to mononucleosides, which are analyzed by reversed-phase high-performance liquid chromatography (RP-HPLC) and detected by monitoring UV absorbance at 260 nm or recording full absorbance spectra with a diode array detector (DAD) (Fig. 2a–c). Alternatively, the mononucleosides can be labeled with ^{32}P (using T4 polynucleotide kinase and γ-^{32}P-ATP), separated by two-dimensional thin layer chromatography (2D–TLC) on cellulose plates and detected using X-ray films or a phosphorimager [15, 16]. Both these detection methods exploit the different physical properties of methylated and unmodified nucleosides, which cause them to migrate differently on either TLC plates or HPLC columns. A detailed protocol for 2D–TLC analysis of modified nucleotides has been published in this series [17]. Here, we focus on RP-HPLC analysis with detection by UV absorbance, which is usually faster than 2D–TLC and also bypasses the requirement for radioactive labeling. In contrast to fluorographic detection of in vitro methylation that can be performed on total RNA, analysis of RNA methylation by RP-HPLC is only applicable when the target RNA is already known and unmodified RNAs should always be analyzed in parallel (Fig. 2b, c). Furthermore, due to the fragmentation of the RNA substrate prior to nucleoside composition analysis, synthetic RNA oligonucleotides (<50 nt) or short in vitro transcripts (<120 nt) are preferred substrates. Methylated nucleosides can be identified by comparison of retention times and UV absorbance profiles with reference standards of known chemical composition. However, as this is only possible for a limited number of methylated nucleosides, HPLC can also be coupled to electrospray ionization mass spectrometry (ESI-MS) to directly identify the methylation products by determination of molecular mass and characteristic molecular fragmentation patterns [18].

Taken together, in vitro methylation coupled to the complementary detection approaches of fluorography and RP-HPLC can provide valuable insights into the target sites, specificity, and function of RNA methyltransferases.

2 Materials

2.1 Chemicals and Other Consumables

1. Amplify™ Fluorographic Reagent (GE Healthcare).
2. Calf intestinal alkaline phosphatase: 1 unit/μL.
3. Gel drying films.
4. Ethidium bromide:10 mg/mL in water.
5. Proteinase K, recombinant PCR grade: 50 units/mL.
6. Nuclease P1: 3 units/μL.
7. Ribolock RNase inhibitor: 40 units/μL.
8. RP-HPLC column, Nucleosil 100–5, 2 × 125 mm (Macherey Nagel).
9. [^{3}H]-S-adenosylmethionine, 2.22 TBq/mmol: 17 μM.
10. Snake venom phosphodiesterase: 1 unit/μL.
11. Whatman filter paper (3 mm).

2.2 Buffers and Solutions

1. 10× methylation buffer: 500 mM Tris–HCl pH 7.4, 50 mM $MgCl_2$, 10 mM DTT, 500 mM NaCl; prepare with RNase-free water and store aliquots at −20 °C until use (*see* **Note 1**).
2. RNA loading dye: 95 % formamide, 5 mM ETDA pH 7.4, 0.03 % bromophenol blue.
3. 1× TBE buffer: 89 mM Tris-borate pH 8.3, 2 mM EDTA.
4. 12 % denaturing gel solution: 7 M urea, 1× TBE buffer, 12 % polyacrylamide (40 %; 19:1); add 0.1 % TEMED and 0.1 % ammonium persulfate before pouring gel.
5. Fixing solution: 10 % methanol, 10 % acetic acid (prepare always fresh).
6. NaCl solution: 3 M in water.
7. 10× RNase P1 buffer: 0.2 M NH_4OAc pH 5.3, 1 mM $ZnCl_2$.

8. 10× Alkaline phosphatase buffer: 100 mM Tris–HCl pH 8.0, 50 mM $MgCl_2$, 1 M KCl (adjust pH at 37 °C).
9. RP-HPLC buffer A: 20 mM NH_4OAc pH 6.5.
10. RP-HPLC solvent B: acetonitrile (HPLC grade).

3 Methods

3.1 *In Vitro* Methylation

1. The methylation reaction is carried out in a 20 μL reaction volume. Calculate the volumes of recombinant protein and RNA (in vitro transcript, synthetic RNA, or total RNA) needed for a final concentration of 1–3 μM of recombinant protein and 1 μM of in vitro transcript or synthetic RNA, or 10 μg of total RNA (*see* **Note 2**).
2. Thaw the required number of ^{3}H–SAM aliquots, vortex shortly, and centrifuge briefly (*see* **Notes 3** and **4**).
3. Mix the following components in the listed order: required volume of H_2O for 20 μL total volume, 2 μL 10× methylation buffer, 1 μL Ribolock and recombinant protein (1–3 μM final concentration).
4. Add 2 μL of ^{3}H–SAM to each sample. Vortex shortly and centrifuge briefly to collect the reaction at the bottom of the tube (*see* **Note 5**).
5. Add the required amount of RNA to each sample. Vortex shortly and centrifuge briefly to collect the reaction at the bottom of the tube, then incubate for 2 h at 22–37 °C (*see* **Note 6**).
6. Centrifuge briefly to collect the reaction at the bottom of the tube, add 1 μL of proteinase K and incubate for 30 min at 22–37 °C to stop the reaction.
7. Add 20 μL of RNA loading dye (*see* **Note 7**).

3.2 Detection of Methylated Target RNAs by Fluorography

1. Cast a denaturing 12 % polyacrylamide gel (*see* **Note 8**).
2. Prerun the gel in 1× TBE for 20–30 min at 25 mA.
3. Heat samples to 95 °C for 3 min and then load each sample (approximately 41 μL) onto the gel after thoroughly rinsing the wells (*see* **Note 9**).
4. Load RNA ladder or 5 μL RNA loading dye with bromophenol blue and xylene cyanol in a free lane to estimate migration of RNAs during electrophoresis (*see* **Note 10**).
5. Run the gel for approximately 1–2 h at 30 mA depending on the size of the target RNA (*see* **Note 11**).
6. Disassemble the gel chamber (*see* **Note 12**) and transfer the gel to 1× TBE containing 0.2–1 μg/mL ethidium bromide. Stain the gel for approximately 5 min then remove the ethidium

bromide solution. Visualize the RNA with UV light and record the image with a gel documentation system.

7. Transfer the gel to 250 mL of Fixing solution. Incubate for 15 min in a fume hood. Repeat this step once.
8. Incubate the gel for 45–60 min in Amplify™ Fluorographic Reagent (*see* **Note 13**).
9. Gel drying: Pre-wet a piece of Whatman filter paper that is approximately 3–4 cm larger than the gel on each side. Place this in the middle of the gel dryer and lay the gel on top. Cover completely with a pre-wetted gel drying film, ensuring that there are no air bubbles between the Whatman filter paper, gel, and gel-drying film. Cover with the lid of the gel dryer, activate the vacuum pump, and dry the gel at 65 °C for at least 2 h.
10. After drying, expose the gel to an X-ray film for between 12 h and 2 weeks at −80 °C before developing.

3.3 Detection of Methylated Products Using RP-HPLC

1. Perform in vitro methylation reactions with unlabeled SAM at 5× the scale described in Subheading 3.1, **steps 1–3**. When using short synthetic nucleotides, perform reaction with approximately 100 pmol synthetic RNA in a 100 μL reaction mixture. Omit the Ribolock RNase inhibitor.
2. Heat the sample for 2 min at 90 °C to denature the MTase (*see* **Note 14**).
3. Recover the RNA by precipitation: Add 1/10 volume of 3 M NaCl and 3 volumes of cold 100 % ethanol. Mix well by inverting the tube several times and store in dry ice for 30 min (*see* **Note 15**). Centrifuge at 16,000 rcf for 30 min at 4 °C, remove the supernatant, and wash the pellet twice with 50 μL of 70 % ethanol.
4. Dissolve the pellet in 14 μL water, add 2 μL of 10× RNase P1 buffer, 2 μL of Nuclease P1, and 2 μL of snake venom phosphodiesterase. Prepare a control sample containing 100 pmol of non-methylated RNA substrate in parallel. Incubate both reactions for 4 h at 37 °C.
5. Add 2.5 μL of 10× alkaline phosphatase buffer and 2.5 μL alkaline phosphatase and incubate for 2 h at 37 °C.
6. Heat the sample for 2 min at 90 °C and allow to slowly cool to room temperature.
7. Add 50 μL of chloroform and vortex vigorously. Centrifuge at 20,000 rcf for 2 min (*see* **Note 16**).
8. Load 15 μL of the upper aqueous phase into a glass vial suitable for the HPLC autosampler.
9. Analyze the digestion products of methylated and unmethylated samples on an RP-C18 HPLC column at 30 °C (*see* **Note 17**),

using a gradient of 0–20 % B in 10 column volumes at a flow rate of 0.5 mL/min (*see* **Note 18**). Record UV absorbance at at least 2 wavelengths (e.g., 260 nm and 280 nm) or record spectra from 220 to 400 nm using a DAD.

10. Analyze reference standards of expected methylated nucleosides in parallel to allow for comparison of retention times and UV spectra.

4 Notes

1. Magnesium ions are important for folding of in vitro transcripts or synthetic RNAs.
2. The catalytic activity of recombinant RNA MTases varies between different proteins and preparations. The optimal amount of protein should be experimentally determined by a dilution series.
3. Tritium (^{3}H) is a weak beta emitter with a physical half-life of approximately 11 years. ^{3}H cannot be detected by a "Geiger"-counter and can only be monitored by scintillation counting ("wipe"-tests).
4. ^{3}H–SAM is unstable at physiological pH at room temperature; therefore, ^{3}H–SAM is delivered in an acidic solution (pH 4.0). Store ^{3}H–SAM at −20 °C in small aliquots and avoid multiple freeze-thaw cycles. Ensure safe handling of radioactive samples according to the corresponding radiation safety rules.
5. If the fluorographic detection method is not being used, the reaction can be performed with nonradioactive SAM (final concentration 1 μM).
6. The methylation reaction can be carried out between 22 and 37 °C. Proteins that tend to aggregate or are instable at higher temperature should be incubated at 22 °C.
7. Samples can be stored at −20 °C until use.
8. Recommended gel size is 18 × 24 cm.
9. Rinse the wells with a syringe at least twice. If urea stays in the wells of the gel it will be difficult to load the whole sample and the RNA might not be properly resolved.
10. The xylene cyanol band migrates similar to a single-stranded RNA of approximately 50 nucleotides and the bromophenol blue band similar to approximately 10 nucleotides under these conditions.
11. Recommendation: use a metal plate clamped to the glass plate to distribute heat equally.
12. ^{3}H–SAM is positively charged and therefore all ^{3}H–SAM that is not incorporated into RNA by the methylation reaction will

stay in the upper bath. The upper bath can therefore contain a significant amount of radioactivity after the gel run.

13. Treatment of the gel with Amplify™ Fluorographic Reagent will increase signal intensity by approximately 1000-fold.
14. Alternatively, remove the MTase protein by PCI (phenol/chloroform/isoamyl-alcohol; 25:24:1) extraction prior to ethanol precipitation.
15. Precipitation can also be performed at −80 °C for 1 h, or at −20 °C for 16 h.
16. Chloroform extraction will only partially remove proteins; however, phenol:chloroform extraction should be avoided as residual phenol often interferes with HPLC analysis. Alternatively, solid-phase extraction or ultrafiltration can also be used to partially remove the proteins. Regardless of which purification method is applied, it is recommended to use a suitable guard column.
17. Retention times vary with temperature. Use of a thermoregulated column oven is recommended for reproducible results.
18. The gradient may need to be adjusted for optimal separation of the desired nucleosides. A slope of 0.5 % per column volume is recommended as an initial condition.

Acknowledgments

We thank Ahmed Warda for comments on the manuscript. This work was supported by the Deutsche Forschungsgemeinschaft (SPP 1784: BO3442/2-1 to M.T.B.; HO4436/2-1 to C.H.) and the Faculty of Medicine, Georg-August-University Göttingen (M.T.B. and "Startförderung" to S.H.).

References

1. Cantara WA, Crain PF, Rozenski J, McCloskey JA, Harris KA, Zhang X et al (2011) The RNA modification database, RNAMDB: 2011 update. Nucleic Acids Res 39:D195–D201
2. Motorin Y, Helm M (2011) RNA nucleotide methylation. Wiley Interdiscip Rev RNA 2: 611–631
3. Jurkowski TP, Meusburger M, Phalke S, Helm M, Nellen W, Reuter G et al (2008) Human DNMT2 methylates tRNA(Asp) molecules using a DNA methyltransferase-like catalytic mechanism. RNA 14:1663–1670
4. Müller S, Windhof IM, Maximov V, Jurkowski TP, Jeltsch A, Förstner KU et al (2013) Target recognition, RNA methylation activity and transcriptional regulation of the Dictyostelium discoideum Dnmt2-homologue (DnmA). Nucleic Acids Res 41:8615–8627
5. Haag S, Warda AS, Kretschmer J, Günnigmann MA, Höbartner C, Bohnsack MT (2015) NSUN6 is a human RNA methyltransferase that catalyzes formation of m5C72 in specific tRNAs. RNA 21:1532–1543
6. Hendricks CL, Ross JR, Pichersky E, Noel JP, Zhou ZS (2004) An enzyme-coupled colorimetric assay for S-adenosylmethionine-dependent methyltransferases. Anal Biochem 326:100–105

7. Schulz D, Rentmeister A (2012) An enzyme-coupled high-throughput assay for screening RNA methyltransferase activity in E.coli cell lysate. RNA Biol 9:577–586
8. Bohnsack MT, Tollervey D, Granneman S (2012) Identification of RNA helicase target sites by UV cross-linking and analysis of cDNA. Methods Enzymol 511:275–288
9. Haag S, Kretschmer J, Sloan KE, and Bohnsack MT (2016) Crosslinking methods to identify RNA methyltransferase targets *in vivo*. Methods Mol Biol, in this issue
10. Schürer H, Lang K, Schuster J, Mörl M (2002) A universal method to produce in vitro transcripts with homogeneous 3′ ends. Nucleic Acids Res 30:e56
11. Graber D, Trappl K, Steger J, Geiermann A-S, Rigger L, Moroder H, Polacek N, Micura R (2012) Deoxyribozyme-base, semisynthetic access to stable peptidyl-tRNAs exemplified by tRNAVal carrying a macrolide antibiotic resistance peptide. Methods Mol Biol 848:201–213
12. Höbartner C, Wachowius F (2010) Chemical synthesis of modified RNA. In: Mayer G (ed) The chemical biology of nucleic acids. Wiley, Chichester, pp 1–37
13. Haag S, Kretschmer J, Bohnsack MT (2015) WBSCR22/Merm1 is required for late nuclear pre-ribosomal RNA processing and mediates N7-methylation of G1639 in human 18S rRNA. RNA 21:1532.1543
14. Baker MR, Zarubica T, Wright HT, Rife JP (2009) Scintillation proximity assay for measurement of RNA methylation. Nucleic Acids Res 37:e32
15. Grosjean H, Droogmans L, Roovers M, Keith G (2007) Detection of enzymatic activity of transfer RNA modification enzymes using radiolabeled tRNA substrates. Methods Enzymol 425:55–101
16. Roovers M, Kaminska KH, Tkaczuk KL, Gigot D, Droogmans L, Bujnicki JM (2008) The YqfN protein of Bacillus subtilis is the tRNA: m1A22 methyltransferase (TrmK). Nucleic Acids Res 36:3252–3262
17. Grosjean H, Keith G, Groogmans L (2004) Detection and quantification of modified nucleotides in RNA using thin-layer chromatography. Methods Mol Biol 265:357–391
18. Kellner S, Burhenne J, Helm M (2010) Detection of RNA modifications. RNA Biol 7:237–247

Chapter 18

Crosslinking Methods to Identify RNA Methyltransferase Targets In Vivo

Sara Haag, Jens Kretschmer, Katherine E. Sloan, and Markus T. Bohnsack

Abstract

Several crosslinking methods have been developed to identify interacting RNAs for proteins of interest. Here, we describe variants of the UV crosslinking and analysis of cDNA (CRAC) method that allow target identification of RNA methyltransferases on a genome-wide scale. We present a detailed protocol for the application of CRAC in human cells that stably express the protein of interest fused to a tandem affinity tag. After the introduction of a covalent link between the protein and its target RNAs, protein-RNA complexes are purified and bound RNAs trimmed, ligated to adapters, reverse transcribed, and amplified. Sequences obtained from next-generation sequencing are then mapped onto the human genome allowing the identification of possible substrates. For some RNA methyltransferases, e.g., m^5C MTases, their catalytic mechanism can be exploited for chemical crosslinking approaches instead of UV based crosslinking.

Key words UV crosslinking, CRAC, Methyltransferase, CLIP, 5-Methylcytosine, 5-Azacytidine

1 Introduction

The enzymatic transfer reactions of methyl groups to ribonucleosides represent important steps during the biogenesis and maturation of many types of cellular RNAs and the modifications can modulate the biogenesis, structure, function, or stability of the cognate RNAs or RNA-protein complexes. Using computational approaches, approximately 1 % of all protein coding genes in yeast and humans were predicted to encode methyltransferases [1, 2], among them many uncharacterized RNA methyltransferases (RNA MTases). Almost all RNA methyltransferases that have been characterized so far specifically catalyze a single type of modification (e.g., m^6A, m^7G, m^5C), but many of them modify multiple target RNAs that can even belong to different classes of RNA. The functional characterization of the modification enzymes therefore requires genome-wide approaches to identify the complete spectrum of their RNA binding and modification sites.

Alexandra Lusser (ed.), *RNA Methylation: Methods and Protocols*, Methods in Molecular Biology, vol. 1562, DOI 10.1007/978-1-4939-6807-7_18,

During the last decade, several UV crosslinking-based methods have been developed to study protein-RNA interactions on a genome-wide scale. UV treatment of cells at 254 nm introduces covalent bonds between the protein of interest and its bound target RNAs, most frequently at uridines. The initially developed UV crosslinking and immunoprecipitation (CLIP) approach utilized specific antibodies against the protein of interest for immunoprecipitation of covalently linked protein-RNA complexes followed by cDNA synthesis and sequencing of associated RNA fragments [3]. Subsequently, combination of this approach with high-throughput sequencing improved genome coverage (HITS-CLIP) [4, 5], and an alternative library preparation method was developed to circumvent the problem of premature stops of reverse transcriptases at crosslinked nucleotides by replacing an intermolecular RNA ligation step with a more efficient intramolecular cDNA circularization step (iCLIP) [6, 7].

In contrast to the original CLIP protocol, the UV crosslinking and analysis of cDNA method (CRAC) does not depend on specific antibodies against the protein of interest, but instead is based on expression of a tagged variant of the protein [8], thereby avoiding the requirement to optimize or develop individual antibodies for each protein of interest. CRAC was initially developed for the identification of protein-RNA interactions in yeast [9–12]. Here, we now present a detailed protocol for its application in human cells (Fig. 1) [13, 14]. This requires the expression of a tagged (e.g., 2xFLAG-PreScission protease cleavage site-His_6-tag) protein of interest, which can either be expressed from a plasmid or after genomic insertion in stable cell lines, as described below for Flp-In T-REx 293 cells (Thermo Fisher Scientific). The combination of genomic insertion of the transgene with regulated expression, e.g., via a tetracycline-regulatable promoter, allows the expression of the tagged protein to be titrated to endogenous levels. A further advantage of the CRAC method is the inclusion of a denaturing affinity chromatography step, which reduces nonspecific interactions that might otherwise lead to increased background [8]. Trimming of the bound RNA to leave a specific "footprint" is carried out prior to this second purification step and transfer of the sample into the denaturing buffer inactivates the RNase. Furthermore, the ligation of the adapter molecules to the RNA fragments takes place while the protein-RNA complexes are immobilized on beads (Fig. 1a). Finally, the RNA is reverse transcribed, amplified, and sequenced by Illumina next-generation sequencing. The obtained sequence reads are mapped onto the human genome allowing the identification of target RNAs and putative modification sites.

More recently, alternative approaches for the in vivo crosslinking of proteins to their cognate RNAs have also been developed, such as the incorporation of photoactivatable ribonucleosides like 4-thiouridine to improve crosslinking specificity (PAR-CLIP) [15].

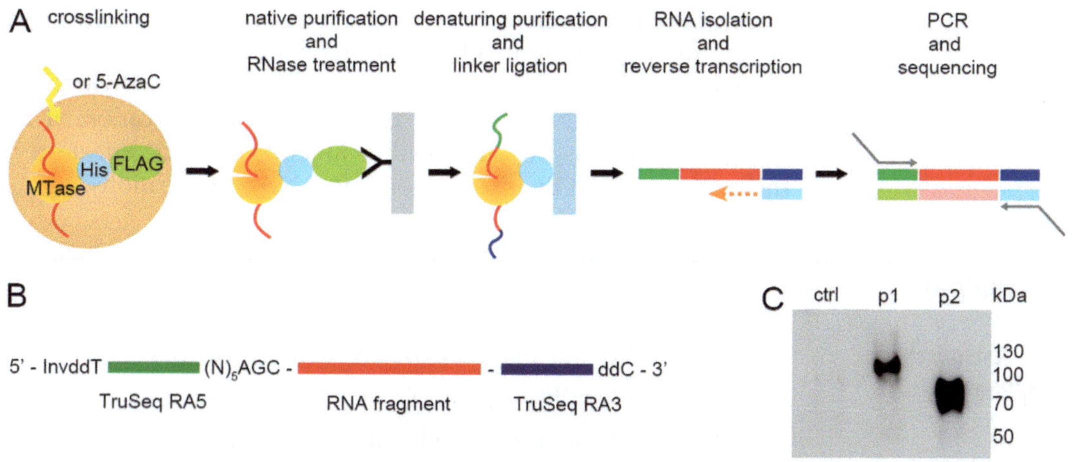

Fig. 1 Overview of UV or 5-azacytidine crosslinking and analysis of cDNA (CRAC). (**a**) Schematic outline of the CRAC method. Cells expressing an RNA MTase fused to a HisPrcFLAG-tag are treated with 5-azacytidine (5-AzaC) or crosslinked using UV light (254 nm). The crosslinked protein-RNA-complexes are affinity purified using anti-FLAG magnetic beads and Ni-NTA agarose. The RNA is trimmed and linkers are ligated to the RNA. After RNA isolation and reverse transcription, the cDNA library is amplified and subjected to Illumina next-generation sequencing. The obtained sequence reads are mapped onto the human genome. (**b**) Schematic view of the product of ligation of the TruSeq RA5 and RA3 adaptors to an isolated RNA fragment. InvddT, inverted dideoxythymidine; ddC, dideoxycytosine. (**c**) Example of an autoradiography of ^{32}P–labeled RNA-protein complexes after separation by SDS-PAGE and transfer to a nitrocellulose membrane. No specific signal is detected after UV crosslinking of cells expressing only the HisPrcFLAG-tag (ctrl), whereas crosslinking of cells expressing RNA-binding proteins such as RNA MTases (p1, p2) leads to radioactive signals approx corresponding to the molecular weights of the proteins, indicating their specific crosslinking to cellular RNAs

Interestingly, for some methyltransferases the specificity of the crosslinking reaction can also be improved by exploiting their catalytic mechanism. RNA MTases, such as 5-methylcytosine (m^{5}C) MTases, initiate the methylation reaction by a nucleophilic attack on the pyrimidine ring of the target cytosine by a cysteine residue that is part of the conserved TC motif, resulting in a covalent intermediate between the protein and its target RNA [16]. After methyl group transfer, release of the target RNA depends on a second cysteine located in a conserved PC motif [17]. Specific crosslinking without UV treatment can be achieved by generating mutants in which the cysteine residue of the PC motif is exchanged for alanine; these mutants can specifically bind their targets but remain in the covalent transition state since the mutation impairs the release of the protein from its target RNA (miCLIP; methylation-iCLIP) [18]. Alternatively, incorporation of the cytosine derivate 5-azacytidine into nascent RNA also traps m^{5}C MTases in the covalent intermediate state resulting in a chemically induced crosslink between the protein and its target RNA (5-AzaC-CLIP/CRAC) [14, 19]. In both cases, the specificity of the crosslinking reaction can be confirmed by generation of catalytically inactive mutants that carry a cysteine to alanine mutation in the TC motif and are therefore impeded in establishing the covalent bond to the target nucleotide. The cross-linking methods presented here thus represent powerful and highly flexible tools for revealing the binding sites of RNA methyltransferases on cellular RNAs, thereby enabling the identification of the target nucleotides [20] and further functional characterization of these enzymes and their corresponding RNA modifications.

2 Materials

2.1 Consumables (See Note 1*)*

1. Anti-FLAG M2 magnetic beads (Sigma-Aldrich).
2. γ-^{32}P-ATP (6000 Ci/mmol).
3. DMEM (Dulbecco's Modified Eagle Medium).
4. Flp-In T-REx 293 Cell Line (Thermo Fisher Scientific).
5. Gel loading dye: 6× (Qiagen).
6. Hybond-C nitrocellulose membrane (GE Healthcare).
7. Low range DNA ladder.
8. Metaphor agarose (Lonza).
9. Micro Bio-Spin columns (Biorad).
10. Ni-NTA agarose.
11. NuPAGE gel (4–12 % polyacrylamide Bis-Tris precast gels; Thermo Fisher Scientific).
12. NuPAGE LDS sample buffer: 4× (supplemented with 50 mM DTT (dithiothreitol); Thermo Fisher Scientific).
13. NuPAGE MES SDS running buffer: 20× (Thermo Fisher Scientific).
14. NuPAGE transfer buffer: 20× (Thermo Fisher Scientific).
15. Prestained protein ladder.
16. pcDNA5/FRT/TO vector (Thermo Fisher Scientific).
17. PreScission protease (approx. 100 μM).
18. Protease inhibitor complete-mini, EDTA free (Roche).
19. Proteinase K, recombinant PCR grade (Roche).
20. QG buffer (Qiagen).
21. RNace-IT ribonuclease cocktail (mixture of RNase A and RNase T1; Agilent Technologies).
22. RNeasy MinElute Cleanup kit (Qiagen).
23. RNasin (RNase inhibitor): 20–40 units/μL (Promega).
24. RNase H: 5 units/μL (New England Biolabs).
25. SuperScript III kit (reverse transcriptase; Invitrogen).
26. T4 PNK (polynucleotide kinase): 5–10 units/μL (New England Biolabs).
27. T4 RNA ligase 1: 10 units/μL (New England Biolabs).

28. T4 RNA ligase 2 deletion mutant: 200 units/μL (Epicentre).
29. TaKaRa LA Taq DNA polymerase and 10× LA Taq buffer (+Mg^{2+}) (Clontech).
30. TruSeq small RNA adapters and oligonucleotides:10 μM RA3 3′ adapter, 100 μM RA5 (N5) 5′ adapter, 10 μM RTP primer, 10 μM RP1 primer, 10 μM RPI primer (Illumina).
31. TSAP (Thermosensitive alkaline phosphatase; Promega).

2.2 Buffers and Solutions

Prepare all buffers with fresh ultrapure water and use only sterile filter tips for pipetting. Adjust the pH of all solutions at the temperature and dilution they will be used at. Store all buffers at 4 °C unless otherwise stated.

1. 1× PBS: 137 mM NaCl, 2.7 mM KCl, 10 mM Na_2HPO_4, 1.8 mM KH_2PO_4 (pH 7.4).
2. TNM150 buffer: 50 mM Tris–HCl pH 7.8, 150 mM NaCl, 1.5 mM $MgCl_2$ 0.1 % NP-40, 5 mM β-mercaptoethanol.
3. TNM1000 buffer: 50 mM Tris–HCl pH 7.8, 1 M NaCl, 1.5 mM $MgCl_2$, 0.1 % NP-40, 5 mM β-mercaptoethanol.
4. 3 M NaCl.
5. 2.5 M imidazole pH 8.0 (adjust pH with HCl).
6. 1× PNK buffer: 50 mM Tris–HCl pH 7.8, 10 mM $MgCl_2$, 0.5 % NP-40, 5 mM β-mercaptoethanol.
7. 5× PNK buffer: 250 mM Tris–HCl pH 7.8, 50 mM $MgCl_2$, 25 mM β-mercaptoethanol (store at −20 °C).
8. Wash buffer I: 50 mM Tris–HCl pH 7.8, 300 mM NaCl, 10 mM imidazole, 6 M guanidine hydrochloride, 0.1 % NP-40, 5 mM β-mercaptoethanol (always prepare fresh).
9. Wash buffer II: 50 mM Tris–HCl pH 7.8,50 mM NaCl, 10 mM imidazole, 0.1 % NP-40, 5 mM β-mercaptoethanol.
10. Elution buffer: 50 mM Tris pH 7.8, 50 mM NaCl, 200 mM imidazole, 0.1 % NP-40, 5 mM β-mercaptoethanol.
11. 1× TBE buffer: 89 mM Tris-borate pH 8.3, 2 mM EDTA.
12. ATP lithium salt solution: 100 mM.
13. ATP solution: 1 mM.
14. BSA solution: 1 μg/μL.
15. SDS solution: 10 %.
16. EDTA: 0.5 M.
17. Sodium acetate (NaAc): 3 M, pH 5.2.
18. Phenol/Chloroform/Isoamylalcohol (25:24:1; PCI).
19. Glycogen: 20 μg/μL.
20. dNTP solution: 5 mM dATP, 5 mM dTTP, 5 mM dGTP, 5 mM dCTP.

3 Method

3.1 Cell Culture

1. Generate a cell line expressing the protein of interest with either a C-terminal His_6-PreScission protease cleavage site (Prc)-2xFLAG tag or a N-terminal 2xFLAG-Prc-His_6 tag under the control of an inducible promoter (*see* **Note 2**).
2. Expand cells in 1× DMEM (37 °C, 5 % CO_2) to approximately 80 % confluence on 3–12 15 cm² dishes.
3. Induce the expression of the tagged RNA methyltransferase with 0.1–1 μM tetracycline 16–36 h prior to UV crosslinking (*see* **Note 3**).

3.2 UV Crosslinking and Cell Lysis

1. Remove DMEM from cells and wash once with 1× PBS.
2. Remove the lid, add 1× PBS to each dish, and irradiate the cells three times with 800 mJ/cm² (254 nm) in a Stratalinker.
3. Remove the PBS, harvest cells by scraping in 150 μL of TMN150 buffer containing protease inhibitors, and transfer cell suspension to a 15 mL tube.
4. Sonicate the cells on ice three times for 15 s (50 % amplitude and 0.3 duty cycle).
5. Centrifuge lysate at 20,000 rcf for 15 min at 4 °C.

3.3 Anti-FLAG Affinity Purification (Native Purification Step)

1. Prepare the anti-FLAG magnetic beads by transferring 100 μL slurry to a 2 mL tube and equilibrating three times with 500 μL TMN150 buffer using a magnetic rack.
2. Transfer the clarified lysate (supernatant of Subheading 3.2, **step 5**) to the magnetic beads and incubate on a rotating wheel at 4 °C for 2 h.
3. Wash the anti-FLAG magnetic beads once with 500 μL TNM150 buffer, twice with 500 μL TNM1000 buffer, and then twice with 500 μL TNM150 buffer as in **step 1**.
4. Elute the complexes by resuspending the anti-FLAG magnetic beads in 600 μL of TMN150, adding 3 μL PreScission protease and incubating overnight at 4 °C (*see* **Note 4**).

3.4 RNase Treatment

1. Transfer the eluate to a fresh tube.
2. Add 0.1–1 unit RNace-It Ribonuclease Cocktail to the sample and incubate for 30–300 s at 37 °C (*see* **Note 5**).
3. Stop the RNase digestion by transferring solution to the tubes containing 0.4 g guanidine hydrochloride, 45 μL 3 M NaCl, and 3 μL 2.5 M imidazole, and vortex well.

3.5 Ni-NTA Affinity Purification (Denaturing Purification Step)

1. Equilibrate 100 μL of Ni-NTA agarose slurry three times with Wash buffer I.
2. Transfer the sample to the equilibrated Ni-NTA agarose and incubate on a rotating wheel for at least 2 h at 4 °C.
3. Centrifuge for 10 s at 500 rcf.
4. Wash the beads twice with 750 μL Wash buffer I and then three times with 750 μL 1× PNK buffer. Centrifuge for 10 s at 500 rcf between washing steps.
5. Transfer the Ni-NTA agarose, in the buffer of the last washing step, to a Micro Bio-Spin column.

3.6 Dephosphorylation of Co-precipitated RNA

1. Centrifuge columns at 500 rcf for 10 s to spin out the remaining buffer, plug the bottom of the column, and add a mixture containing 54 μL water, 16 μL 5× PNK buffer, 2 μL RNasin, and 8 μL TSAP. Cover the top of the column loosely, mix gently, and incubate for 30 min at 37 °C (*see* **Notes 6** and 7).
2. Wash the beads once with 400 μL Wash buffer I and then three times with 400 μL 1× PNK buffer. Allow the column to empty by gravity flow between washing steps (*see* **Note 8**).

3.7 On-Bead Ligation of the 3′ Adapter

1. Spin out the remaining buffer, plug the bottom of the column, and add a mixture containing 34 μL water, 16 μL 5× PNK buffer, 8 μL TruSeq RA3 3′-adapter, 16 μL PEG8000, 2 μL RNasin, and 4 μL T4 RNA ligase 2 deletion mutant. Cover the top of the column loosely, mix gently, and incubate the reaction at 16 °C overnight (Fig. 1b; *see* **Note 9**).
2. Wash the beads once with 400 μL of wash buffer I and then three times with 400 μL 1× PNK buffer. Allow the column to empty by gravity flow between washing steps.

3.8 Phosphorylation of the 5′ Ends of the Co-precipitated RNA

1. Spin out the remaining buffer, plug the bottom of the column, and add a mixture containing 50 μL water, 16 μL 5× PNK buffer, 8 μL T4 PNK, 2 μL RNasin, and 4 μL of γ-^{32}P-ATP (*see* **Note 10**).
2. Cover the top of the column loosely and incubate for 40 min at 37 °C.
3. Add 1 μL ATP lithium salt solution and allow the reaction to proceed for a further 20 min (*see* **Note 11**).
4. Wash beads once with 400 μL of Wash buffer I and then three times with 400 μL 1× PNK buffer. Allow the column to empty by gravity flow between washing steps.

3.9 On-Bead Ligation of the 5′ Adapter

1. Spin out the remaining buffer, plug the bottom of the column, and add a mixture containing 49 μL water, 16 μL 5× PNK buffer, 1 μL TruSeq RA5 5′-adapter, 8 μL ATP, 2 μL RNasin, and4 μL T4 RNA ligase 1, and incubate overnight at 16 °C (Fig. 1b; *see* **Note 12**).

2. Wash the beads once with 400 μL of Wash buffer I and three times with 400 μL of Wash buffer II. Allow the column to empty by gravity flow between washing steps.

3.10 Elution from Ni-NTA Agarose and TCA Precipitation

1. Spin out the remaining buffer and incubate the beads in 200 μL elution buffer for 5 min. Collect the eluate in a clean tube by centrifuging at 500 rcf for 10 s. Repeat and pool both elution fractions in one tube.
2. Add 2 μL of BSA solution and 100 μL of 100 % TCA (20 % final conc.), vortex and incubate on ice for 20 min.
3. Centrifuge for 30 min at 4 °C at 20,000 rcf and carefully remove the supernatant with a pipette (*see* **Note 13**).
4. Add 500 μL of ice-cold acetone to the tube containing the pellet, centrifuge for 20 min at 20,000 rcf, and carefully remove supernatant.
5. Air-dry the pellet for at least 5 min at room temperature, resuspend in 30 μL of NuPAGE LDS sample buffer, and incubate at 65 °C for 10 min (*see* **Note 14**).

3.11 SDS-PAGE and RNA Isolation

1. Separate the sample on a 4–12 % Bis-Tris NuPAGE gel alongside 5 μL of prestained protein marker. Leave at least one empty lane in between different samples. Run the gel in 1× NuPAGE MES SDS running buffer at 100 V for approximately 1.5 h (*see* **Note 15**).
2. Transfer the complexes to Hybond-C nitrocellulose membrane using a wet blot system and 1× NuPAGE transfer buffer complemented with 20 % methanol for 2 h at 80 V at 4 °C.
3. Air-dry the membrane and detect crosslinked, radioactively labeled RNAs by autoradiography (Fig. 1c; *see* **Note 16**).
4. Cut out the region of the membrane corresponding to the size of the protein of interest as well as the area directly above that will contain complexes of the protein and its associated RNA.
5. Incubate the membrane slices in 400 μL of Wash buffer II supplemented with 40 μL of 10 % SDS, 4 μL of 0.5 M EDTA, and 6 μL of proteinase K solution for at least 2 h at 55 °C.
6. Add 50 μL of 3 M NaAc and 500 μL of PCI, mix thoroughly, and centrifuge at 20,000 rcf for 5 min. Transfer the upper aqueous phase to a clean tube, add 1 μL glycogen, and precipitate the RNA with 1 mL of 100 % ethanol. Incubate overnight at −20 °C.
7. Pellet the RNA by centrifuging for 30 min at 20,000 rcf at 4 °C, wash the pellet with 500 μL 70 % ethanol, and air-dry for at least 5 min.

3.12 Reverse Transcription and PCR Reaction

1. Resuspend the RNA pellet in 10 μL water, 1 μL RTP primer, and 2 μL of dNTPs.
2. Heat the mixture to 80 °C for 3 min and incubate on ice for 5 min. Collect the contents at the bottom of the tube by brief centrifugation and add 4 μL 5× first strand buffer, 1 μL of 100 mM DTT, and 1 μL RNasin (*see* **Note 17**).
3. Incubate the mixture at 55 °C for 3 min and then add 1 μL of Superscript III reverse transcriptase and incubate the reaction at 55 °C for 1 h.
4. Inactivate Superscipt III reverse transcriptase by incubation at 65 °C for 15 min.
5. Add 2 μL of RNase H and incubate for 30 min at 37 °C (*see* **Note 18**).
6. Set up 3–6 50 μL PCR reactions each containing 40 μL water, 1 μL of the synthesized cDNA, 5 μL 10× LA Taq Buffer (+ Mg^{2+}), 1 μL RP1 primer, 1 μL RPI primer, 1.5 μL dNTPs and 0.5 μL Takara LA Taq polymerase (*see* **Note 19**).
7. Apply the following thermocycling program: denaturing at 95 °C for 2 min; up to 40 cycles of 98 °C for 30 s, 55 °C for 40 s and 68 °C for 40 s; final elongation at 72 °C for 5 min.
8. Pool the PCR reactions, add 1/10 of volume of NaAc, extract the DNA with PCI, and ethanol precipitate it at −20 °C overnight as above (*see* Subheading 3.11, **steps 6–7**).

3.13 Agarose Gel Electrophoresis and Gel Elution

1. Prepare a 3 % Metaphor agarose gel in 1× TBE (*see* **Note 20**).
2. Dissolve the DNA in 15 μL ultrapure water and add 5 μL 6× Gel loading dye.
3. Load 10 μL in two adjacent lanes and load 5 μL of the Gene ruler low range DNA ladder at least five lanes away from the sample.
4. Run the gel in 1× TBE at 100 mA until the Orange G band migrates to the bottom of the gel.
5. Excise gel slices containing fragments of approximately 150–300 bp using a clean scalpel (*see* **Note 21**).
6. To extract the DNA fragments add 500 μL QG buffer to the agarose slice and incubate at 50 °C until agarose is dissolved completely.
7. Add 200 μL isopropanol and transfer onto a column from the RNeasy MinElute Cleanup kit, centrifuge at 10,000 rcf for 1 min, and then transfer the column to a new collection tube.
8. Add 500 μL QG buffer and centrifuge at 10,000 rcf for 1 min.
9. Add 500 μL 80 % ethanol and centrifuge at 10,000 rcf for 1 min, then transfer the column to a clean collection tube.

10. Centrifuge 5 min at 10,000 rcf and transfer the column to a clean tube.
11. Add 20 μL of ultrapure water, incubate for 1 min at room temperature, and elute DNA by centrifuging at 10,000 rcf for 1 min.
12. Determine the DNA concentration using a Qubit fluorometer or a Bioanalyzer according to the manufacturer's instructions.

3.14 Next-Generation Sequencing and Data Analysis

1. Sequence the cDNA library using an Illumina system, such as HiSeq2000 or HiSeq4000 (*see* **Note 22**).
2. Map acquired sequencing reads to the human genome using programs such as Flexbar [21] for quality control and adapter removal, and an alignment software such as Bowtie [22], Bowtie2 [23], Novoalign (www.novocraft.com) or TopHat2 [24] for mapping onto the human genome (*see* **Note 23**).
3. Further analysis of the data can be performed in online environments such as CLIPZ [25] or using custom-made pipelines (*see* **Note 24**).

4 Notes

1. It is strongly recommended to use chemicals and consumables from the suggested suppliers where stated. This is important due to the sensitivity of the protocol and the requirement for specific properties such as high chemical purity and compatibility with the buffers used in this protocol (*see also* **Note 6**).
2. The Flp-In T-REx 293 cell line system is suitable as it enables the integration of the transgene at a single, specific locus in the genome and the inducible expression of the tagged protein via a tetracycline regulatable promoter. The expression level of the tagged protein should be titrated to the level of the endogenous protein. Cells expressing only the tag should be included in CRAC experiments as a control.
3. For m^5C MTases, chemical crosslinking can be done by adding 4 μM 5-azacytdine to the media 14–16 h prior to harvesting. The UV crosslinking (Subheading 3.2, **step 2**) should be omitted for cells that have been previously treated with chemical crosslinkers.
4. The amount of self-made PreScission protease needs to be empirically determined for each batch of protein since the catalytic activity can vary between different preparations.
5. The amount of RNase and the incubation time required to partially digest the RNA need to be empirically determined for each protein-RNA complex. RNase concentrations and treatment times can be titrated and the RNA fragments cloned using the

TOPO-TA kit (Invitrogen) then sequenced in a 96-well plate to identify the optimal RNase digestion conditions. RNA fragments of approximately 20–40 nt in length are desired as they can be sequenced by Illumina deep sequencing and will enable a specific binding site on the target RNA to be identified.

6. Enzymes that are used for the CRAC experiment should not contain His-tags to avoid binding to the Ni-NTA beads. The alkaline phosphatase used in this protocol (TSAP; Promega) works well in 1× PNK buffer.
7. Close top with parafilm or loosely with a lid. Do not press a lid tightly onto the column as the liquid sample may otherwise be forced through the column filter.
8. The washing step with Wash buffer I inactivates the TSAP.
9. The 3′ adapter used in this step has a blocked 3′dideoxycytosine (ddC) end and is pre-adenylated (rApp) at the 5′ end. These adapters can be ligated to the RNA substrate by the T4 RNA ligase 2 deletion mutant in the absence of ATP.
10. Co-precipitated RNAs are radioactively labeled to enable detection and isolation of protein-RNA complexes after SDS-PAGE and autoradiography. γ-^{32}P is a beta emitter with a half-life of approximately 14 days. It should always be handled with caution (e.g., shielding using a plexiglass screen) and can be monitored using a Geiger counter. Ensure safe handling of radioactive samples according to the corresponding radiation safety rules.
11. This step ensures that the large majority of the RNA 5′ ends carry a phosphate group.
12. The 5′-adapter is a DNA-RNA hybrid, which has an inverted dideoxythymidine (InvddT) at the 5′ end to block multimerization of the oligonucleotide during the ligation reaction. Furthermore, a (N_5)AGC sequence is included at the 3′end of the adapter to enable multiple sequence reads generated from the same PCR template to be disregarded during the data analysis. Inclusion of the AGC sequence was found to improve ligation efficiency.
13. The efficiency of the precipitation can be monitored with a "Geiger"-counter as approximately 90–100 % of the radioactive signal should be detectable in the pellet.
14. Samples can be stored at −20 °C until use.
15. The use of a well-buffered gel system such as the NuPAGE gel system is essential since the pH remains close to 7.0 throughout separation whereas with conventional Laemmli-based SDS-PAGE systems the pH can increase to 9.0 leading to hydrolysis of the RNA. Note that for large proteins, NuPAGE MOPS SDS running buffer may provide more optimal separation.

16. The exposure time can vary between 20 min and 20 h due to differences in the crosslinking efficiency and the protein expression level. Use a fluorescent ruler to align the X-ray film on the membrane and enable accurate cutting of the desired region of membrane.
17. These chemicals are taken from the Superscript III reverse transcriptase kit.
18. cDNA can be stored at −80 °C until use.
19. The Illumina TruSeq small RNA index primers (RPI) introduce a defined barcode required for a multiplexed sequencing reaction. In this case, the Illumina pooling guidelines for recommended barcode combinations should be followed.
20. Metaphor agarose is an intermediate melting temperature agarose that is optimal for the fine resolution of small DNA fragments (20–800 bp) and can therefore be used for the visualization and recovery of the amplified cDNA library. To reduce foam formation during Metaphor agarose preparation, soak the agarose in 1× TBE for 30 min prior to heating in a microwave.
21. A band corresponding to an adaptor-adaptor ligation product may occur at approximately 120 bp. Cut bands above only.
22. The Illumina system delivers a good read quality with a high number of short reads. Paired-end sequencing can be performed to further increase the sequence length and quality. The samples can be multiplexed by using barcoded adapters (*see* **Note 19**). The number of samples per lane and the mixing ratio depends on the required number of sequencing reads per sample.
23. Different assemblies of the human genome can be downloaded from genome.ucsc.edu or www.ensembl.org.
24. Custom-made pipelines often consist of individual modules/programs for data mapping and analysis that are linked, e.g., by python scripts. Various such pipelines are currently available (see, for example, [15, 26, 27]).

Acknowledgments

This work was supported by the Deutsche Forschungsgemeinschaft (SPP 1784: BO3442/2-1 to M.T.B.), the Alexander von Humboldt Foundation (K.E.S. and M.T.B.), and the Faculty of Medicine, Georg-August-University Göttingen (M.T.B. and "Startförderung" to S.H.).

References

1. Petrossian T, Clarke S (2009) Bioinformatic identification of novel methyltransferases. Epigenomics 1:163–175
2. Petrossian TC, Clarke SG (2011) Uncovering the human methyltransferasome. Mol Cell Proteomics 10:M110.000976
3. Ule J, Jensen KB, Ruggiu M, Mele A, Ule A, Darnell RB (2003) CLIP identifies Nova-regulated RNA networks in the brain. Science 302:1212–1215
4. Licatalosi DD, Mele A, Fak JJ, Ule J, Kayikci M, Chi SW et al (2008) HITS-CLIP yields genome-wide insights into brain alternative RNA processing. Nature 456:464–469
5. Zhang C, Darnell RB (2011) Mapping in vivo protein-RNA interactions at single-nucleotide resolution from HITS-CLIP data. NatBiotechnol 29:607–614
6. Wang Z, Kayikci M, Briese M, Zarnack K, Luscombe NM, Rot G et al (2010) iCLIP predicts the dual splicing effects of TIA-RNA interactions. PLoSBiol 8:e1000530
7. Konig J, Zarnack K, Rot G, Curk T, Kayikci M, Zupan B et al (2010) iCLIP reveals the function of hnRNP particles in splicing at individual nucleotide resolution. Nat Struct Mol Biol 17:909–915
8. Bohnsack MT, Tollervey D, Granneman S (2012) Identification of RNA helicase target sites by UV cross-linking and analysis of cDNA. Methods Enzymol 511:275–288
9. Grannemann S, Kudla G, Petfalski E, Tollervey D (2009) Identification of protein binding sites on U3 snoRNA and pre-rRNA by UV cross-linking and high-throughput analysis of cDNAs. Proc Natl Acad Sci U S A 106:9613–9618
10. Granneman S, Petfalski E, Swiatkowska A, Tollervey D (2010) Cracking pre-40S ribosomal subunit structure by systematic analyses of RNA-protein cross-linking. EMBO J 29:2026–2036
11. Bohnsack MT, Martin R, Granneman S, Ruprecht M, Schleiff E, Tollervey D (2009) Prp43 bound at different sites on the pre-rRNA performs distinct functions in ribosome synthesis. Mol Cell 36:583–592
12. Martin R, Hackert P, Ruprecht M, Simm S, Brüning L, Mirus O et al (2014) A pre-ribosomal RNA interaction network involving snoRNAs and the Rok1 helicase. RNA 20:1173–1182
13. Sloan KE, Leisegang MS, Doebele C, Ramirez AS, Simm S, Safferthal C et al (2015) The association of late-acting snoRNPs with human pre-ribosomal complexes requires the RNA helicase DDX21. Nucleic Acids Res 43, 553–564.
14. Haag S, Warda AS, Kretschmer J, Günnigmann MA, Höbartner C, Bohnsack MT (2015) NSUN6 is a human RNA methyltransferase that catalyzes formation of m5C72 in specific tRNAs. RNA 21:1532–1543
15. Hafner M, Landthaler M, Burger L, Khorshid M, Hausser J, Berninger P et al (2010) Transcriptome-wide identification of RNA-binding protein and microRNA target sites by PAR-CLIP. Cell 141:129–141
16. Liu Y, Santi DV (2000) m5C RNA and m5C DNA methyl transferases use different cysteine residues as catalysts. Proc Natl Acad Sci U S A 97:8263–8265
17. King MY, Redman KL (2002) RNA methyltransferases utilize two cysteine residues in the formation of 5-methylcytosine. Biochemistry 41:11218–11225
18. Hussain S, Sajini AA, Blanco S, Dietmann S, Lombard P, Sugimoto Y et al (2013) NSun2-mediated cytosine-5 methylation of vault non-coding RNA determines its processing into regulatory small RNAs. Cell Rep 4:255–261
19. Khoddami V, Cairns BR (2013) Identification of direct targets and modified bases of RNA cytosine methyltransferases. Nat Biotechnol 31:458–464
20. Haag S, Höbartner C, and Bohnsack MT (2016) In vitro assays for RNA methyltransferase activity. Methods Mol Biol, in this issue.
21. Dodt M, Roehr JT, Ahmed R, Dieterich C (2012) Flexbar - flexible barcode and adapter processing for next-generation sequencing platforms. Biology 1:895–905
22. Langmead B, Trapnell C, Pop M, Salzberg SL (2009) Ultrafast and memory-efficient alignment of short DNA sequences to the human genome. GenomeBiol 10:R25
23. Langmead B, Salzberg S (2012) Fast gapped-read alignment with Bowtie 2. Nat Methods 9:357–359
24. Kim D, Pertea G, Trapnell C, Pimentel H, Kelley R, Salzberg SL (2013) TopHat2: accurate alignment of transcriptomes in the presence of insertions, deletions and gene fusions. Genome Biol 14:R36
25. Khorshid M, Rodak C, Zavolan M (2011) CLIPZ: a database and analysis environment for experimentally determined binding sites of RNA-binding proteins. Nucleic Acids Res 39:D245–D252
26. Webb S, Hector RD, Kudla G, Granneman S (2014) PAR-CLIP data indicate that Nrd1-Nab3-dependent transcription termination regulates expression of hundreds of protein coding genes in yeast. Genome Biol 15:R8
27. Chi SW, Zang JB, Mele A, Darnell RB (2009) Ago HITS-CLIP decodes miRNA-mRNA interaction maps. Nature 460:479–486

Chapter 19

Methylated mRNA Nucleotides as Regulators for Ribosomal Translation

Thomas P. Hoernes and Matthias D. Erlacher

Abstract

Methylated RNA nucleotides were recently discovered to be highly abundant in RNAs. The effects of these methylations were mainly attributed to altered mRNA stabilities, protein-binding affinities, or RNA structures. The direct impact of RNA modifications on the performance of the ribosome has not been investigated so far. In this chapter, we describe an approach that allows introducing RNA modifications site-specifically into coding sequences of mRNAs and determining their effect on the translation machinery in a well-defined bacterial in vitro system.

Key words Splinted ligation, In vitro translation, RNA purification, Peptide purification

1 Introduction

Ribosomal translation is an essential process for every organism in all three domains of life. A cell dedicates tremendous amounts of resources to provide fast and accurate protein biosynthesis. Modified RNA nucleotides are absolutely crucial for translation and are present in all involved classes of RNA. The largest number and the greatest variety of RNA modifications are located in transfer RNAs (tRNAs). Some of these very complex RNA nucleotide derivatives were shown to be pivotal for the optimal speed and the accuracy of protein synthesis [1, 2]. Ribosomal RNAs (rRNAs) as well were found to carry nucleotide modifications, probably involved in fine-tuning of structural and functional aspects during translation [3]. Recently, the class of mRNAs was revealed to contain far more nucleotide modifications than previously assumed. Using transcriptome-wide profiling for edited RNA nucleotides, methylated adenosines (m^6A) and cytosines (m^5C) as well as pseudouridines (Ψ) were detected in untranslated and translated regions of mRNAs [4–9]. During revision of this protocol, also the presence of m^1A within mRNAs has been unveiled [10, 11].

Alexandra Lusser (ed.), *RNA Methylation: Methods and Protocols*, Methods in Molecular Biology, vol. 1562,
DOI 10.1007/978-1-4939-6807-7_19, © Springer Science+Business Media LLC 2017

m^6A, the most abundant modification within mRNAs, is present in mRNAs from yeast to mammals [4]. Thereby, m^6A modifications are located in conserved mRNA regions and are enriched in the vicinity of stop codons, potentially influencing mRNA stability or its regulation through miRNAs [12]. The discovery of m^6A demethylases revealed a reversible methylation process, implying a rather dynamic role of m^6A during gene expression [13, 14]. Very recently, also bacterial mRNAs were reported to harbor numerous m^6A nucleotides within their coding sequences, emphasizing their importance not only in eukaryotes [15].

m^5C was described to be highly abundant in eukaryal mRNAs. Through transcriptome analysis, more than 10,000 m^5C sites have been identified in mRNAs [6]. This modification was also reported to be present in archaea, but has not been identified in bacteria so far [5].

The 2′-O-methylation (2′-OCH_3) of riboses might also be employed for translation regulation. So far, this modification has not been identified within coding sequences of mRNAs. However, methylases and furthermore distinct RNA-protein complexes (small nucleolar ribonucleoprotein complexes) are present within eukaryal cells, which possibly introduce this modification into open reading frames of mRNAs [16].

Whereas various publications reported that these mRNA nucleotide derivatives can alter stability, protein-binding affinities, or splicing [17–19], only very little is known about the immediate effect on the translation process. In former studies mRNAs harboring randomly incorporated modifications were employed, which only provided limited insights into the direct effect on protein synthesis [20–23].

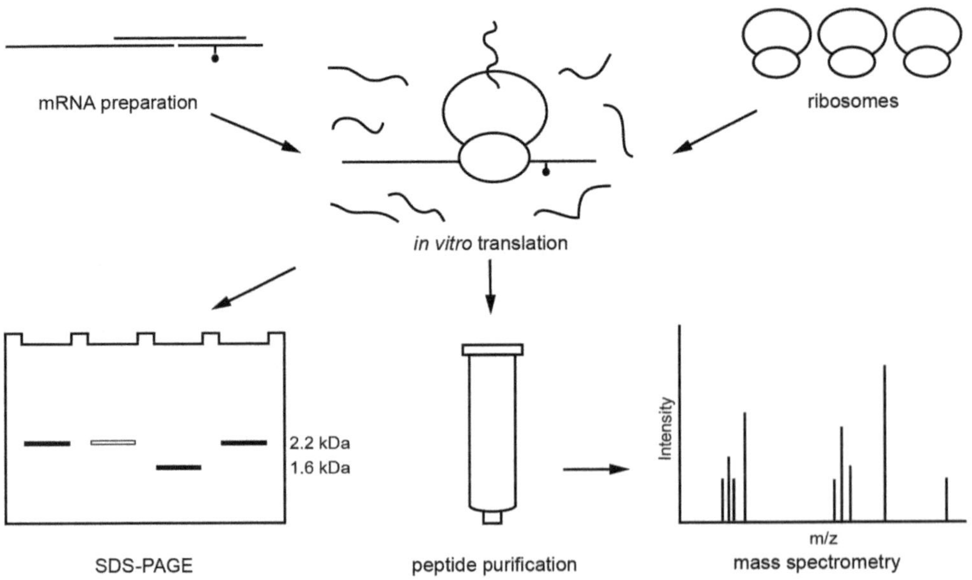

Fig. 1 Experimental strategy to investigate the immediate impact of mRNA modifications on ribosomal translation. mRNAs carrying modifications (*black dot*) are generated by splinted ligation. Subsequently, the mRNA is translated in vitro employing purified *E. coli* 70S ribosomes and the product is quantitatively analyzed by SDS-PAGE. The amino acid composition can also be determined by MS analysis after an additional purification step

By using an RNA ligation approach (Fig. 1), it is possible to introduce a defined number of modifications strictly site-specifically into coding and noncoding regions of mRNAs [24–26]. The products of the subsequent in vitro translation reactions employing these modified mRNAs can then be analyzed for their abundance and amino acid composition. This enables the discovery of potential novel regulatory roles of mRNA modifications during protein biosynthesis.

In the following chapter, we describe an experimental setup allowing the analysis of the impact of mRNA modifications on the translation process and the resulting peptide product.

2 Materials

All the solutions are prepared with ultrapure water and analytical grade reagents. Consult your mass spectrometry (MS) facility to assure the compatibility of the reagents for subsequent MS analysis of the peptide. Peculiar waste disposal of employed materials is indicated in Subheading 4.

2.1 RNA Ligation

1. T7 RNA in vitro transcription kit: HiScribe™ T7 Quick High Yield RNA Synthesis kit (New England Biolabs).
2. Ligases for splinted RNA ligation: T4 DNA ligase (Fermentas) or RNA ligase 2 (New England Biolabs) with the corresponding 10× ligation buffers.
3. PEG 4000.
4. RNase Inhibitor.
5. DNase I.
6. Roti®-Phenol/Chloroform/Isoamylalkohol.
7. GlycoBlue.
8. 30 % Acrylamide/Bisacrylamide (29:1).
9. Ammonium persulfate (APS): 10 % in ddH_2O. Store at 4 °C (*see* **Note 1**).
10. N,N,N′,N′-tetramethyl-ethylenediamine (TEMED). Store at 4 °C.
11. RNA elution buffer: 300 mM NaCl, 60 mM NaOAc (pH 5.2), 0.2 % SDS. Store at room temperature (RT).
12. 2× RNA loading dye: 0.5 mM EDTA, 95 % formamide, 0.025 % SDS, 0.25 % bromophenol blue.

2.2 In Vitro Translation

1. PURExpress® in vitro protein synthesis kit (New England Biolabs).
2. [^{35}S]-Methionine/[^{35}S]-Cysteine or [^{35}S]-Methionine, 31.75 TBq/mmol.

2.3 Peptide Analysis

1. 4× Laemmli loading buffer: 200 mM Tris–HCl (pH 6.8), 8 % SDS, 40 % glycerol, 4 % β-mercaptoethanol, 50 mM EDTA, 0.08 % bromophenol blue. Store at −20 °C.
2. 10× Cathode buffer [27]: 1 M Tris, 1 M Tricine, 1 % SDS. pH should be 8.25, do not adjust the pH. Store at RT.
3. 10× Anode buffer [27]: 1 M Tris, 0.225 HCl. Adjust pH to 8.9. Store at RT.
4. SDS-Tricine-PAGE: Novex™ 16 % Tricine protein gels (Thermo Fisher Scientific).
5. Staining solution: 50 % ethanol, 5 % acetic acid, 1.5 g Comassie Brilliant Blue. Store at RT.
6. Destaining solution: 20 % ethanol, 7.5 % acetic acid. Store at RT.
7. Detection of ^{35}S–labeled peptides: storage phosphor screen. Alternatively, high sensitivity films (BioMax MS films (Kodak)) can be used.
8. Centrifugal concentrator: Vivaspin 2 Hydrosart® membrane, 5000 MWCO (VIVAproducts).

3 Methods

3.1 Preparation of Site-Specifically Modified mRNAs Employing Splinted RNA Ligation

The approach to introduce site-specific modifications into mRNAs is based on a splinted ligation technique (Fig. 1). The ErmCL mRNA [28] represents an ideal candidate for this strategy and is used as an example in the following section.

In general, the mRNAs used for this protocol have to meet several criteria to be suitable for this experimental setup. (1) The

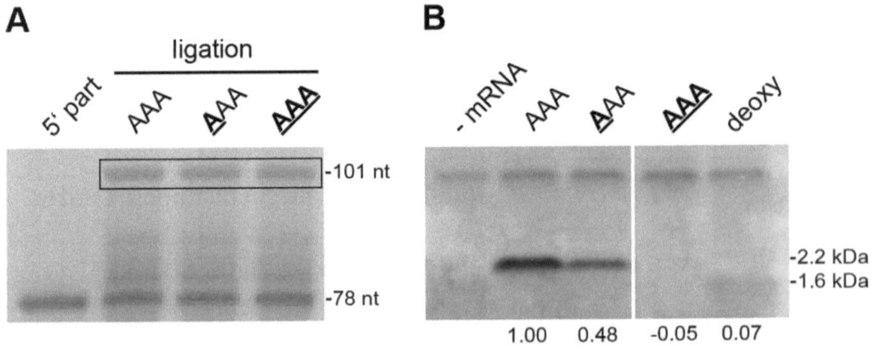

Fig. 2 Preparation and in vitro translation of ErmCL mRNAs modified at defined codon positions. (**a**) Chemically synthesized m^{6}A modified oligos (23 nt) are ligated to T7 tanscripts (78 nt) yielding the full-length ErmCL reporter mRNA (101 nt, frame). Modifications are introduced at the indicated codon positions (*bold* and *underlined*). (**b**) SDS-PAGE analysis of in vitro translated mRNA constructs, carrying modifications at the indicated positions (*bold* and *underlined*); mRNAs containing a codon, harboring three deoxyriboses at codon position 15 of the ErmCL reporter, serve as control (deoxy). SDS-PAGE allows the discrimination between full-length (i.e., 2.2 kDa) and truncated peptides (i.e., 1.6 kDa)

mRNA has to be designed to provide an efficient in vitro translation. (2) It is crucial that the length of the mRNA allows a separation of the unligated and ligated RNA fragments by gel electrophoresis (Fig. 2a). In case of the described ErmCL mRNA the full-length mRNA has a length of 101 nucleotides (nts) and the unligated educts are 78 and 23 nts in length, respectively. This allowed the application of gel electrophoresis to efficiently isolate and purify the ligated product. (3) It is also essential that the size of the translated full-length peptide can be distinguished from the truncated product by SDS-PAGE analysis (Figs. 1 and 2b).

To introduce a modification site specifically into the mRNA chemically synthesized oligonucleotides are used. This enables placing a variety of modifications at any position within the coding sequence of the RNA. Also, the positioning of multiple modified RNA nucleotides is feasible. Because the length of these synthesized RNA fragments is limited to about 70 nts, we use splinted ligations to generate suitable full-length mRNAs. The two RNAs, which need to be ligated, are positioned and bridged by a splint oligonucleotide, base pairing with both fragments during the ligation reaction (Fig. 1). The unmodified RNA fragment can be either generated by T7 transcription or by solid-phase synthesis (*see* **Notes 2** and **3**). The only prerequisite for the 3′ RNA part is a 5′monophosphate essential for the ligation reaction (*see* **Note 4**).

This instruction is applicable to the ligation of various mRNAs; however, different constructs might need minor modifications concerning the ligation point and the employed DNA-splint (*see* **Notes 5** and **6**).

1. Splinted mRNA ligations are generally performed as described [29]. Mix 60 pmol of the transcripts encoding the 5′ part of the ErmCL mRNA with 90 pmol of the synthesized RNA oligonucleotide with a 5′ monophosphate (5′-P-AUUAU<u>XXX</u> CCAAACAAAAAAUAA-3′; the modified codon is underlined) and 60 pmol of the DNA splint (5′-TTTGTTTGG<u>YYY</u> ATAATGAACTGTG-3′) spanning the two RNA fragments (*see* **Note 7**). The volume should be adjusted to 16.2 μl with ddH_2O (*see* **Note 8**).
2. Denature the nucleic acids at 94 °C for 3 min and cool them down to RT for 15 min to allow annealing of the DNA/RNA hybrids. These temperature steps can be performed in a PCR cycler (94 °C, 3 min; 75 °C, 3 min; 50 °C, 3 min; 22 °C, 9 min).
3. Supplement the annealed nucleic acids with the remaining ligation components (*see* **Note 9**): 2.4 μl 10× ligation buffer, 2.4 μl PEG 4000, 40 U RiboLock™ RNase Inhibitor, and 10 U T4 DNA ligase or RNA ligase 2 to a final volume of 24 μl and incubate at 35 °C (T4 DNA ligase) or 37 °C (RNA ligase 2) for 4 h (*see* **Note 10**).
4. Usually, 8–10 ligation reactions are set up in parallel to increase the efficiency of the subsequent purification. Pool all ligation

reactions and digest the DNA splint by the addition of 10 U DNase I at 37 °C for 30 min (*see* **Note 11**).

5. The reactions are purified from proteins by employing phenol-chloroform-isoamylalcohol (PCI) extractions (*see* **Note 12**). Add ddH_2O to a volume of 400 µl to the DNase-treated ligation mix and supplement with the same volume of PCI. Vortex for 30 s and centrifuge at 13,000 × *g* for 10 min. Proteins will separate in the lower organic phase, while RNA remains in the upper aqueous phase (*see* **Note 13**).
6. Transfer the upper aqueous phase to a fresh reaction tube und wash with 400 µl chloroform. Vortex the solution again for 30 s and centrifuge at 13,000 × *g* for 10 min to allow phase separation.
7. Carefully transfer the aqueous phase again to a fresh tube. To precipitate the RNA add 1 ml 100 % ethanol and 40 µl 3 M sodium acetate (pH 5.2) (*see* **Note 14**). Incubate the solution at −80 °C for 20–30 min or precipitate overnight at −20 °C.
8. Pellet the RNA by a centrifugation step at 13,000 × *g* at 4 °C for 30 min. Discard the supernatant without disrupting the RNA pellet and subsequently wash the pellet with 1 ml precooled 70 % ethanol (*see* **Note 15**).
9. Centrifuge again at 13,000 × *g* at 4 °C for 10 min. Remove the supernatant and let the RNA pellet air dry for 2–3 min.
10. Dissolve the RNA in 15 µl ddH_2O and add 15 µl 2× RNA loading dye.
11. Denature the sample at 95 °C for 3 min, transfer the reaction tube to ice, and immediately load on a denaturing 8 % PAA gel (*see* **Notes 16** and **17**).
12. Let the sample enter the gel at 150–180 V and increase the voltage after 15 min to 250–400 V (*see* **Note 18**).
13. Remove the PAA gel from the gel chamber and stain the gel in a plastic tub with ethidium bromide (*see* **Note 19**). We usually add 3 µl 1 % ethidium bromide to 100 ml ddH_2O and stain the gel for 10 min under agitation.
14. Visualize the RNA by employing an UV gel documentation system (Fig. 2a). Excise the ligation product from the gel with a clean scalpel and transfer the gel slice to a fresh reaction tube.
15. Cover the gel slice with elution buffer in the reaction tube. Usually, 100–150 µl elution buffer is sufficient. Shock-freeze the reaction tube in liquid nitrogen and elute the RNA at 4 °C overnight under agitation.
16. Incubate the gel slice for 60 min at 60 °C under agitation. Transfer the solution to a fresh reaction tube and discard the gel. Precipitate the eluted RNA from the solution as described in **steps 7–9**.

17. Resuspend the RNA pellet in ddH_2O, typical volumes range from 8 to 15 μl, depending on the number of ligation reactions and size of the pellet. Store the RNA at −80 °C for downstream assays.

3.2 In-Vitro Translation Employing a Recombinant Bacterial System

To pinpoint immediate effects of modified mRNA nucleotides on ribosomal translation process per se and to exclude indirect effects on translation mediated by proteins present in complex cell lysates, the recombinant prokaryal PURExpress in vitro protein synthesis system is employed [30]. This system is based on a minimal set of purified recombinant protein components essential for translation and thereby considerably reduces the complexity of the reaction. However, the ligated mRNA is suitable for various *E. coli* based translation systems as well [31].

1. Mix the following components in the indicated order in a PCR reaction tube and fill up to 7.5 μl with ddH_2O: 3 μl Solution A, 2.25 μl Solution B, 0.5 μl RiboLock™, 0.5 μl [^{35}S]-Met or [^{35}S]-Met/[^{35}S]-Cys, 200 ng ErmCL mRNA (*see* **Notes 20** and **21**).
2. Mix gently and incubate the reaction at 37 °C for 2 h in a PCR cycler with a heated lid (*see* **Note 22**).
3. Freeze the samples at −20 °C for use at a later time or directly continue with separation of the peptides employing SDS-PAGE.

3.3 Analysis of Translation Products and Preparation for Mass Spectrometry Analysis

Depending on your experimental approach it might be of interest to analyze the quantity and the size of your translated peptide as well as the quality of the translation product, i.e., the exact amino acid sequence. In the former case, conventional sodium dodecyl sulfate polyacrylamide gel electrophoresis (SDS-PAGE) can be employed. In the latter case, mass spectrometry offers a powerful technique to analyze the amino acid sequence. Here, we provide straightforward instructions to prepare the reaction product to be applied for both techniques.

By employing Tricine-SDS-PAGE short peptides can be separated with high resolution [27]:

1. Add 2.5 μl 4× Laemmli buffer to the translation reaction. Denature your sample at 94 °C for 2 min (*see* **Note 23**).
2. Add 1× cathode buffer to the cathode buffer reservoir (upper chamber) and 1× anode buffer to the anode buffer reservoir (lower chamber) of your electrophoresis system.
3. Load the whole denatured PURExpress® reaction to a 16 % Tricine-SDS gel and run it at 100 V. (*see* **Note 24**).
4. Stop electrophoresis shortly before the bromophenol blue dye runs out of the gel. Fix the gel in staining solution for 15 min under agitation.

5. Discard the solution and bath the gel in destaining solution for 15 min (*see* **Note 25**).
6. Move the gel to a Whatman paper, cover it with clingfilm, and dry the gel in a slab gel dryer at 80 °C for 45 min.
7. Expose your dried gel to a phosphorimager screen or a film with high sensitivity (e.g., BioMax MS films, Kodak) overnight.
8. Scan the phosphorimager screen with an appropriate scanner (e.g., a STORM 840 scanner) or develop your film, respectively (Fig. 2b).

Alternatively, the ErmCL peptide can be prepared for subsequent MS analysis. Therefore, the peptide is separated from the protein components of the PURExpress® reaction employing centrifugal concentrators (e.g., Vivaspin 2, Hydrosart® membrane, 5000 MWCO). This step significantly increases the sensitivity in the subsequent MS analysis:

1. Prerinse the centrifugal concentrator with 200 μl ddH_2O to remove storage chemicals (e.g., sodium azide) interfering with MS analysis. Centrifuge the column at 8000 × *g* for 15 min. Discard the flow through.
2. Adjust the PURExpress® reaction to 70 % acetonitrile and 0.1 % formic acid in a final volume of 200 μl (*see* **Note 26**). Load the sample to the centrifugal concentrator und centrifuge at 8000 × *g*, 4 °C for 50 min.
3. Collect the flow through and rinse the centrifugal concentrator with 100 μl 70 % acetonitrile/0.1 % formic acid. Centrifuge again at 8000 × g, 4 °C for 10 min and pool the flow through with the primary flow through.
4. By the addition of 1.2 ml precooled acetone proteins are precipitated. Incubate at −20 °C for 30 min and pellet the ErmCL peptide at 13,000 g, 4 °C for 30 min.
5. Discard the supernatant, let the pellet air dry and dissolve it in 10 μl ddH_2O. Store the sample at −80 °C or submit it directly to MS analysis.

4 Notes

1. APS should not be stored longer than 4 weeks at 4 °C.
2. A high-quality T7-transcript is essential. Assay the length and the purity of the transcripts on 8 % denaturing polyacrylamide (PAA) gels and gel purify the transcripts if necessary.
3. Commercial RNA oligonucleotides significantly differ in quality and purity [32]. In our experience Dharmacon as well as IDT provide a reasonable oligonucleotide quality.

4. T7 transcripts exhibit a 5′-triphosphate group. When a T7 transcript is employed as 3′ part in the splinted ligation the triphosphate has to be converted into a monophosphate to be recognized by DNA/RNA ligases as a substrate. We usually employ FastAP (Thermo Fisher Scientific), CIAP (Promega), or antarctic phosphatase (New England Biolabs) to remove the initial triphosphate according to the manufacturers' instructions. Subsequently, the monophosphate is attached to the RNA by T4 PNK (New England Biolabs) according to the provided protocol.
5. The position of the ligation sites clearly influences the efficiency of the ligation itself. Different ligation sites should be tested to optimize the reaction.
6. For high ligation efficiency different splinter oligonucleotides varying in length should be tested. For most sequences a splint base pairing to 8–15 nts on both sides of the ligation site might be sufficient. The splinter oligonucleotide can be either DNA or 2′O-methylated RNA. 2′O-methylated RNA oligos have been reported to more efficiently break up intramolecular base pairs of the ligation educts resulting in higher ligation efficiencies [29]. In our hands both splint variants work equally efficient.
7. The sequence of the full-length ErmCL reporter construct:
   GGATCC*TAATACGACTCACTATA*GGGCTCTTTAACA
   ATTTATCAGATCCAAT**AGGAGG**AACAAT
   ATGGGCATTTTTAGTATTTTTGTAATCAGCA
   CAGTTCATTATCAACCAAACAAAAAA**TAA.**

   (T7 promoter is depicted in italics. The Shine-Dalgarno sequence as well as the start and stop codons are indicated in bold, the modified codon is underlined).
8. The ligation reaction is assembled on ice.
9. Ligations can be performed with either DNA or RNA ligases. We made good experiences with T4 DNA ligase (Fermentas) and RNA ligase 2 (New England Biolabs). RNA ligase 2 yields higher ligation rates, but in some cases unspecific ligation products can be observed. RNA ligase 2 is especially recommended for the ligation of two chemically synthesized oligonucleotides of high quality. T4 DNA ligase is not as efficient as RNA ligase 2. In any case, it is recommended to pool several ligation reactions for higher yields.
10. Incubation is performed in a PCR cycler with a heated lid.
11. This step can be omitted, if the ligation reaction is purified afterward by a denaturing PAA gel.
12. Phenol is carcinogenic. Perform all PCI work under a fume hood and follow the guidelines for waste disposal.

13. The pH of the PCI reagent determines which nucleic acids partition in the aqueous phase and which in the organic phase. At a neutral pH DNA as well as RNA is found in the aqueous phase. When acidic PCI is used, DNA separates to the organic phase und solely RNA is present in the aqueous phase.
14. It can be helpful to add GlycoBlue to the solution. GlycoBlue facilitates the precipitation of small amounts of RNA and colors the RNA pellet for better visibility. Dilute GlycoBlue (15 mg/ml) 1:500–1:1000 in 100 % ethanol.
15. To avoid disruption of the RNA pellet, we recommend using flattened gel loading tips to remove the supernatant.
16. Add 300 μl 10 % APS and 25 μl TEMED to 60 ml 8 % PAA to start the polymerization process and pour the gel. The size of the employed gels was 150 × 200 mm.
17. It is beneficial to apply the unligated RNA as size marker on the gel.
18. We run the PAA gels until the bromophenol blue dye front reaches the last third of the gel.
19. Ethidium bromide is under suspicion to be mutagenic, carcinogenic, and teratogenic. Wear nitrile gloves that are certified for ethidium bromide work when handling the solutions and the stained gel. Follow the waste disposal regulations for ethidium bromide waste.
20. Optionally 5–10 pmol of 70S ribosomes can be added to the PURExpress® reaction to increase the translation efficiency. Purify ribosomes as described [33]. New England Biolabs recommends 25 μl PURExpress® in vitro translation reactions. However, the reaction can be scaled down to 7.5 μl or even 5 μl. The amount of mRNA, which was scaled down accordingly, should be at least equimolar to the 70S ribosomes.
21. Stick to the general rules for working with radioactive material and disposal of radioactive waste. If your laboratory is not equipped for radioactive experiments omit [^{35}S]-Met from the reaction and visualize ErmCL by silver staining.
22. You will detect low amounts of your translation product after 2–5 min of incubation. In most experiments, 45 min of incubation is sufficient to reach the end point.
23. Boiling your sample can force proteins to aggregate with the SDS in the Laemmli buffer. Alternatively, the sample can be incubated at 37 °C for 15 min.
24. We usually run precast Novex 16 % Tricine gels (Thermo Fisher Scientific), but you can cast tricine gels easily yourself as described [27].

25. After incubating the gel in the staining solution the solution exhibits considerable radioactivity. Discard it to the radioactive liquid waste.

26. 70 % acetonitrile/0.1 % formic acid in solution prevents the ErmCL peptide to stick to the plastic surface of the centrifugal concentrator.

Acknowledgment

This work was supported by the FWF (P 22658-B12 and P 28494-BBL to M.E.)

References

1. El Yacoubi B, Bailly M, de Crecy-Lagard V (2012) Biosynthesis and function of posttranscriptional modifications of transfer RNAs. Annu Rev Genet 46:69–95
2. Nedialkova DD, Leidel SA (2015) Optimization of codon translation rates via tRNA modifications maintains proteome integrity. Cell 161(7):1606–1618
3. Wilson DN, Nierhaus KH (2007) The weird and wonderful world of bacterial ribosome regulation. Crit Rev Biochem Mol Biol 42(3):187–219
4. Pan T (2013) N6-methyl-adenosine modification in messenger and long non-coding RNA. Trends Biochem Sci 38(4):204–209
5. Edelheit S, Schwartz S, Mumbach MR, Wurtzel O, Sorek R (2013) Transcriptome-wide mapping of 5-methylcytidine RNA modifications in bacteria, archaea, and yeast reveals m5C within archaeal mRNAs. PLoS Genet 9(6):e1003602
6. Squires JE, Patel HR, Nousch M, Sibbritt T, Humphreys DT, Parker BJ, Suter CM, Preiss T (2012) Widespread occurrence of 5-methylcytosine in human coding and non-coding RNA. Nucleic Acids Res 40(11):5023–5033
7. Carlile TM, Rojas-Duran MF, Zinshteyn B, Shin H, Bartoli KM, Gilbert WV (2014) Pseudouridine profiling reveals regulated mRNA pseudouridylation in yeast and human cells. Nature 515:143–146
8. Lovejoy AF, Riordan DP, Brown PO (2014) Transcriptome-wide mapping of pseudouridines: pseudouridine synthases modify specific mRNAs in *S. cerevisiae*. PLoS One 9(10):e110799
9. Hoernes TP, Erlacher MD (2016) Translating the epitranscriptome. WIREs RNA 2016. doi:10.1002/wrna.1375
10. Li X, Xiong X, Wang K, Wang L, Shu X, Ma S, Yi C (2016) Transcriptome-wide mapping reveals reversible and dynamic N(1)-methyladenosine methylome. Nat Chem Biol 12(5):311–316
11. Dominissini D, Nachtergaele S, Moshitch-Moshkovitz S, Peer E, Kol N, Ben-Haim MS, Dai Q, Di Segni A, Salmon-Divon M, Clark WC et al (2016) The dynamic N(1)-methyladenosine methylome in eukaryotic messenger RNA. Nature 530(7591):441–446
12. Meyer KD, Saletore Y, Zumbo P, Elemento O, Mason CE, Jaffrey SR (2012) Comprehensive analysis of mRNA methylation reveals enrichment in 3′ UTRs and near stop codons. Cell 149(7):1635–1646
13. Fu Y, Dominissini D, Rechavi G, He C (2014) Gene expression regulation mediated through reversible m(6)A RNA methylation. Nat Rev Genet 15(5):293–306
14. Jia G, Fu Y, Zhao X, Dai Q, Zheng G, Yang Y, Yi C, Lindahl T, Pan T, Yang YG, He C (2011) N6-methyladenosine in nuclear RNA is a major substrate of the obesity-associated FTO. Nat Chem Biol 7(12):885–887
15. Deng X, Chen K, Luo GZ, Weng X, Ji Q, Zhou T, He C (2015) Widespread occurrence of N6-methyladenosine in bacterial mRNA. Nucleic Acids Res 43:6557–6567
16. Vitali P, Basyuk E, Le Meur E, Bertrand E, Muscatelli F, Cavaille J, Huttenhofer A (2005) ADAR2-mediated editing of RNA substrates in the nucleolus is inhibited by C/D small nucleolar RNAs. J Cell Biol 169(5):745–753
17. Wang Y, Li Y, Toth JI, Petroski MD, Zhang Z, Zhao JC (2014) N6-methyladenosine modification destabilizes developmental regulators in embryonic stem cells. Nat Cell Biol 16(2):191–198

18. Wang X, Zhao BS, Roundtree IA, Lu Z, Han D, Ma H, Weng X, Chen K, Shi H, He C (2015) N(6)-methyladenosine modulates messenger RNA translation efficiency. Cell 161(6):1388–1399
19. Liu N, Dai Q, Zheng G, He C, Parisien M, Pan T (2015) N(6)-methyladenosine-dependent RNA structural switches regulate RNA-protein interactions. Nature 518(7540):560–564
20. Kariko K, Muramatsu H, Keller JM, Weissman D (2012) Increased erythropoiesis in mice injected with submicrogram quantities of pseudouridine-containing mRNA encoding erythropoietin. Mol Ther 20(5):948–953
21. Kariko K, Muramatsu H, Ludwig J, Weissman D (2011) Generating the optimal mRNA for therapy: HPLC purification eliminates immune activation and improves translation of nucleoside-modified, protein-encoding mRNA. Nucleic Acids Res 39(21):e142
22. Kariko K, Muramatsu H, Welsh FA, Ludwig J, Kato H, Akira S, Weissman D (2008) Incorporation of pseudouridine into mRNA yields superior nonimmunogenic vector with increased translational capacity and biological stability. Mol Ther 16(11):1833–1840
23. Heilman KL, Leach RA, Tuck MT (1996) Internal 6-methyladenine residues increase the in vitro translation efficiency of dihydrofolate reductase messenger RNA. Int J Biochem Cell Biol 28(7):823–829
24. Simms CL, Hudson BH, Mosior JW, Rangwala AS, Zaher HS (2014) An active role for the ribosome in determining the fate of oxidized mRNA. Cell Rep 9(4):1256–1264
25. Hudson BH, Zaher HS (2015) O6-Methylguanosine leads to position-dependent effects on ribosome speed and fidelity. RNA 21(9):1648–1659
26. Hoernes TP, Clementi N, Faserl K, Glasner H, Breuker K, Lindner H, Hüttenhofer H, Erlacher MD (2016) Nucleotide modifications within bacterial messenger RNAs regulate their translation and are able to rewire the genetic code. Nucleic Acids Res 44(2):852–862
27. Schägger H (2006) Tricine-SDS-PAGE. Nat Protoc 1(1):16–22 doi:nprot.2006.4
28. Vazquez-Laslop N, Thum C, Mankin AS (2008) Molecular mechanism of drug-dependent ribosome stalling. Mol Cell 30(2): 190–202
29. Lang K, Micura R (2008) The preparation of site-specifically modified riboswitch domains as an example for enzymatic ligation of chemically synthesized RNA fragments. Nat Protoc 3(9):1457–1466
30. Shimizu Y, Inoue A, Tomari Y, Suzuki T, Yokogawa T, Nishikawa K, Ueda T (2001) Cell-free translation reconstituted with purified components. Nat Biotechnol 19(8): 751–755
31. Erlacher MD, Chirkova A, Voegele P, Polacek N (2011) Generation of chemically engineered ribosomes for atomic mutagenesis studies on protein biosynthesis. Nat Protoc 6(5):580–592
32. Schmid K, Thuring K, Keller P, Ochel A, Kellner S, Helm M (2015) Variable presence of 5-methylcytosine in commercial RNA and DNA. RNA Biol 12(10):1152–1158
33. Bommer U, Burkhardt N, Jünemann R, Spahn CM, Triana-Alonso FJ, Nierhaus KH (1997) Ribosomes and polysomes. Subcellular fractionation—a practical approach. IRL Press, Washington, DC, pp 271–301

Chapter 20

Automated Chemical Solid-Phase Synthesis and Deprotection of 5-Hydroxymethylcytosine-Containing RNA

Christian Riml and Ronald Micura

Abstract

5-Hydroxymethylcytosine is an epigenetic base modification that is part of the demethylation pathway of 5-methylcytosine in DNA. 5-Methylcytosine is iteratively oxidized to 5-hydroxymethylcytosine, 5-formylcytosine, and 5-carboxycytosine by enzymes of the TET protein family. Since the discovery of 5-hydroxymethylcytosine also in RNA its role in regulatory processes and metabolism remains elusive. To gain more insight into the function of RNA containing 5-hydroxymethylcytidine, innovative and interdisciplinary approaches are required. In this context, synthetic oligoribonucleotides containing 5-hyroxymethylcytidine are an inevitable tool. Therefore, in this chapter, we present the efficient synthesis of RNA oligonucleotides containing 5-hydroxymethylcytosine by solid-phase synthesis. The incorporation of the modified cytosine derivative into RNA is compatible with standard phosphoramidite-based synthesis procedures of oligoribonucleotides.

Key words Solid-phase synthesis, RNA, Methylation, Epigenetics, Cytosine

1 Introduction

5-Methylcytosine (5-mC) is one of the most abundant epigenetic base modifications in eukaryotic organisms and is well conserved among most plant, animal, and fungal models [1]. In genomic DNA the dynamic regulatory process of methylation and demethylation of cytosine has been extensively studied [2]. Methylation is conducted by well-known methyltransferases that are dependent on common cofactors like cobalamin and *S*-adenosylmethionine [3, 4]. 5-Hydroxymethyldeoxycytidine (5-hmdC), 5-formyldeoxycytidine (5-fdC), and 5-carboxymethyldeoxycytidine (5-cadC) are the products of the iterative oxidation by the enzymes of the Ten-Eleven-Translocation (TET) protein family. Recent studies have proposed these oxidized cytosine derivatives as intermediates of active

Alexandra Lusser (ed.), *RNA Methylation: Methods and Protocols*, Methods in Molecular Biology, vol. 1562,
DOI 10.1007/978-1-4939-6807-7_20, © Springer Science+Business Media LLC 2017

cytosine demethylation in DNA [5]. Thymine-DNA-Glycosylase (TDG)-mediated base excision repair (BER) has been shown to reverse cytosine methylation by excision of 5-fC or 5-caC and subsequent replacement of the abasic site by a cytosine [6, 7]. While the abundance of 5-methylcytidine (5-mrC) in tRNA has been known for several decades, only recently its occurrence in human coding and noncoding RNAs has been discovered [8, 9]. In addition to 5-mrC, also 5-hydroxymethylcytidine (5-hmrC) has been found to be present in RNA and the ability of TET enzymes to oxidize methylated cytidine to 5-hmrC in RNA has been shown [10]. The possible involvement of 5-mrC in "RNA epigenetics" [11] led to increased interest in 5-methylcytosine derivatives in RNA. There are already a variety of methods available for the investigation of DNA modifications. Oxidative and reduced bisulfite sequencing gives detailed sequence information of epigenetic base modifications [12, 13] and enzyme-based methods enable an enrichment of 5-hmdC from biological samples [14]. However, the majority of the existing methods are to the best of our knowledge not directly transferable to RNA for different reasons like stability of RNA samples or enzyme selectivity towards DNA. Toward the aim to create suitable tools for RNA epigenetics research, synthetic oligoribonucleotides containing 5-hydroxymethylcytidine (5-hmrC) would represent an invaluable one. While DNA oligonucleotides including 5-hmdC are commercially available, this is not the case for RNA. As the deprotection of 5-hydroxymethylated cytosine DNA oligonucleotides has been reported to be troublesome and in general requires nonstandard conditions [15–18], our motivation was to provide a straightforward protocol as close as possible to standard RNA deprotection, so that it becomes broadly accessible and easy-to-apply, also for the nonsynthetic chemist.

The following protocol describes all necessary details to synthesize 5-hmrC containing oligoribonucleotides by solid-phase synthesis based on a recently introduced nucleoside building block [19]. Thereby, the 4-amino and 5-hydroxymethyl functionalities at the nucleobase are masked by acetyl groups while the 2′-hydroxylgroup is masked by a *tert*.butyldimethylsilyl (*t*BDMS) group. The ribose 3′-hydroxylgroup is functionalized as cyanoethyl phosphoramidite (*see* Fig. 1). This derivatization pattern allows in combination with the usage of standard (RNA) 2′-*O*-[(triisopropylsilyl) oxy]methyl(TOM)-protected phosphoramidites [20] the automated solid-phase synthesis of oligoribonucleotides in a straightforward manner. The subsequent removal of the base-labile and fluoride-labile protecting groups and the cleavage from the solid support follow standard conditions for RNA deprotection using methylamine-ammonia conditions.

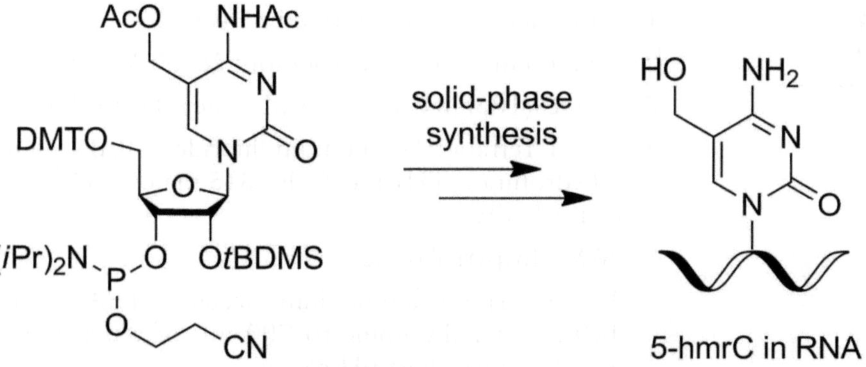

Fig. 1 The incorporation of 5-hydroxymethylated cytosine in oligoribonucleotides (*right*) via solid-phase synthesis and N^4-acetyl-5-acetoxymethyl-5′-*O*-(4,4′-dimethoxytrityl)-2′-*O*-*tert*-butyldimethylsilylcytidine 3′-*O*-2-cyanoethyl-*N*,*N*-diisopropylphosphoramidite (*left*)

2 Materials

All fine chemicals are available at commercial suppliers. For all buffers use ultrapure deionized water.

2.1 RNA Solid-Phase Synthesis

1. Instrumental equipment: Nucleic Acid Synthesizer.
2. Solid Support: Polystyrene solid support PS 200, degree of nucleoside substitution 80 μmol/g.
3. Phosphoramidites: 2′-*O*-TOM-cyanoethyl-phosphoramidites for standard nucleosides (*N*-acetyl protected rC, rG and rA), and N^4-acetyl-5-acetoxymethyl-5′-*O*-(4,4′-dimethoxytrityl)-2′-*O*-*tert*-butyldimethylsilylcytidine 3′-*O*-2-cyanoethyl-*N*,*N*-diisopropylphosphoramidite synthesized according to ref. [19] (*see* Fig. 1). Per coupled phosphoramidite use 130 μL of a 0.1 M solution in acetonitrile and dry it overnight over activated molecular sieve (3 Å) prior to synthesis.
4. Detritylation solution: 4 % Dicholoroacetic acid in 1,2-dichloroethane.
5. Activator solution: Per coupled phosphoramidite use 360 μL of a 0.3 M solution of 5-benzylthio-1*H*-tetrazole (BTT) in acetonitrile and dry it overnight over activated molecular sieve (3 Å) prior to synthesis.
6. Capping solutions: Cap A-solution: 0.5 M 4-(dimethylamino)-pyridine in acetonitrile. Cap B-solution: Acetic anhydride/sym-collidine/acetonitrile (2/3/5). Prepare 1 mL per coupling for both solutions.
7. Oxidation solution: 20 mM iodine in THF/pyridine/H_2O (7/2/1). Prepare 1 mL per coupling.

2.2 RNA Deprotection and Purification

1. 40 % Methylamine solution in H_2O.
2. Ammonium hydroxide solution: 28–30 % (w/w).
3. Funnel (10 mm diameter) with sintered glass disc for filtration.
4. 1.0 M Tetrabutylammonium fluoride (TBAF) solution in tetrahydrofuran (THF). Dissolve 315 mg of TBAF monohydrate in 1 mL THF.
5. *N*-Methylpyrrolidone.
6. 1.0 M Triethylammonium acetate (TEAA) buffer: Add 140 mL Triethylamine to 700 mL of water and slowly add acetic acid to adjust pH to 7.4.
7. Size exclusion column for desalting: Sephadex G25, GE HiPrep 26/10 Desalting.
8. HPLC system.
9. Anion exchange column DNAPac PA-100, 4 × 250 mm (Dionex).
10. Anion exchange column DNAPac PA-100, 9 × 250 mm (Dionex).
11. 0.1 M Triethylammoniumbicarbonate (TEAB) buffer (pH = 8.0): Add 140 mL Et_3N to 860 mL H_2O and bubble CO_2 through the solution until pH reaches 8.0. Then dilute 1:10.
12. Reversed Phase (C18) cartridge for desalting.
13. Buffers for anion-exchange chromatography: Eluant A: 25 mM Tris–HCl, pH 8.0, 6 M urea. Eluant B: 25 mM Tris–HCl, pH 8.0, 0.5 M $NaClO_4$, 6 M urea. Prepare the eluants from a stock solution of 250 mM Tris–HCl, pH 8.0. Filtration of the eluants through a cellulose acetate filter, 0.2 μm pore size, is strongly recommended.

2.3 Mass Spectrometry

1. 40 mM EDTA solution. Dissolve 13 mg ethylenediaminetetraacetic acid disodium salt dihydrate in 1 mL H_2O.
2. Electrospray ionization mass spectrometer (Ion trap) coupled to an microHPLC system (μLC-ESI-MS).
3. Reversed-phase column for desalting: C18, 50 × 4.6 mm, 5 μm.
4. Eluents: Buffer A: 8.6 mM Et_3N, 100 mM 1,1,1,3,3,3-hexafluoroisopropanol (HFIP), pH 8.3. To 598 μL of triethylamine in H_2O add 5.23 mL HFIP. Buffer B: HPLC grade methanol.

3 Methods

3.1 RNA Solid-Phase Synthesis

1. For a 1.5 μmol scale synthesis, use 19 mg of the polystyrene solid support and place it in the synthesis cartridge.
2. Following standard methods of automated RNA strand elongation, perform the synthesis under the following conditions: Detritylation 120 s (detritylation solution), coupling 180 s

for standard phosphoramidites and 360 s for modified phosphoramidites (activator solution and phosphoramidite solution), capping 2 × 20 s (capping A and capping B solution) and oxidation 60 s (oxidation solution) (*see* **Note 1**).

3. Dry the solid-support for 30 min under high vacuum in the synthesis cartridge (*see* **Note 2**).

3.2 Deprotection

1. Place the solid support in a screw-cap 1.5 mL tube, add 650 μL of aqueous methylamine solution, and 650 μL of ammonium hydroxide solution and close the tube.
2. Vortex the tube.
3. Put the tube on a preheated heating block at 65 °C for 30 min.
4. Let the tube cool down. Do not open before it reaches room temperature.
5. Centrifuge the tube at 15,600 × *g* for 15 s.
6. Decant the supernatant that contains the synthesized 2′-*O*-silylated RNA and push it through a sintered glass filter with a syringe to remove remaining traces of the solid support. Wash the residual solid support in the tube twice by adding 1 mL of H_2O, vortexing, centrifuging, and filtering. Using a 25 mL flask to collect the supernatant and the washes is advisable, to control steaming in the next step.
7. Evaporate the solvent under vacuum. The synthesized 2′-*O*-silylated RNA will remain in the flask as an off-white barely visible residue.
8. Dilute the synthesized 2′-*O*-silylated RNA in 1 mL 1 M TBAF solution in THF. In case of low solubility, add 100 μL *N*-methylpyrrolidone.
9. Keep the flask at 37 °C for 16 h to ensure complete deprotection of the 2′-*O*-silyl groups.
10. Neutralize the solution with 1 mL of TEAA buffer (pH 7.4) and evaporate the THF under vacuum.
11. Desalt the RNA sample by running it through a Sephadex size exclusion column with water as eluent. Use a flow of 1 mL/min.
12. Evaporate the solvent under vacuum.
13. Dissolve the crude RNA in 1 mL of H_2O to make a stock solution (*see* **Note 3**).

3.3 Purification

1. Analyze the crude product RNA by anion-exchange chromatography on a HPLC system to document the quality of the synthesis and to estimate the appropriate solvent gradient for the purification. Inject 2 μL of the stock solution. We recommend the use of a Dionex DNAPac PA-100 column (4 × 250 mm) at 80 °C, a flow of 1 mL/min, and a gradient of 0–60 % eluant B in eluant A within 45 min; UV detection at 260 nm (*see* Fig. 2).

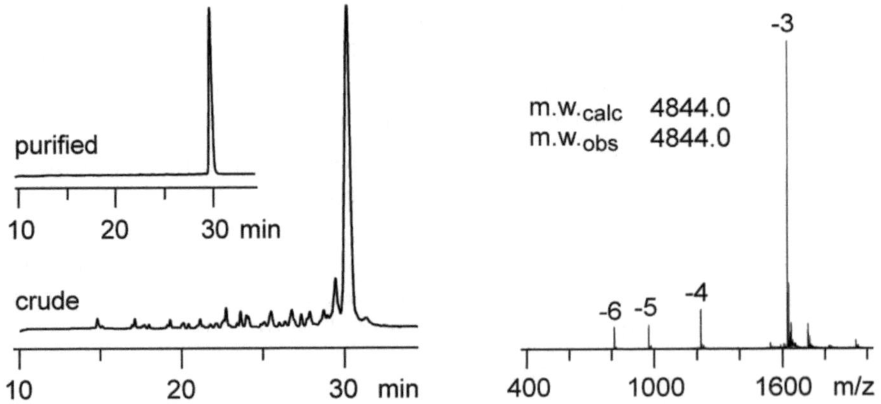

Fig. 2 Characterization of a 15 nt oligoribonucleotide with the sequence 5′-GAAGGGCAAC(5hmrC)UUCG-3′. Anion-exchange HPLC traces of the crude and purified product RNA (*left*) and the corresponding LC-ESI mass spectra (*right*). HPLC conditions: Dionex DNAPac column (4 × 250 mm), 80 °C, 1 mL/min, 0–60 % buffer B in buffer A within 45 min

2. Purify the crude product RNA by anion-exchange chromatography on a HPLC system. Inject 70–150 μL of the stock solution and combine the product-containing fractions. We recommend the use of a Dionex DNAPac PA-100 column (9 × 250 mm) at 80 °C, a flow of 2 mL/min, and a gradient of approximately 30–50 % eluant B in eluant A within 20 min; UV detection at 260 nm.
3. Dilute the product-containing fractions 1:1 with 0.1 M TEAB buffer.
4. Pretreat the C18 cartridge consecutively with 20 mL acetonitrile, 30 mL acetonitrile/H_2O, 30 mL H_2O, and 30 mL of 0.1 M TEAB buffer.
5. Load the product-containing solution on the C18 cartridge.
6. Wash with 50 mL of H_2O.
7. Elute the product RNA with acetonitrile/H_2O (1/1).
8. Evaporate the solvents to dryness.
9. Dissolve the product RNA in 1 mL of H_2O and make a stock solution of the purified RNA (*see* **Note 3**).
10. Quantify the RNA by UV absorbance at 260 nm.
11. Analyze the purified product RNA by anion-exchange chromatography under the same conditions as in **step 1** (*see* Fig. 2). The pure RNA appears as a single symmetric peak in the HPLC profile and in general, has an estimated purity of >90–95 %.

3.4 Mass Spectrometry

To analyze the integrity of the synthesized RNA, the molecular weight is determined by liquid chromatography-electrospray ionization mass spectrometry (μLC-ESI MS).

1. Prepare a sample of 200 pmol RNA in 30 μL 20 mM EDTA solution.
2. Analyze by μLC-ESI-MS in the negative ion mode: Desalting is achieved by the RP-column chromatography at a flow rate of 30 μL/min and 21 °C; gradient: 0–100 % buffer B in buffer A within 30 min. UV detection at 254 °C (*see* Fig. 2).

4 Notes

1. Depending on the type of nucleic acid synthesizer, the times can vary due to the different device properties.
2. At this point it is possible to split the solid-support into several portions and store a back-up at −20 °C.
3. At this stage the sample can be stored at −20 °C.

Acknowledgments

We thank the Austrian Science Fund FWF (I1040 and P27947) for funding.

References

1. Smith ZD, Meissner A (2013) DNA methylation: roles in mammalian development. Nat Rev Genet 14(3):204–220. doi:10.1038/nrg3354
2. Shen L, Song C, He C et al (2014) Mechanism and function of oxidative reversal of DNA and RNA methylation. Annu Rev Biochem 83(1):585–614. doi:10.1146/annurev-biochem-060713-035513
3. Goll MG, Bestor TH (2005) Eukaryotic cytosine methyltransferases. Annu Rev Biochem 74(1):481–514. doi:10.1146/annurev.biochem.74.010904.153721
4. Kumar S, Cheng X, Klimasauskas S et al (1994) The DNA (cytosine-5) methyltransferases. Nucleic Acids Res 22(1):1–10. doi:10.1093/nar/22.1.1
5. Fu Y, He C (2012) Nucleic acid modifications with epigenetic significance. Curr Opin Chem Biol 16(516):–524. doi:10.1016/j.cbpa.2012.10.002
6. He Y, Li B, Li Z et al (2011) Tet-mediated formation of 5-carboxylcytosine and its excision by TDG in mammalian DNA. Science 333(6047):1303–1307. doi:10.1126/science.1210944
7. Maiti A, Drohat AC (2011) Thymine DNA glycosylase can rapidly excise 5-formylcytosine and 5-carboxylcytosine: potential Implications for active demethylation of CpG sites. J Biol Chem 286(41):35334–35338. doi:10.1074/jbc.C111.284620
8. Squires JE, Patel HR, Nousch M et al (2012) Widespread occurrence of 5-methylcytosine in human coding and non-coding RNA. Nucleic Acids Res 40(11):5023–5033. doi:10.1093/nar/gks144
9. Amort T, Soulière MF, Wille A et al (2013) Long non-coding RNAs as targets for cytosine methylation. RNA Biol 10(6):1003–1008. doi:10.4161/rna.24454
10. Fu L, Guerrero CR, Zhong N et al (2014) Tet-mediated formation of 5-hydroxymethylcytosine in RNA. J Am Chem Soc 136(33):11582–11585. doi:10.1021/ja505305z
11. He C (2010) Grand challenge commentary: RNA epigenetics? Nat Chem Biol 6(12):863–865. doi:10.1038/nchembio.482
12. Booth MJ, Ost TWB, Beraldi D et al (2013) Oxidative bisulfite sequencing of 5-methylcytosine and 5-hydroxymethylcytosine. Nat Protocols 8(10):1841–1851

13. Booth MJ, Marsico G, Bachman M et al (2014) Quantitative sequencing of 5-formylcytosine in DNA at single-base resolution. Nat Chem 6(5):435–440
14. Song C, Szulwach KE, Fu Y et al (2011) Selective chemical labeling reveals the genome-wide distribution of 5-hydroxymethylcytosine. Nat Biotech 29(1):68–72. doi:10.1038/nbt.1732
15. Tardy-Planechaud S, Fujimoto J, Lin SS et al (1997) Solid phase synthesis and restriction endonuclease cleavage of oligodeoxynucleotides containing 5-(Hydroxymethyl)-cytosine. Nucleic Acids Res 25(3):553–558. doi:10.1093/nar/25.3.553
16. Tan L, Shi YG (2012) Tet family proteins and 5-hydroxymethylcytosine in development and disease. Development 139(11):1895–1902. doi:10.1242/dev.070771
17. Dai Q, Song C, Pan T et al (2011) Syntheses of two 5-hydroxymethyl-2′-deoxycytidine phosphoramidites with TBDMS as the 5-hydroxyl protecting group and their incorporation into DNA. J Org Chem 76(10):4182–4188. doi:10.1021/jo200566d
18. Münzel M, Globisch D, Trindler C et al (2010) Efficient synthesis of 5-hydroxymethylcytosine containing DNA. Org Lett 12(24):5671–5673. doi:10.1021/ol102408t
19. Riml C, Micura R (2016) Synthesis of 5-hydroxymethylcytidine- and 5-hydroxymethyluridine-modified RNA. Synthesis (Stuttg) 48:1108–1116. doi:10.1055/s-0035-1561220
20. Pitsch S, Weiss PA, Jenny L et al (2001) Reliable chemical synthesis of oligoribonucleotides (RNA) with 2′-O-[(Triisopropylsilyl)oxy]methyl(2′-O-tom)-protected phosphoramidites. HCA 84(12):3773–3795. doi:10.1002/1522-2675(20011219)84:12<3773:AID-HLCA3773>3.0.CO;2-E

INDEX

Alexandra Lusser (ed.), *RNA Methylation: Methods and Protocols*, Methods in Molecular Biology, vol. 1562, DOI 10.1007/978-1-4939-6807-7,

MIX
Papier aus verantwortungsvollen Quellen
Paper from responsible sources
FSC® C105338

If you have any concerns about our products,
you can contact us on
ProductSafety@springernature.com

In case Publisher is established outside the EU,
the EU authorized representative is:
Springer Nature Customer Service Center GmbH
Europaplatz 3, 69115 Heidelberg, Germany

Printed by Libri Plureos GmbH
in Hamburg, Germany